Old Waterfront Walls

CIRIA, the Construction Industry Research and Information Association, is an independent, non-profit-distributing body which initiates, organizes and manages research and provides information on all topics of concern to professionals in the construction industry.

Organizations participating in CIRIA's activities represent all sides of the construction industry. Through the elected Council, programme advisory committees and project steering groups, they control the selection and direction of all CIRIA's research projects, ensuring that CIRIA's work is practical and directly useful as well as independent and authoritative. It is through these steering and working groups that the essential spirit of CIRIA – the exchange of ideas and experience to solve common problems – is realized.

CIRIA projects bring together knowledge, experience and information from a wide range of sources and organizations, to form the basis of well-considered recommendations for best practice.

The British Ports Federation is the only trade association representing the majority of port interests. Within its London headquarters, representative ports large and small, in competition and partnership, meet to review, discuss and deal with matters of concern to the industry.

The British Ports Federation acts as a watchdog for the industry, scrutinizing proposed new legislation which might affect the statutory and commercial role of its member ports. It also advises Government when consulted, and when not, upon proposed legislation and other important policy matters.

Behind the scenes, the British Ports Federation is responsible for developing and maintaining research and training programmes, safe operating practices and the industry's statistics.

It is the central voice of the industry representing more than 100 port authorities, including all those of any commercial significance.

Old Waterfront Walls

Management, maintenance and rehabilitation

R.N. BRAY
Consulting Engineer
(formerly Project Director,
Livesey Henderson Ltd)

P.F.B. TATHAM
Livesey Henderson Ltd

Construction Industry Research
and Information Association
British Ports Federation

Published by E & FN Spon, an imprint of Chapman & Hall, 2–6 Boundary Row, London SE1 8HN

Chapman & Hall, 2–6 Boundary Row, London SE1 8HN, UK

Blackie Academic & Professional, Wester Cleddens Road, Bishopbriggs, Glasgow G64 2NZ, UK

Van Nostrand Reinhold Inc., 115 5th Avenue, New York, NY10003, USA

Chapman & Hall Japan, Thomson Publishing Japan, Hirakawacho Nemoto Building, 6F, 1-7-11 Hirakawa-cho, Chiyoda-ku, Tokyo 102, Japan

Chapman & Hall Australia, Thomas Nelson Australia, 102 Dodds Street, South Melbourne, Victoria 3205, Australia

Chapman & Hall India, R. Seshadri, 32 Second Main Road, CIT East, Madras 600 035, India

---

First edition 1992

Typeset in 10½/12pt Sabon by Falcon Graphic Art Ltd, Wallington, Surrey
Printed in Great Britain at the Alden Press, Oxford

ISBN 0 419 17640 3 0 442 31554 6 (USA)

A catalogue record for this book is available from the British Library

Library of Congress Cataloging-in-Publication data available

Bray, R.N. (Richard Nicholas)
Old waterfront walls: management, maintenance and rehabilitation
R.N. Bray, P.F.B. Tatham
p. cm.
Includes bibliographical references and index.
ISBN 0–442–31554–6
1. Bulkheads—Maintenance and repair. 2. Sea-walls–Maintenace and repair. I. Tatham, P.F.B. II. Title
TA760.B73 1992
627'.58—dc20 91–33375
CIP

Contents

Acknowledgements

This Report was written by R.N. Bray, an independent consulting engineer, and P.F.B. Tatham of Livesey Henderson Ltd, (part of the Binnie & Partners Group) under contract to CIRIA. Supporting contributions were made by Dr D.M. McCann of the British Geological Survey (geophysics), A.D.M. Bellis, Consulting Engineer (technical review), and C.W. Frith of Binnie & Partners (part of Chapters 4 and 5).

The CIRIA project was carried out with the guidance and assistance of the steering group listed below:

P.S. Lucas (Chairman)	Mersey Docks & Harbour Company
A.P. Barwell	British Rail
J.J. Bell	Sealink Harbours Ltd
M.E. Bramley (to July 1989)	CIRIA
G.M. Gray (from July 1989)	CIRIA
D.H. Cooper	Associated British Ports
R. Gardener	Engineering Geophysicist
G. Haider	British Waterways
J. Horne	Property Services Agency
A.B. Hughes	Ministry of Agriculture, Fisheries and Food
P. Kinsey	Department of Transport
J.V. Levy	Thames Water (now National Rivers Authority)
D.J. Palmer	Tees & Hartlepool Port Authority
H.R. Payne	Welsh Office
P.J. Quarton	Tarmac Structural Repairs Ltd
I.W. Stickland	Posford Duvivier
P. Treadgold	Flynn & Rothwell (previously with LDDC)
P. Wright	Scottish Development Agency

Corresponding members

I. Whittle	National Rivers Authority
A.J. Allison	Ministry of Agriculture, Fisheries and Food

The project was funded by:

Ministry of Agriculture, Fisheries and Food
Department of Transport
British Ports Federation
British Waterways
British Rail
Property Services Agency

The assistance of the following individuals and organizations in providing information on design practice and experience with respect to old waterfront walls in the UK is acknowledged:

Aberconwy Borough Council
Aberdeen Harbour Board
ABP Research and Consultancy Ltd
T.M. Adie and Sons
Adur District Council
Alta Geophysics
Ampthill Geophysical Consultants
Applied Geophysics Ltd
Arfon Borough Council
Ove Arup and Partners, Ireland
Associated British Ports, Goole
Associated British Ports, Grimsby
Associated British Ports, Port of Hull
Associated British Ports, Swansea
W.S. Atkins and Partners

Babtie Shaw and Morton
Beckett Rankine Partnership
Belfast Harbour Commissioners
Bembridge Harbour Improvements Co Ltd
Berkshire County Council
Blackpool Borough Council
Blyth Valley Borough Council
Bournemouth Borough Council
Peter Brett Associates
Brighton Borough Council
Bristol City Council
British Cement Association
British Ceramic Research Ltd
British Rail
British Waterways
Sir Bruce White, Wolfe Barry and Partners
Building Research Establishment
Buro Happold
Bury Metropolitan Borough Council
Bylander Group

Calderdale Metropolitan Borough Council
Cambridgeshire County Council
Carmarthen District Council
Cementation Construction Ltd
Charcon Tunnels Ltd
Cheshire County Council
Cleethorpes Borough Council
Cleveland County Council
Clyde Port Authority
S.P. Collins and Associates
Colwyn Borough Council
Concrete Repairs Ltd
Cork Harbour Commissioners
Council for British Archaeology
County Surveyors' Society
Coventry City Council
R.J. Crocker and Partners, Bromley

Darpor Engineering & Associates
Dean and Dyball Construction
Dee-Cee Group Services Ltd
Demex Ground Engineering Ltd
DHV Burrow-Crocker Consulting
Dobbie & Partners
Dr I.G. Doran and Partners
Dow Construction Chemicals Ltd
Dover Harbour Board
Dudley Metropolitan Borough Council
Dumfries and Galloway Regional Council

East Yorkshire Borough Council
East Devon District Council
East Sussex County Council
Eastbourne Borough Council
ELS Land Consultants Ltd
English Heritage
Essex County Council Highways Department

Fairclough Civil Engineering Ltd
Falmouth Harbour Commissioners
Flynn & Rothwell
Forth Ports Authority
Fosroc CCD Ltd
Fylde Borough Council

Geotechnical Consulting Group
GKN Colcrete Ltd
Vale of Glamorgan Borough Council
Frank Graham & Partners
Grampian Regional Council
Great Yarmouth Port & Haven Commissioners
Gwynedd County Council

Harbour & General Works Ltd
Hastings Borough Council
Hereford and Worcester County Council
Hertfordshire County Council
Highland Regional Council
Holderness Borough Council
Hove Borough Council
Humberside County Council
Hydraulics Research Ltd

Imperial College Library
Institution of Civil Engineers Library

R.T. James and Partners

P. Kalaugher, Exeter University
Kent County Council
Knowsley Metropolitan Borough Council

Lancaster City Council
Langbaurgh Borough Council
Ronald Leach & Associates Ltd
John Lelliott (Structural Renovation) Ltd
Lerwick Harbour Trust
Lewes District Council
Port of Liverpool
London Borough of Lambeth
London Docklands Development Corporation
London Docklands Museum
London Borough of Ealing
Lovells Wharf Ltd
Loy Surveys Ltd

Sir M. MacDonald and Partners
Sir Robert McAlpine and Sons Ltd
Maldon District Council
Manchester Ship Canal Company
G. Maunsell and Partners
Medina Borough Council
Medway Ports Authority
Mevagissey Harbour Trustees
Ministry of Agriculture, Fisheries and Food (Regional Offices)
Mobbs and English
Mobell Marine Ltd
A. Monk and Co Ltd
Mott Hay & Anderson
Mowlem Civil Engineering

National Rivers Authority
National Trust
Newcastle University
Newcastle upon Tyne City Council
New Forest District Council
Newport Borough Council
North East London Polytechnic
North Norfolk District Council
Northern Ireland Fishery Harbour Authority

Oceanfix International Ltd
Ogwr Borough Council
Olympus Industrial
Orkney Islands Council
Geoffrey Osborne Ltd
Osiris Seaway Ltd
Oxfordshire County Council

Parkman Consulting Engineers
Patent Library
Peterhead Bay Authority

Plymouth City Council
Port Talbot Borough Council
Posford Duvivier
Powys County Council
Pozament Ltd
Prakla-Seismos AG
Preston Borough Council
Property Services Agency
Proton PCS Ltd

Ram Services Ltd
Rendel Palmer & Tritton
Rhuddlan Borough Council
Robertson Group PLC
Rochester-upon-Medway City Council

Scarborough Borough Council
Scottish Development Department
Scrabster Harbour Trust
Sea Fish Industry Authority
Sealink Harbours Ltd
Sealocrete (UK) Ltd
Sefton Metropolitan Borough Council
Shoreline Holdings
Shropshire County Council
Simrad Hydrospace Ltd
Sir Frederick Snow & Partners
Solex International
South Hams District Council
A.M. Sowden
Staffordshire County Council
Stonbury Ltd
Strathclyde Regional Council
Stroud District Council
Structure Testing Services (UK) Ltd
Subsurface Geotechnical Services
Suffolk Coastal District Council
Suffolk County Council
Port of Sunderland
Sunderland Borough Council
Sutton Harbour Company

Tameside Borough Council
Tees and Hartlepool Port Authority
Thanet District Council
Tilt Measurement Ltd
Torridge District Council
Transport and Road Research Laboratory
Port of Tyne Authority

UML Ltd

Wallace Evans & Partners
Walsall Metropolitan Borough Council
Welsh Office
Robert West & Partners
West Sussex County Council
West Somerset District Council
Westcrete Specialist Contracts Ltd
Wigan Metropolitan Borough Council
James Williamson & Partners
Wimpey Laboratories
Wirral Borough Council

Foreword

This book is the final report of a CIRIA research project on old waterfront walls. It seeks to provide an overview of the types of wall in existence, their strengths and failings, the management of their maintenance and, where appropriate, the methods available for their rehabilitation.

Information for the research project was gleaned from a variety of sources, including a literature review, a questionnaire to engineers and organizations responsible for the maintenance of waterfront walls, discussions with a large selection of interested parties, and a particularly well-informed and encouraging steering group. The authors are aware that there is still a wealth of untapped information on this subject waiting to be collated, but time and budgetary constraints prevail.

The history of the subject spans at least 2000 years. During the project, many interesting points were uncovered. In particular it was noted that over this period engineering skills were lost and found again, and that engineering failure at 'the sharp end of technology' was frequently accepted as being a necessary evil.

That the construction of waterfront walls has never been considered easy is of little doubt. Sir Benjamin Baker, in his well-known 1881 ICE paper 'On the actual lateral pressure of earthwork', stated 'If an engineer has not had some failures with retaining walls, it is merely evidence that his practice has not been sufficiently extensive: for the attempt to guard against every contingency in all instances would lead to ruinous and unjustifiable extravagance . . .' He also pointed out that, with respect to the failure of dock walls on soft ground, a movement to a greater or lesser extent was the rule. So much so that Voisin Bey, the distinguished engineer-in-chief of the Suez Canal, could name no exception to it 'since he had failed to find any long line of quay wall, which on further inspection proved to be perfectly straight in line and free from indications of movement'.

Here, then, is a review of those walls which proved so troublesome to construct, but whose longevity is a measure of man's persistence to succeed. Here also is a guide to how we may, by observation, investigation and application of modern construction techniques, manage the substantial national asset which these walls represent.

Glossary

Abutment — End support of a bridge which also connects the structure to the ground

Apron
1. An area of open land adjacent to a berth (immediately behind the quay face)
2. A layer of rubble, stone or a concrete slab, with or without toe piles, to protect the toe of a wall against scour

Ashlar
1. A square-hewn stone
2. Masonry consisting of blocks of stone, finely square dressed to given dimensions, and laid in courses with thin joints

Bag joggle — A bag filled with mortar which is inserted in a preformed keyway in vertical masonry or concrete blockwork joints to provide resistance to sliding

Bagwork — A revetment to protect walls from scour, consisting of dry concrete in bags

Belting — Horizontal projection around the hull of a ship at or above water level to give increased impact protection when berthing.

Berth — A place where a ship can tie up and load or unload

Blockage — The ratio of the cross-section area of the immersed portion of a vessel to the cross-section area of the waterway in which it floats

Bond
1. An interlocking arrangement of stones or blocks within a wall to provide stability
2. Adhesion between mortar and masonry or concrete block, in a structure composed of these materials

Cofferdam A dam (usually temporary) to give access to an area which is ordinarily submerged

Cope The top edge of a quay adjacent to a berth

Counterfort A strengthening pier at right angles to a retaining wall on the side of the retained material

Dead man A large buried concrete or masonry block to which a tie rod is anchored

Drawdown Lowering of a water level

Dry stone wall A stone wall constructed without mortar

Geophysical Of or pertaining to the physics of the earth, as applied to investigation methods

Gravity wall A retaining wall of broad cross-section which depends for its stability on its own weight

Hearting The in-filling of broken stone and other granular materials within a wall

Heel The landward projection of a wall base

Liquefaction A condition affecting saturated loose sand when subjected to shock, vibration, flowing water or sudden loading

Littoral drift The sedimentary material moved in the **littoral zone** under the influence of waves and currents

Littoral transport The movement of material in the **littoral zone** under the action of waves and currents

Littoral zone The coastal zone in which material on the seabed is transported by the action of waves and currents

Overburden drilling A drilling technique for penetrating a mixture of soils and rocks

Oversails Horizontal projections from the back face of a wall intended to increase stability

Overtopping The flow of water over the top of a structure as a result of wave run-up or surge action

Pointing The filling with mortar of the joints in a wall from which the bedding or jointing mortar has been raked out

Pozzolana Originally a volcanic dust used as a hydraulic cement when mixed with lime. It can be made artificially by burning and grinding clay or shale

Puddle clay	A plastic mixture of clay and water which is used to form an impermeable layer
Quay	A berthing structure backing on to the shore or reclaimed land
Rip-rap	A layer, facing or protective mound of randomly placed stones providing protection against scour or erosion
Rubble	Stone of irregular shape and size
Rubble masonry	Masonry formed from rubble
Scour	The underwater removal of bed material by waves or currents
Sea wall	A shoreline structure primarily designed to prevent flooding, erosion and other damage caused by wave action
Tsunami	A large wave induced by the effects of an earthquake or other underwater disturbance
Translation	The lateral movement of a whole structure
Wedge	A thin masonry slice driven into an external joint in a rubble wall to prevent movement of the facing stones
Wharf	See **Quay**

Introduction

Definitions

This book is concerned with the maintenance and rehabilitation of old waterfront walls. In this context maintenance is defined as:

> routine inspection, structural evaluation and periodic small scale works to repair parts of a wall. Such works are usually classified as 'revenue works' and exclude reconstruction.

Rehabilitation is defined as:

> major works to restore a wall to its original state or to upgrade it to a new standard or function. Such works are usually classified as 'capital works'.

Old waterfront walls are defined as:

> gravity walls of masonry, concrete blockwork, brickwork or mass concrete with vertical or near vertical exposed faces fronting on to the sea, rivers, canals, lakes or docks. They may be used or were formerly used as quays, locks, flood defences, coast protection, retaining walls or bridge abutments. Such walls include thin masonry facings to rock or cohesive soils but exclude revetments, sloping sea defences, dams, sheet piled walls and reinforced concrete structures.

Background

Old waterfront walls, as defined above, date in the main from about 1700, although some examples go back as far as the Romans. An indication of the extent of these walls may be gained from the following:

1. there are over 300 harbours in the UK with old breakwaters or quay walls;
2. there are thousands of miles of canal which have many quays, locks, retaining walls and bridge abutments;
3. many sea defence walls come within the definition of old waterfront walls;

4. roads and railways have many retaining walls and bridge abutments fronting on to water;
5. rivers and estuaries, in built-up areas, have many miles of walls on their banks for river training, flood protection and building foundations.

There is a great variety in the design of walls. In some cases the use of a wall has changed and the wall may have been neglected. In others a once-neglected wall is now seen as an attractive part of a new waterfront development scheme in which it will be subjected to a change of service conditions.

Authorities responsible for the walls vary from organizations with considerable engineering expertise and resources to those with no in-house engineering knowledge. In many cases they will be unfamiliar with the problems associated with this kind of structure and the effects of water behind, under, in front of and inside the wall.

Inspection and assessment of the structure and stability of the walls can be difficult and costly, particularly where there is no information on the original design or much of the wall is permanently submerged. Assessment involves consideration of ground conditions in the foundations and behind the wall, as well as the integrity of the wall itself. Investigation requires specialist knowledge and equipment, and the selection of appropriate techniques.

The methods of maintenance and rehabilitation need to take account of engineering, environmental and financial factors and must be practicable, effective and appropriate to the future use and life required of the structure. Schemes involving major rehabilitation also require careful consideration of future maintenance costs and the preservation of the appearance for aesthetic reasons. Particular skill is thus needed to ensure that problem areas are correctly understood and identified, and that the most cost-effective course of inspection/assessment, repair and rehabilitation is undertaken.

This book is mostly based on UK data and practice but will also be applicable to similar structures in other parts of the world.

Objectives

The objectives of this book are as follows:

1. to review the types of old waterfront walls which exist, and their performance;
2. to identify and describe the most suitable techniques available for the inspection, structural evaluation, repair and rehabilitation of old waterfront walls.

Structure of the book

Chapters 1 and 2 contain a review of the types and forms of old waterfront walls and the circumstances which affect their performance. The review is based on information collected by the study team from questionnaires, literature research and direct contact with those responsible for walls and their maintenance. Chapter 3 sets out the procedure for managing the maintenance and rehabilitation of walls. It explains how the information contained in the book may be used to improve the efficiency of the investigation, assessment and repair stages of maintenance and rehabilitation. It refers extensively to the other chapters. Chapters 4 and 5 contain a review of the methods currently used for inspecting and monitoring walls and the techniques available for detailed investigation. Chapter 6 sets out the philosophy behind the process of evaluating the condition of a wall and assessing its stability. Chapter 7 contains a review of the methods currently available for repairing and rehabilitating walls.

The book also includes three appendices. These contain detailed information on subjects which are considered to be particularly relevant to the project but whose inclusion in the main text would have upset the balance and made reading more difficult.

It is suggested that the most effective way of using this book is as follows. Read Chapters 1 and 2 to obtain an insight into the nature of old waterfront walls and the circumstances which affect their behaviour. Read Chapter 3 to obtain a broad appreciation of the way in which the maintenance and rehabilitation process is addressed. This chapter is the key to the whole book and should be used as the main guide to implementation.

The remaining chapters and appendices may be referred to at the appropriate time. Although each of Chapters 4 to 7 covers a distinct area of interest the full significance of each chapter cannot be appreciated unless Chapters 1 and 2 have been read.

Relationship with other publications

Reference is made at various points in this publication to the book *Maintenance of Brick and Stone Masonry Structures* edited by A.M. Sowden and published by E & FN Spon. This covers all types of structure but excludes those of concrete construction. Both this book and Sowden's are stand-alone publications but in many respects are complementary.

Chapter one

Types of wall

Walls have been categorized into the following main groups:

1. quays, docks and locks;
2. breakwaters;
3. sea walls for coastal defence;
4. retaining walls and flood defences;
5. skin walls;
6. bridge piers and abutments.

The descriptions of the features of these walls, which follow, are generally based on the first group since it contains great variety. The other groups are described in relation to the first.

1.1 General wall characteristics

1.1.1 Geographical distribution

Waterfront walls are well distributed over the whole country, with the exception of hilly or mountainous regions, where they are generally only found as lakeside retaining walls or in the abutments of bridges.

In the distant past, in places such as East Anglia with poor or non-existent local sources of rock, timber structures were erected. As these decayed they were replaced with masonry and concrete walls. Other regions, such as south-west England, Wales and the highlands of Scotland, had ample supplies of good quality rock nearby and hence construction with masonry was feasible from the outset.

The first significant constructors of waterfront walls in the British Isles were the Romans. Figure 1.1 shows the main settlements developed during the Roman occupation[1] and Figure 1.2 is a recent photograph of the quayside wall of the Roman waterfront in Chester. Many of these sites are now obscured by later construction but the original Roman foundations are often still in place.

Subsequent development of ports and harbours was dictated by trade patterns and the defence of the country. Figure 1.3 shows the British customs ports in England and Wales in the eighteenth and early nineteenth centuries.[2] The figure also shows the principal inland navigation routes in the middle of the eighteenth century. This was before the major canal building era, which began in 1761. Inland navigation was vastly improved by the development of the canal system up to around 1820. The canal network in 1789 is shown in Figure 1.4.[3] Details of the canals built after this date may be found in Hadfield.[3] Many of the port and canal structures were built in masonry and brickwork, and a considerable number of waterfront walls from this period are extant today.

In more recent times waterfront walls have been constructed for the railway and road networks, both in coastal locations, and alongside and over rivers and estuaries. They have also been built for coastal and flood defence purposes. Many examples of this work are recorded

Figure 1.1 Roman Britain.□, large fortresses; ■, small fortresses; ○, large town; ● small towns. *Source:* Haverfield, F., *The Roman Occupation of Britain,* Clarendon Press, 1924.

Figure 1.2 The quayside wall of the Roman waterfront at Chester
Source: The Archaeological Unit of the Department of Leisure Services, Chester City Council.

in the Proceedings of the Institution of Civil Engineers and other publications of the period. The possible sources of information relating to these walls are given in Chapter 5.

It is particularly difficult to quantify the different types of wall in existence at the present time, or the number in a specific area, because of the number and variety of types of owner. For example, in the Highland region of Scotland alone there are over 100 harbours, piers and ferry terminals of various size and type, many kilometres of coastal walls, and in excess of 1800 bridges.

1.1.2 Designers

The results of a questionnaire sent to wall owners and other responsible authorities showed that there was no relationship between a wall's location and its designer (see Appendix C). In spite of the difficulties in travelling long distances, some of the early engineers designed and supervised works in many different parts of the country and overseas. For instance, Captain John Perry (1670–1733) worked in Dagenham, King's Lynn, Dublin, Dover and Russia.

Although relationships between designers and their areas of work are unforthcoming, there is considerable benefit in identifying the designer of a wall since this might lead to additional information about the wall and its construction.

Swann[4] and Skempton[5] give a considerable amount of information on the engineers of English port improvements between 1640 and 1840. After this date it is possible to obtain information which relates engineers to places from the library of the Institution of Civil Engineers, London. The library has a computerized database relating names, places and subjects. (See also Chapter 5.)

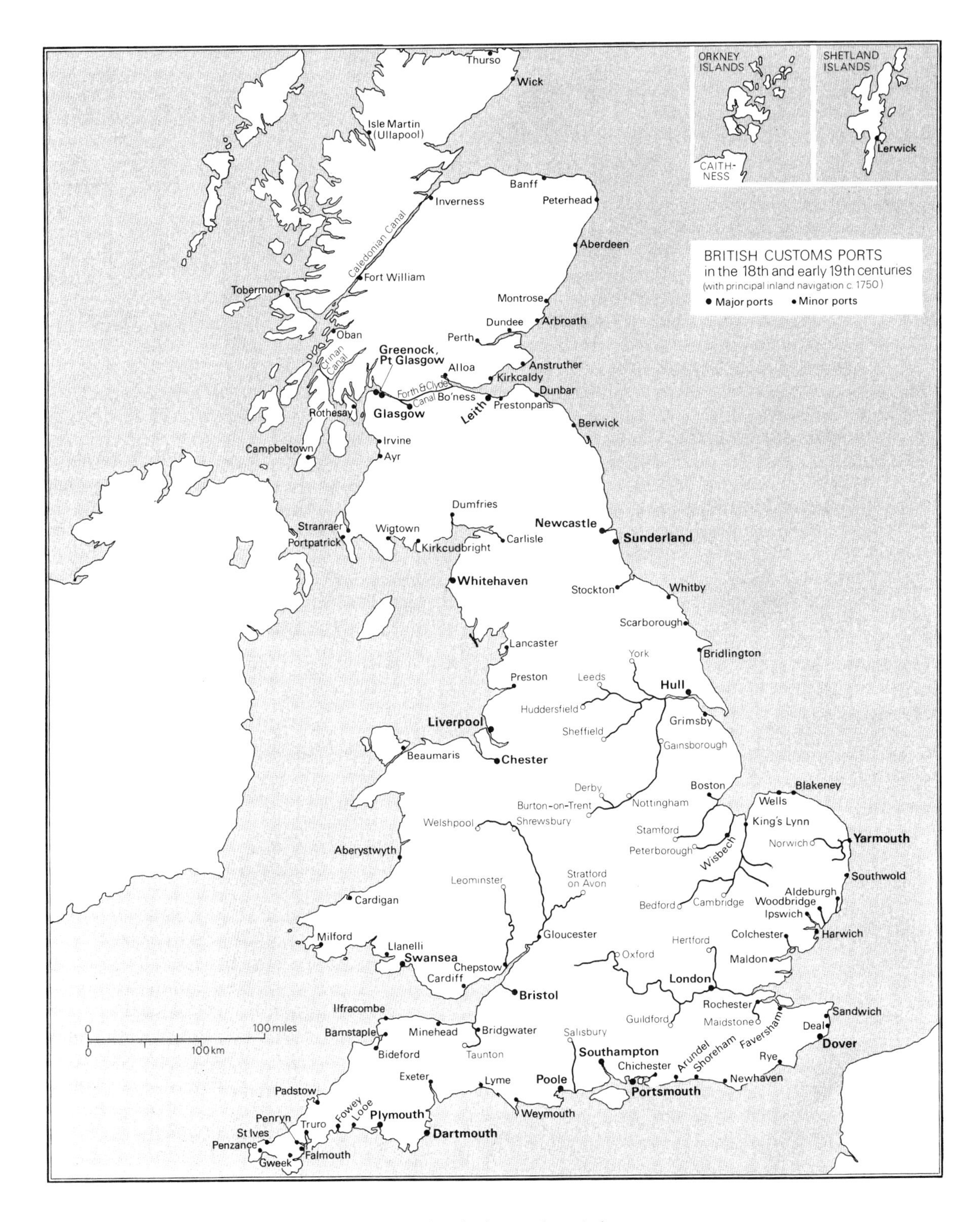

Figure 1.3 British Customs ports in the eighteenth and early nineteenth centuries
Source: Jackson, G., *The History and Archaeology of Ports*, World's Work Ltd, 1983.

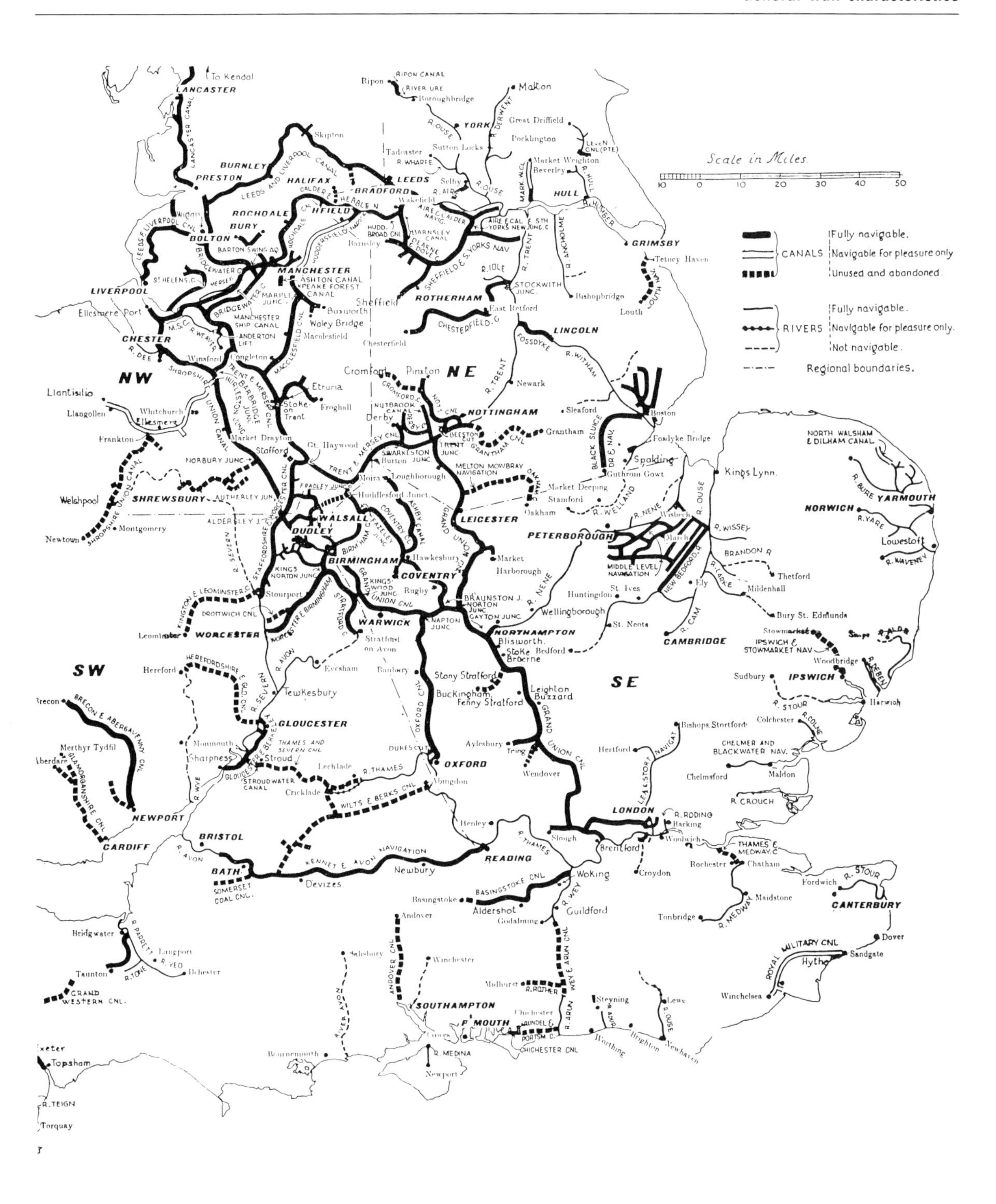

Figure 1.4 The waterway system in England
Source: Maré, E. de., *The Canals of England*, The Architectural Press, London.

Figure 1.5 Trig Lane, London, showing back-braced riverfront revetments behind fifteenth century river wall
Source: Council for British Archaeology Research Report No. 41. Figure 36, page 33. Courtesy of 'Museum in Docklands, PLA Collection'.

1.1.3 Materials and design philosophy

Historical

Although the Romans were competent builders of masonry structures, and, more importantly, of reliable and durable foundations, the art of constructing masonry waterfront walls seems to have lapsed after the Roman occupation until around the fifteenth century. A possible exception to this was the construction of moated castles and similar waterfronted fortifications by the Normans.

Prior to the fifteenth century riverside retaining walls tended to be of timber construction, braced from the river side[6] (see Figure 1.5). This was feasible, since the retaining wall was to form additional riverside land, and the shallow-draught vessels of the time used to beach in the shallow creeks to discharge their cargo. With the advent of deeper-draught vessels, and the need to provide a more substantial and durable structure, the riverside walls became both retaining walls and wharves (see Figure 1.6).

Figure 1.6 Baynard's Castle dock, London, showing timber rubbing posts which protected vessels from damage by the stone wall
Source: Council for British Archaeology Research Report No. 41. Figure 13, page 15. Courtesy of 'Museum in Docklands, PLA Collection'.

The earliest types of masonry quay wall and breakwater were built of blocks. Later, quay walls were formed from rubble masonry faced with ashlar or brick. As the stones in rubble masonry are irregular, the performance of the structure was dependent on the quality of the mortar, which was much thicker than in an ashlar or brick wall.

Properties required

The main properties required of materials to be used in gravity waterfront walls are durability, density and the ability to bind together. High tensile and compressive strengths are generally not important due to the design of such walls. An important exception to this is the design of the counterfort (see below).

A gravity wall is usually designed so that under normal loading there is no tension in the structure and the bearing stress diagram throughout the depth of the wall is triangular (see Figure 1.7). Thus the maximum compressive stress for a uniform width is only twice the average stress arising from the weight of the wall materials above the level considered. For a wall with a height of 10 metres the maximum compressive stress would be about 0.50 N/mm^2, or less if much of the wall is underwater. Another factor limiting the maximum compressive stress in the wall itself is often the strength of the soil under the foundations.

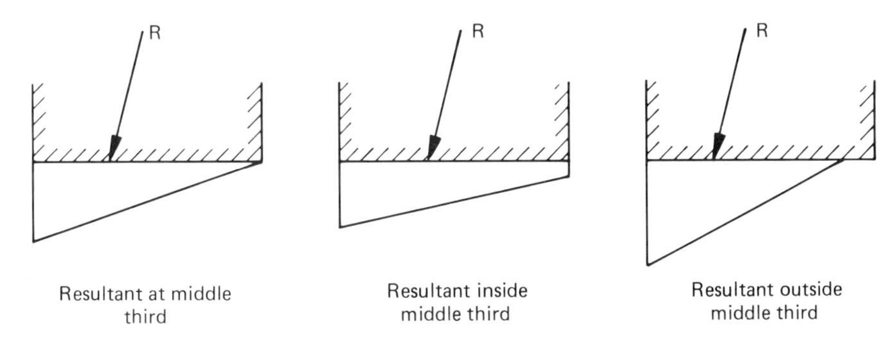

Figure 1.7 Bearing stress distribution under a waterfront wall
Source: Livesey Henderson.

The composite construction

The principal materials used in gravity wall construction are:

1. brick;
2. stone;
3. mortar;
4. concrete;
5. timber.

Steel and iron were occasionally used as inserts in walls in such places as a projecting toe or for anchoring bollards. There are also a number of examples of chains and tie rods being used to tie back temporary works, and then built into the main body of the wall. Materials are frequently used in combination, a typical example being a wall with a stone facing on the outside and rubble concrete on the inside. Thus the outside appearance of the wall is not necessarily a guide to what is inside. Indeed one of the main problems in analysing walls is identifying the materials within which cannot be seen. Facing stonework is often of a different type to that inside the wall and the cope is frequently of a different type again.

The designers of some of the old waterfront walls used materials of different densities in an attempt to improve stability and to reduce maximum working stresses in the structure and on the ground. Materials were also varied or voids left to reduce construction costs.

Brick

Where bricks have been used in the construction of waterfront walls their longevity is mostly determined by their resistance to alternate wetting and drying, freeze/thaw cycles and abrasion. It should be noted that old bricks from the same source will vary considerably in their properties due to variations in the quality of the raw materials and the firing of the bricks. This accounts for the enormous variation in the degree of degradation found in adjacent bricks in old structures. It is interesting to note that the British Standards Institution only started specifying brick durability in the 1970s. A review of the types of brick and stone found in old masonry structures is given in Sowden.[8] The types of brick are also summarized in Section 3.1 of CIRIA Report 111.[9]

Stone

Natural stone is one of the most important materials in waterfront walls and there are few walls which do not have some stone in their composition. Knowledge of the use and behaviour of stone is thus a key factor in the understanding of wall problems and the evaluation of a wall's condition.

A number of points are worth noting here:

1. since stone is a natural material the quality of stone from a particular quarry will vary and thus the rate of deterioration of individual stones will vary;
2. in waterfront structures the types of stone most likely to be durable are those of high density and low porosity. This type of stone is less easy to bond with mortar;
3. stone of high density and low porosity has a high compressive strength but is susceptible to crystallization deterioration and does not necessarily have a high modulus of rupture.

The following notes regarding the use of stone in waterfront walls may be helpful:

1. granite facing was particularly favoured for facing sea walls which were subjected to abrasion and wave attack;
2. in good masonry great care was taken to ensure that ashlar facing blocks were keyed into the structure to prevent stones becoming separated from each other from the core (see Figure 1.8[10]);
3. stone from the excavation was sometimes

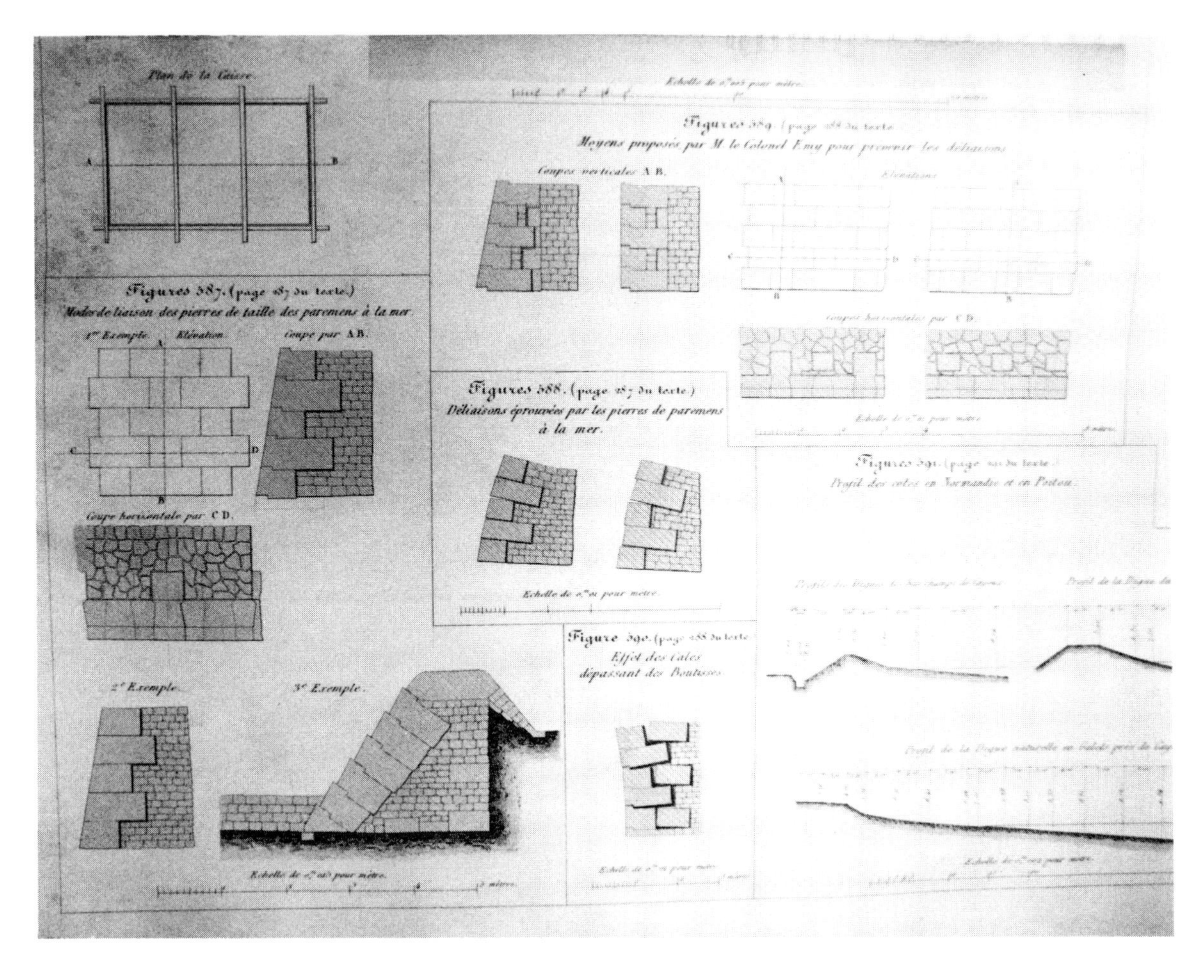

Figure 1.8 Illustrations from a French treatise on civil engineering (1839), showing methods of interlocking blocks and the effects of not doing this
Source: Sganzin, J.M., and Relbell, F., (Programme ou resumé des leçons d'un cours de construction, avec des applications tirées specialement de l'art de l'ingenieur des ponts et chaussées, Paris, 1839.) Plate 115, figure 587 to 590 inclusive.

used for the core of the wall, and in at least one instance[11] was used to form the ashlar facing;

4. because cement was expensive, stone was often used in the core of a wall in place of concrete. There are many examples in which stones were hand-placed (labour was cheap) in the core of a wall and then grout was inserted into the voids;
5. large stones were also used between concrete pours to increase the bond and improve resistance to sliding (see Figure 1.9);
6. where large preformed blocks of stone or concrete were used for the facing to a wall, the internal portions of the block were made slightly under-size to ensure that no open joints were left after construction.

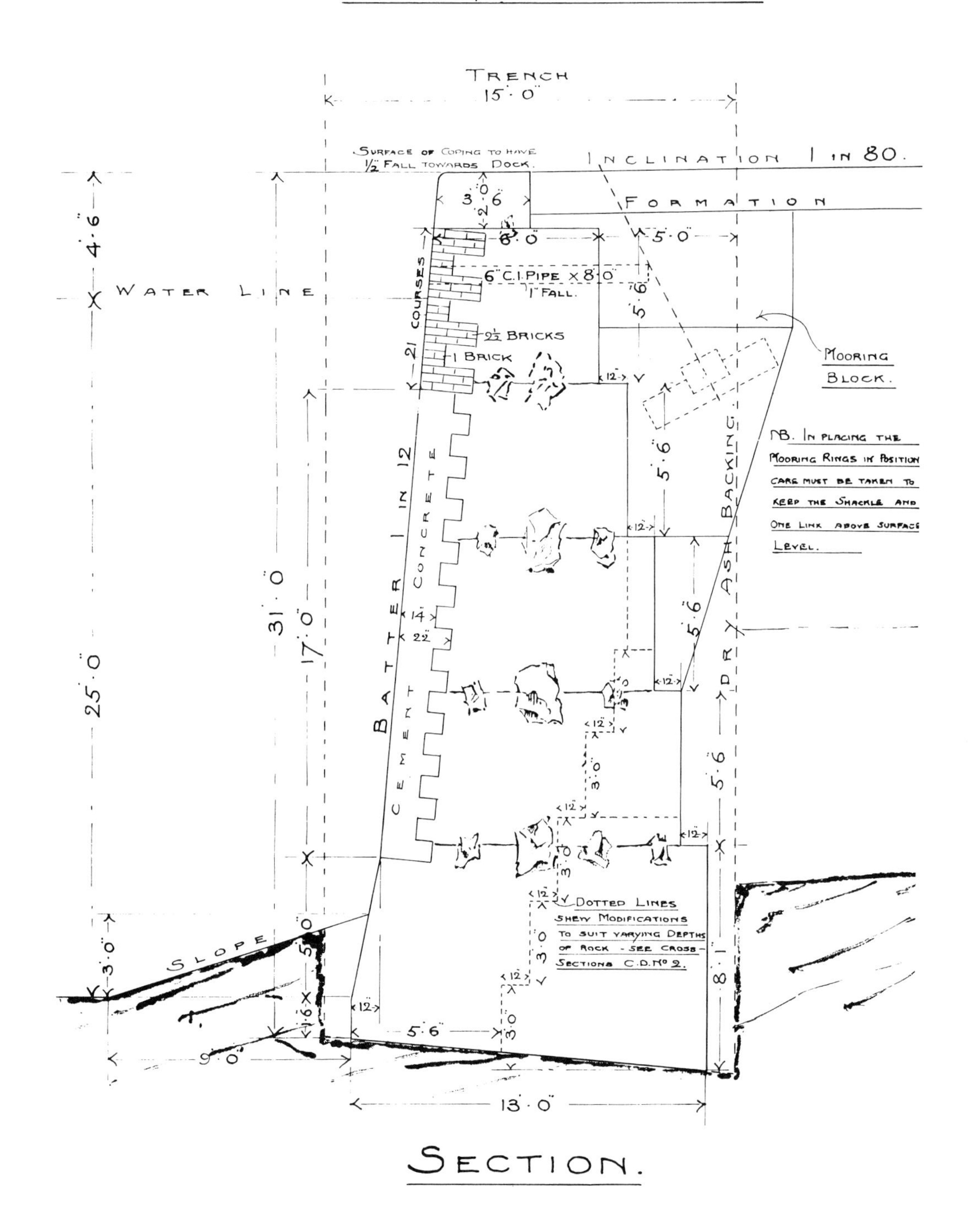

Figure 1.9 Section of dock wall, Sharpness, circa 1880
Source: British Waterways Archives, Gloucester.

Mortar

Although the Romans used lime and pozzalana to make mortar and concrete, reliable methods of cement manufacture for use in structures were later forgotten. John Smeaton improved the performance of mortar in the 1750s. Smeaton used mortar made with lime and pozzalana in the construction of the Eddystone Lighthouse in 1758 and this mortar was still in good condition in 1882 when the lighthouse was dismantled.[12] An example of a mortar specification[13] used in the construction of the Gloucester and Berkeley canal in 1795 is given in the box below.

Portland cement manufacture began in 1824 but lime mortar continued to be the norm until 1900. Between 1900 and 1930, cement mortars became widely used, although lime is still used with Portland cement today to improve the plasticity of mortar.

Mortar specification for the construction of the Gloucester & Berkeley canal (1795), Public Record Office, Kew

'All the Ashlar stone shall be set for six inches in depth from the face inward in Tarris mortar composed of one third part of Aberthaw or Chepstow lime, one third of good pozzolana Earth from Italy or Dutch tarris and one third of clean fine well washed pure River Sand divested of all muddy particles and gravel. And all other parts of the joints of the said Stone work shall be laid in Mortar composed of Clay stone or Bristol Lime one half lime and the other half of well washed clean and pure River or pit sand screened from all gravel or small stones.'

The lowest standard of brickwork with lime mortar in the eighteenth and nineteenth centuries[9] has a permissible compressive stress of 0.42 N/mm^2. Recent tests[14] gave an average strength of 0.96 N/mm^2 for a column of bricks and mortar cut from a building where the working load was 0.45 N/mm^2. Although this gives a factor of safety below that recommended in BS 5628 : Part 1 (*Structural use of unreinforced masonry*),[15] the building showed no signs of distress. (A working stress of 0.45 N/mm^2 is similar to that discussed earlier in this section as likely to exist in a gravity wall 10 metres high.)

Concrete

Concrete was used by the Romans for marine structures and there are reports that they used precast concrete blocks for the construction of breakwaters. However, like many of the other constructional skills which the Romans enjoyed, the knowledge of how to make good concrete was later lost.

Concrete was reintroduced as a material for use in dock works and other maritime structures in the first half of the nineteenth century. It was not considered to be particularly durable, and thus in the early walls it was seldom found in the outer faces. By 1850, although Portland cement was being exported from England to France in large quantities for use in harbour construction, it was not being used on a large scale in the UK.

As cements improved, concrete gained in popularity and by the turn of the century was in wide use in harbour engineering. Extensive details relating to the development of cement and use of concrete may be obtained from the British Cement Association.

Timber

Although this book defines old waterfront walls as being masonry or concrete structures, it is rare to find a wall in which timber does not feature, and the study of its use in wall construction is an important aspect of waterfront wall technology.

Timber is normally found either in the foundations or as an insert in the body of the wall.

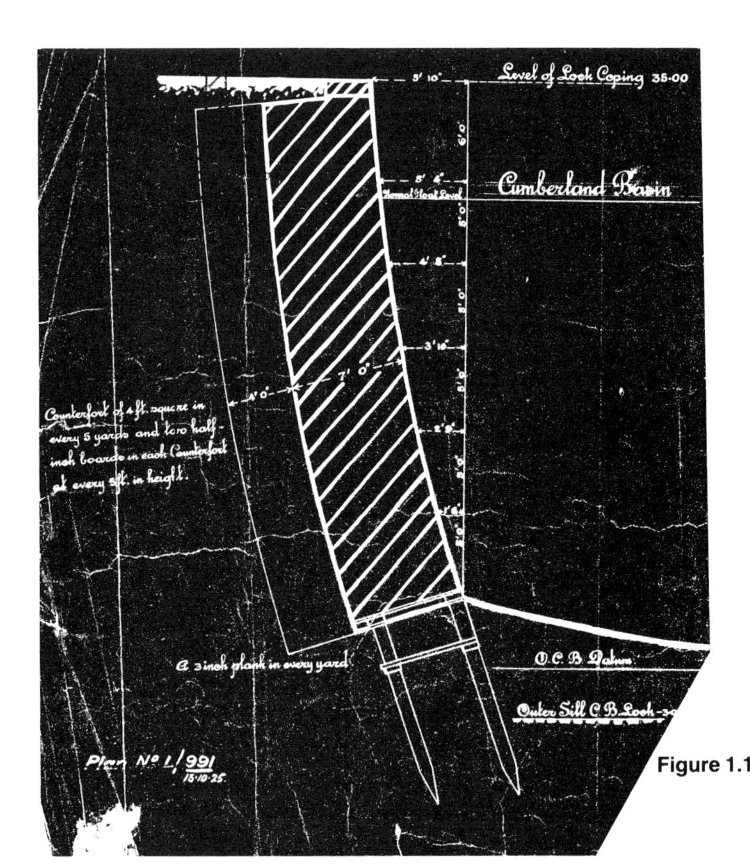

Figure 1.10 Section showing counterfort strengthened by a timber tie, Cumberland Basin, Bristol, (William Jessop) 1793
Source: Bristol City Council.

In some cases the timber was used as a former around which the wall was built and in others it was used as a tension member, such as might be required to strengthen the upper part of a counterfort (see box), or to tie two walls together (see Figure 1.11). It is thus not unusual to find timber in the core of a wall.

Because of the ability of timber to survive for long periods when deprived of oxygen it is timber immersed in water or buried in soft foundations which still plays an important role in wall stability. Chrimes[16] gives a detailed account of the development of piled foundations as would have been used for wall and bridge foundation work.

Example of timber in a dock wall, Bristol

An example of the use of timber in a wall is shown in Figure 1.10 which shows the cross-section of a quay wall designed by William Jessop in 1793 for the Cumberland Basin at Bristol. The wall is supported on timber piles. The note about the counterforts says 'two half inch boards in each counterfort at every 5 ft in height'. This may have been intended to provide tension reinforcement between the counterfort and the body of the wall. The note at the bottom of the figure, 'A 3 inch plank in every yard', may refer to the timbers joining the piles.

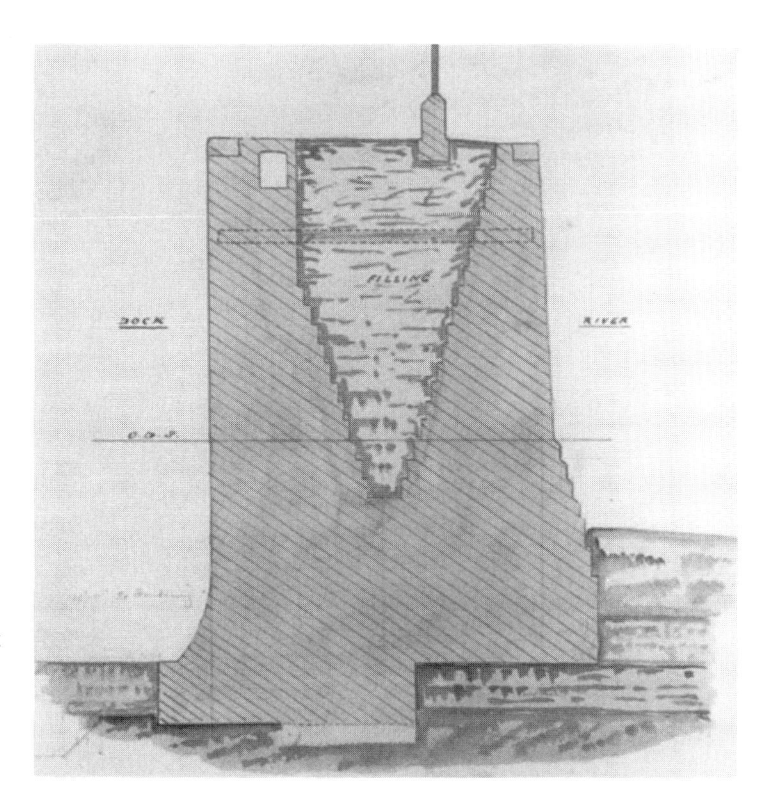

Figure 1.11 Section showing river and dock wall connected by a timber tie, Sandon half-tide dock, Liverpool, 1901
Source: Mersey Docks and Harbours Company.

1.2 Quay, dock and lock walls

1.2.1 Function and design

The main components of a quay wall are illustrated in Figure 1.12. Essentially the structure is designed to provide a vertical, or near vertical, front face against which vessels may lie.

The top surface is normally flat, allowing unimpeded access to the edge of the structure. Immediately to the landward side there is nearly always an open area, the apron, where cargo was handled on to and off the vessels. The apron is supported on the backfilling and usually falls slightly towards the top of the wall to facilitate drainage. The back face of the wall is rarely vertical, the wall usually being widened with depth, and often exhibits features to increase the soil friction on the wall.

The base of the wall may be founded on the natural soils or rocks existing at that level or may rest on other materials which have been placed under the base to spread or transfer the load to a lower founding stratum. Some bases contain features to improve their resistance to sliding and many are extended forwards to provide a toe. The toe may act as an anti-scour device as well as increasing the resistance to overturning.

The whole structure is designed to resist the forces imposed by vessels alongside, the pressure exerted by the backfill, hydraulic pressures, surcharge loads on the top surface and its own self-weight. In many cases the internal portions of the wall are built of a different material from the outside shell, for reasons of economy, and the backfilling material may be specially chosen for its geotechnical properties. These components are described in more detail below.

One point of note is that there may be considerable variety in the designs of dock walls

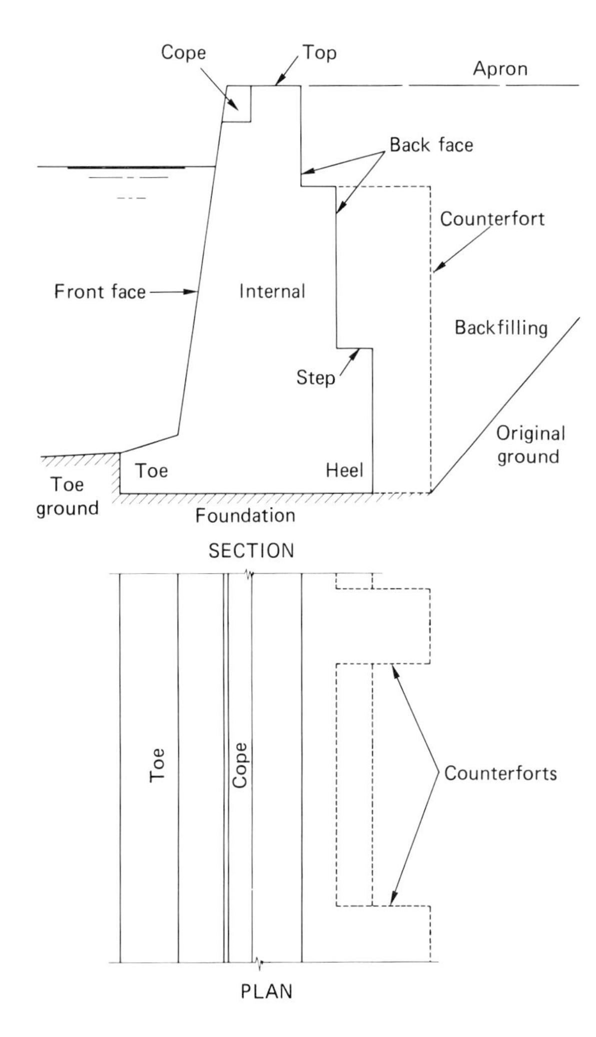

Figure 1.12 The main components of a quay wall
Source: Livesey Henderson.

in one location. Such places as London, Leith, Belfast, Liverpool and Gloucester (see Figure 1.13) each contain examples of many radically different wall configurations. Thus the design of one wall in a port may have little bearing on that of the other walls in the area.

1.2.2 Component characteristics

Front face

The front face of a wall in this category may be vertical (Figure 1.14), slightly inclined (Figure 1.15), or of curved profile (Figures 1.16 and 1.17). The latter is often accentuated towards the toe, the non-vertical shape being functionally acceptable because of the tendency of old vessels to be similarly shaped in cross-section. In locks and narrow docks this curvature is often extended into the base of the lock or dock to form a structurally efficient inverted arch (Figure 1.18) which transfers the hydrostatic uplift to the walls and acts as a strut between the walls. In later years the upper face of a wall was often built of stone and proud of the lower face, to ensure that the berthing, and other damaging forces, were applied to the more durable material (see Figure 1.19).

In a number of cases the front face of the wall is discontinuous at the lower level due to the presence of an arched construction similar to a railway viaduct (Figure 1.20), the arches being closed at the back (sometimes arched in plan – see Figure 1.21). The arches are usually of short span but occasionally longer spans, such as 12 metres (Figure 1.22), are found, and the depth of the arch may be considerable (see Figure 1.23). The use of arches in port construction dates back at least to the Roman era.[17] The front face of a lock wall may also have apertures for sluices, draw shafts and the like.

Back face

The back faces of waterfront walls form the link between the wall and the retained material behind. Being functionally unconstrained with respect to their shape they have been the subject of a variety of methods to improve wall stability.

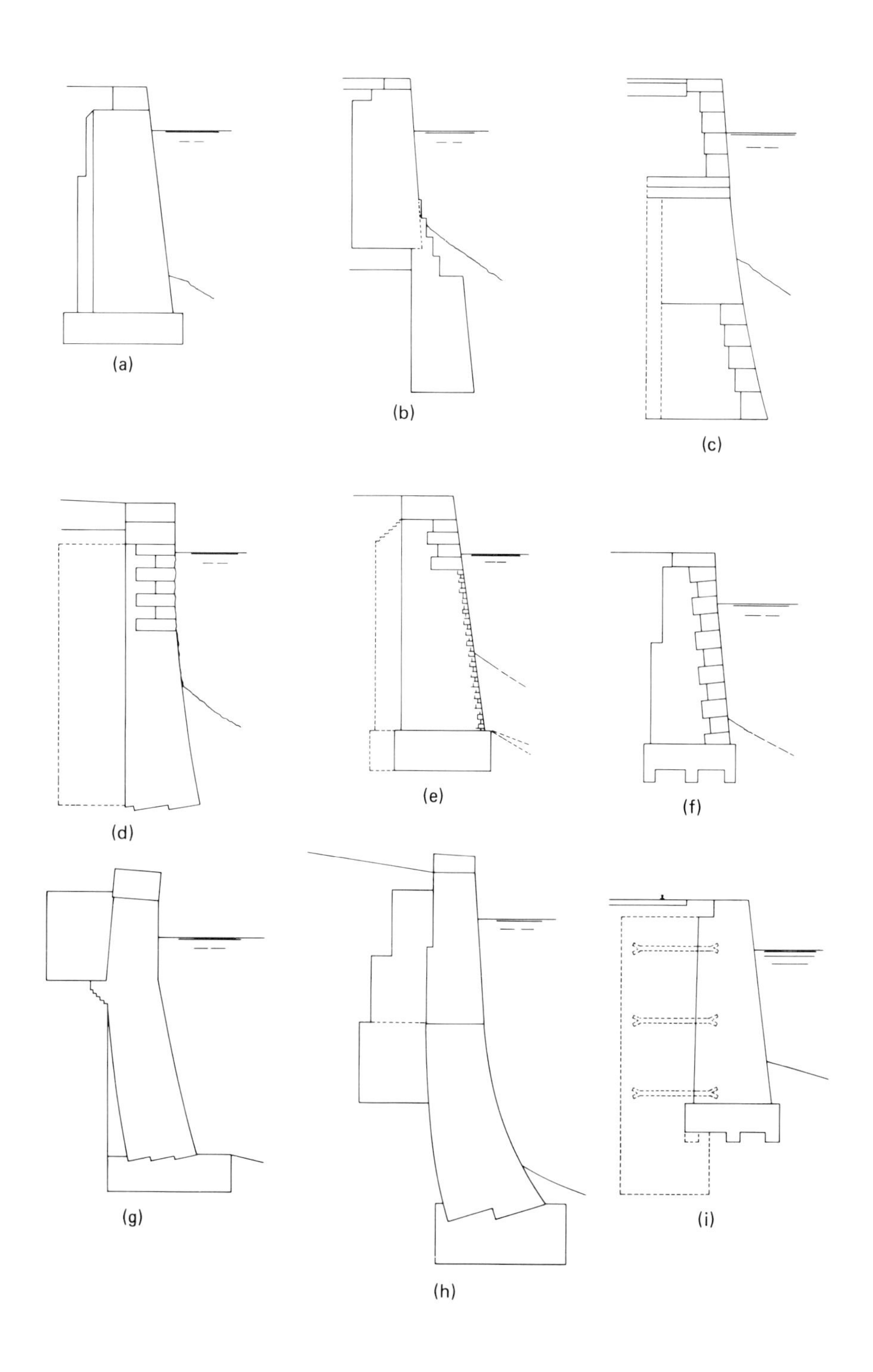

Figure 1.13 Sections of the walls in Gloucester Docks
Source: Livesey Henderson, British Waterways.

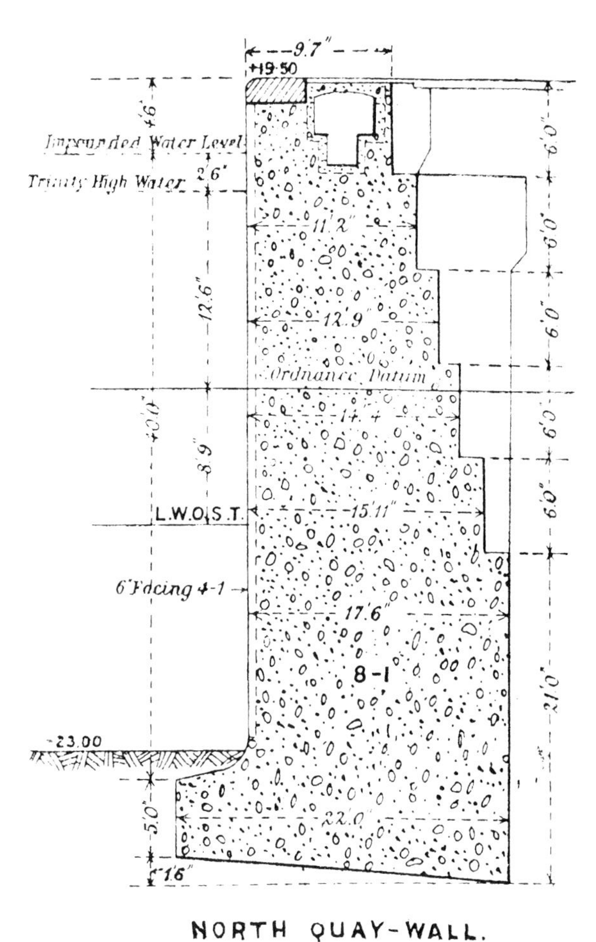

Figure 1.14 King George V Dock, London, 1921
Source: Binns, A., 'The King George V Dock, London', ICE Proc. Vol. 216. April 1923, plate 7, figure 3.

Figure 1.15 Portsmouth Dockyard tidal basin extension, 1864
Source: Vernon-Harcourt, L.F., *Harbours and docks*, Vol. II, 1885, plate 14, figure 9, Clarendon Press.

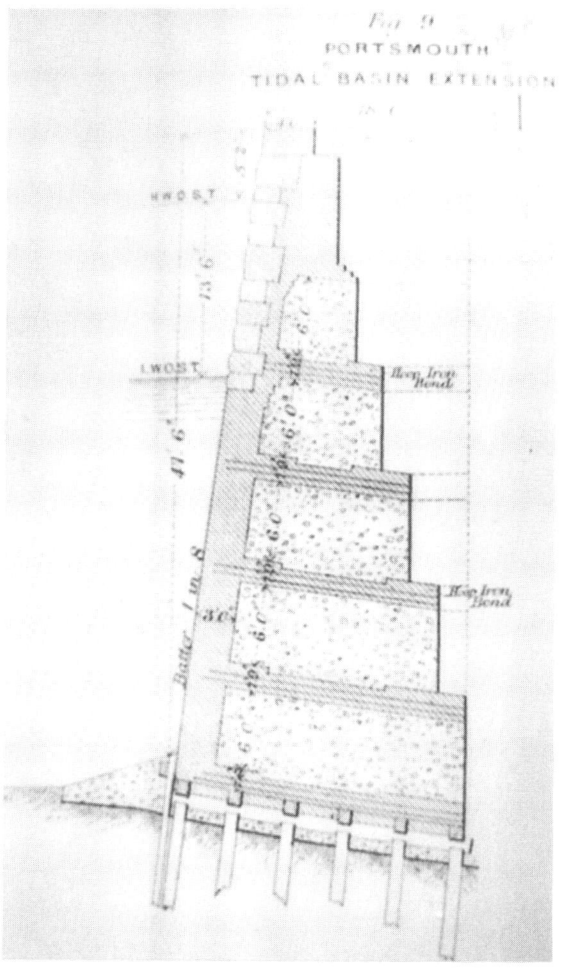

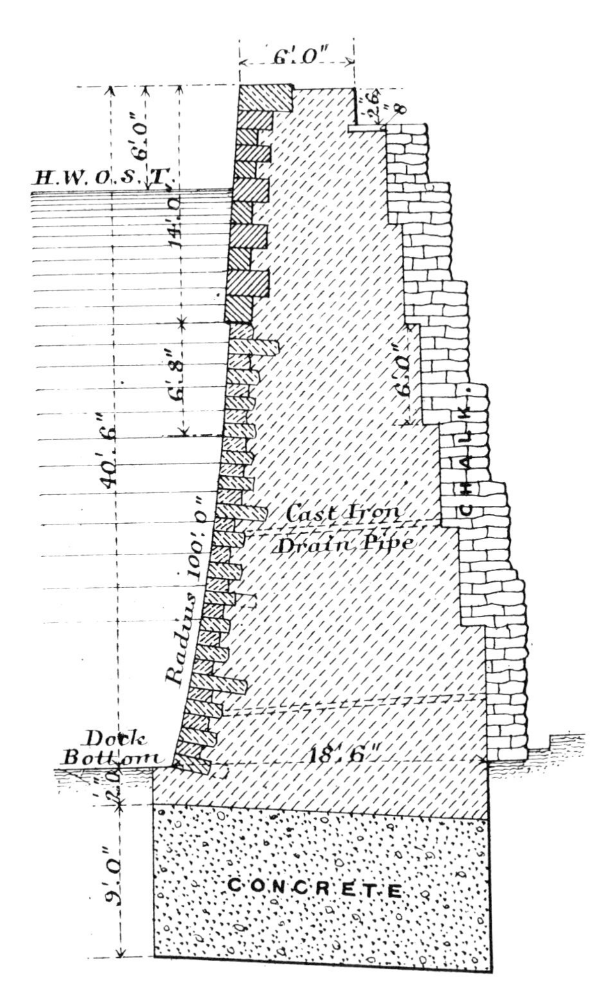

Figure 1.16 The Alexandra Dock, Hull, 1887–88
Source: Hurtzig, A.C., 'The Alexandra Dock, Hull', ICE Proc., Vol. 92, Part III, 1888, plate 2, figure 6.

Figure 1.17 The Royal Albert Dock, London, 1880
Source: Stevenson, T., *Design and Construction of Harbours*, 1886.

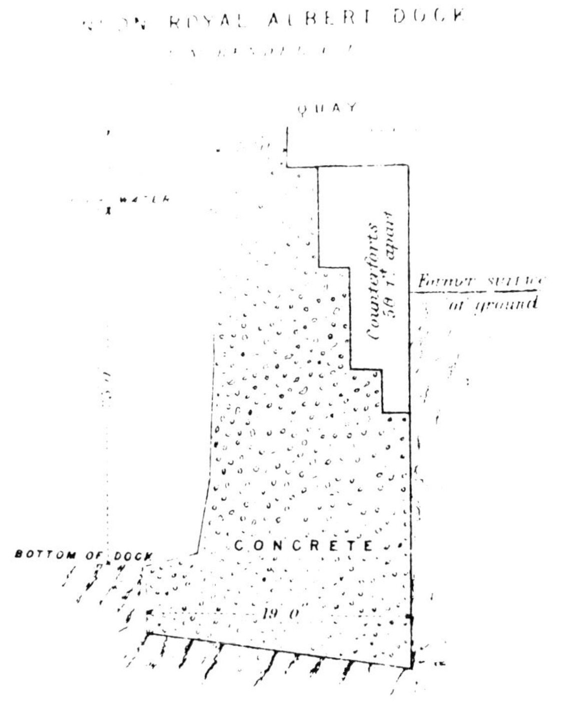

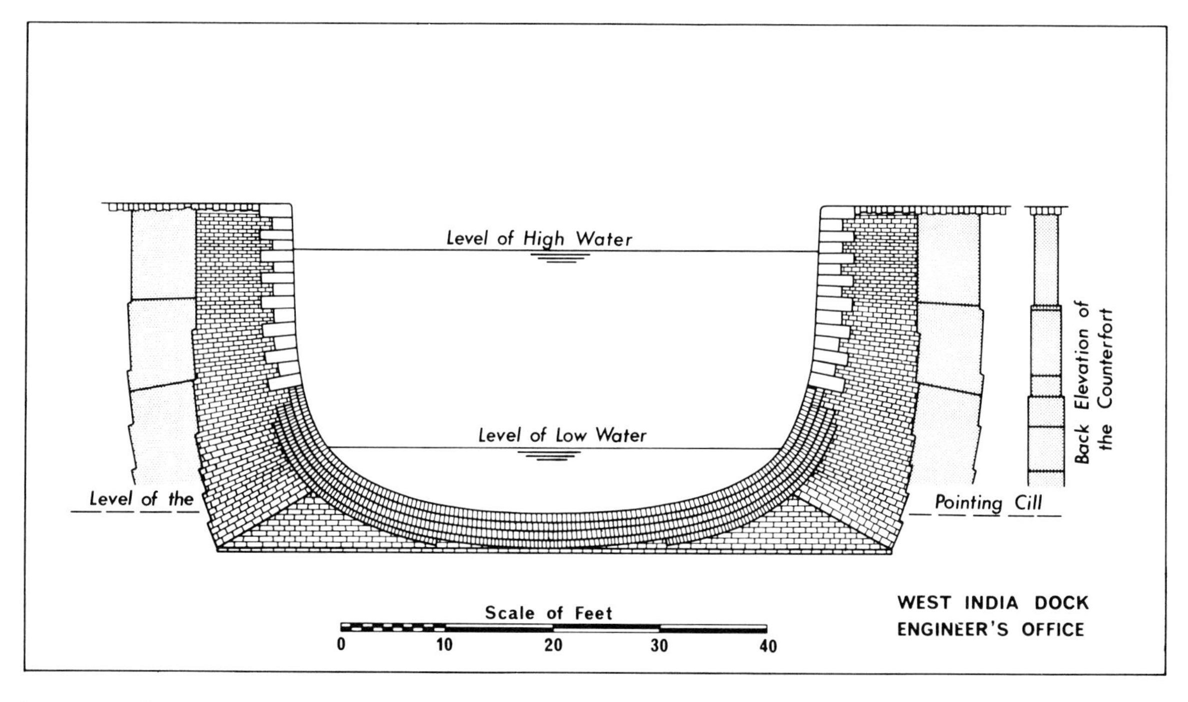

Figure 1.18 Section of the chamber of the entrance lock at Blackwall, West India Dock, 1800–1806
Source: Hadfield, C. and Skempton, A.W., *William Jessop, Engineer*, David and Charles, 1979.

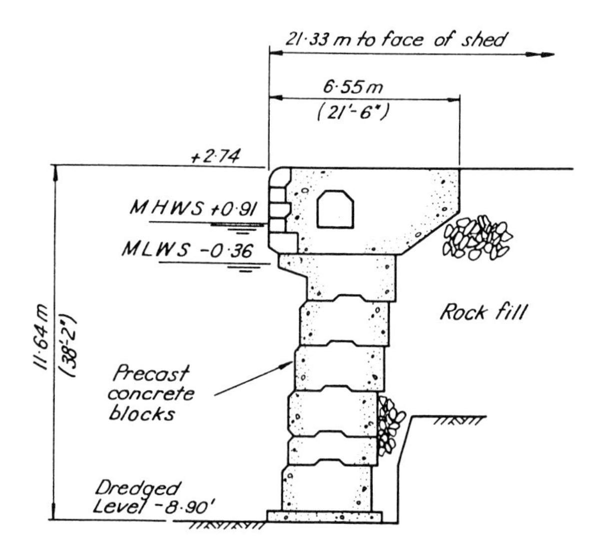

Figure 1.19 Quay wall at Tema, Ghana, 1960
Source: National Ports Council, *Research project on port structures*, Vol. 2, 1969.

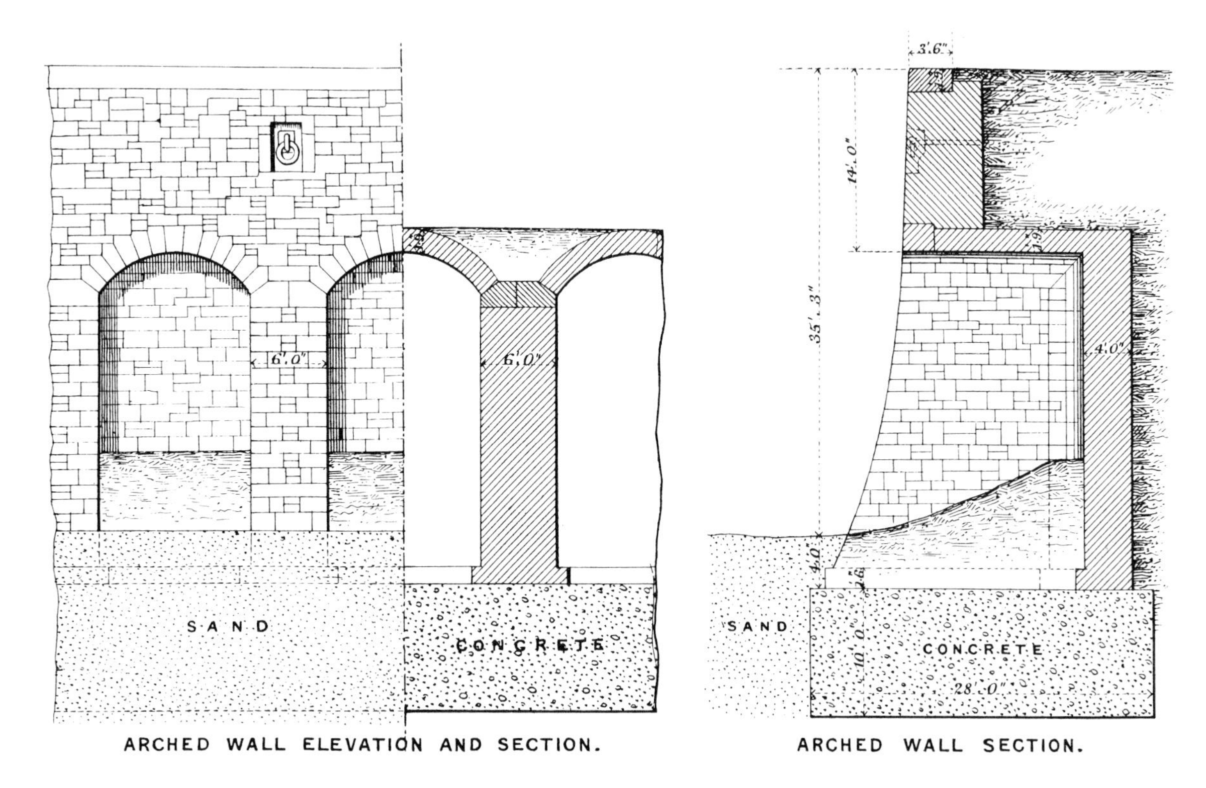

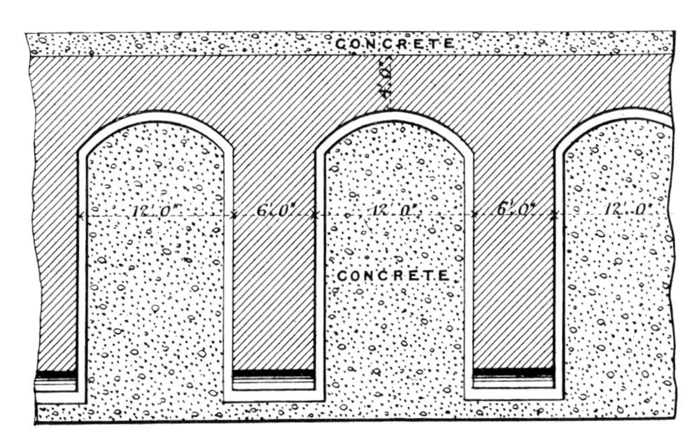

Figure 1.20 The arched wall of the Albert Dock at Hull, 1861
Source: Hawkshaw, J.C., 'The construction of the Albert Dock at Kingston upon Hull', ICE Proc., Vol. 41, 1875, plate 9, figures 1, 2 and 3.

Figure 1.21 Photograph of the arched dock walls in the Canada Dock, London
Source: Courtesy of 'Museum in Docklands, PLA Collection'..

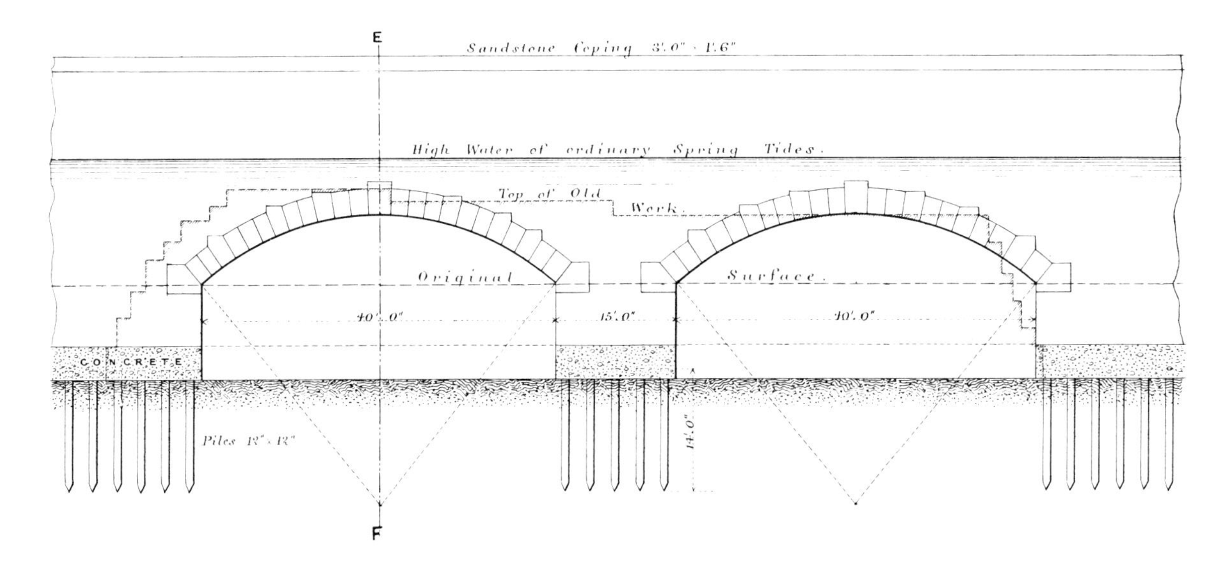

Figure 1.22 The arched dock wall at Whitehaven
Source: Williams, T. E.,'Whitehaven harbour and dock works', ICE Proc., Vol. 55, 1879, plate 4, figure 7.

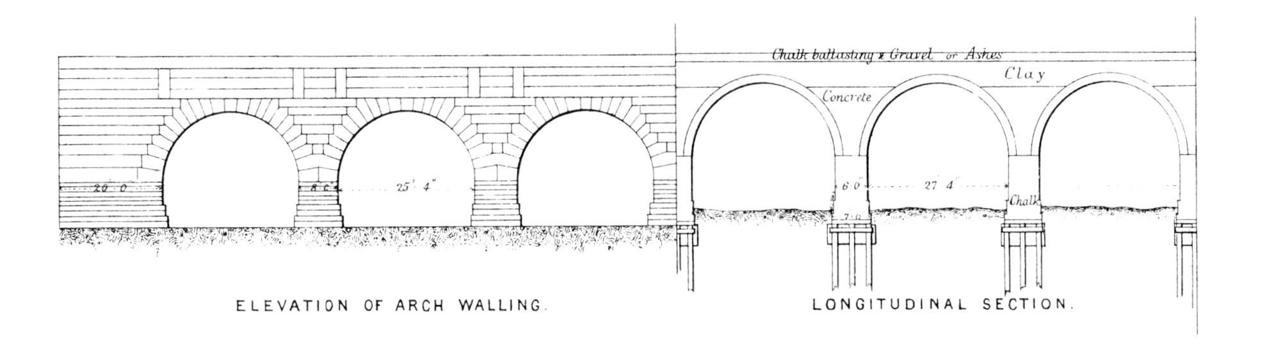

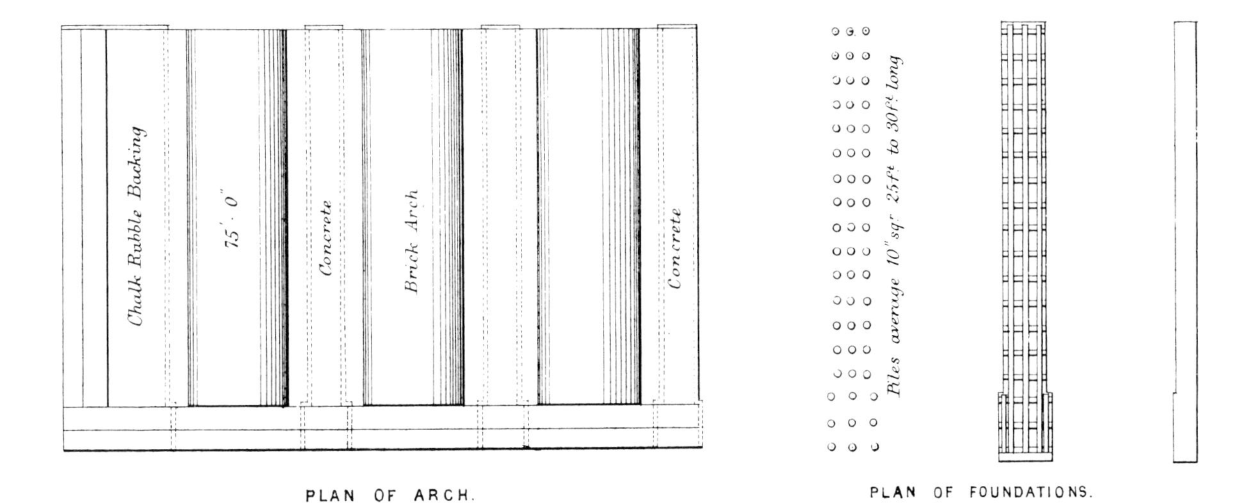

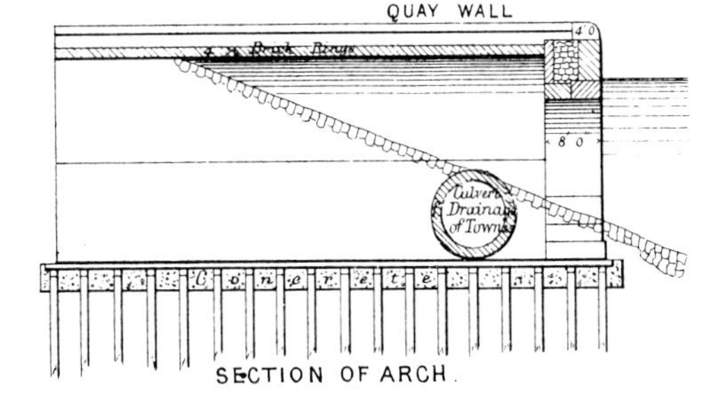

Figure 1.23 Plan and elevations of the arched wall at the Great Grimsby (Royal) Docks, 1864–65
Source: Clark, E.H., 'Great Grimsby (Royal) Docks', ICE Proc., Vol. XXIV, plate 2A, figures 4, 5 and 6.

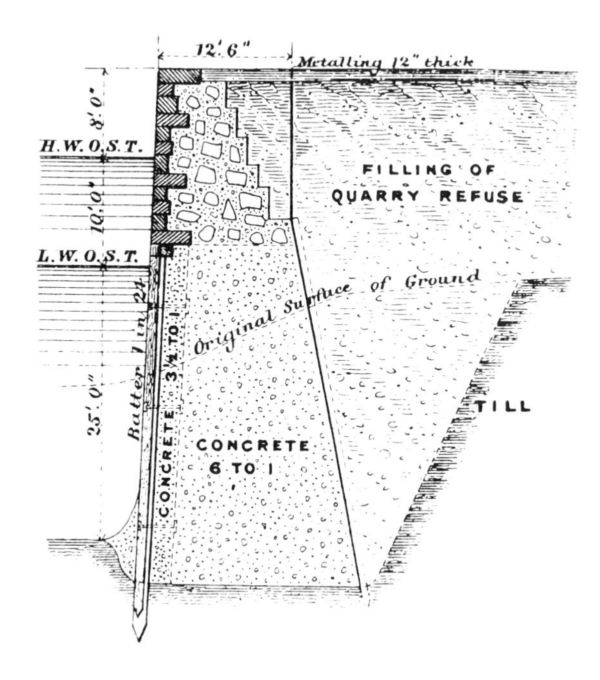

Figure 1.24 Western tidal harbour wall at Greenock, 1878–86
Source: Kinniple, W.R., 'Greenock Harbour', ICE Proc., Vol. 130, plate 7, 1897, figure 4.

A simple method of increasing the resistance of the wall to overturning is to widen the structure towards the base by sloping the back face (Figure 1.24). However, a more common and more effective way of achieving this is to step the back face (Figure 1.25), which has the added advantage of increasing the friction between the wall and the backfilling material. Another method of increasing this friction on vertical or near-vertical back faces is by means of oversails (Figure 1.26), a method much favoured by Brunel.

A common method of improving the stability of relatively slender walls is the counterfort. This is a section of wall which projects perpendicularly backwards into the backfilling material. Counterforts are placed at regular intervals along the back of the wall. They may be the same height as the wall (Figure 1.27) or smaller (Figure 1.17) and may be stepped or sloping (Figures 1.28 and 1.29). Some counterforts were used as piers to support relieving arches.

The counterfort not only acts as a counterweight against the overturning forces of the backfill, but is also intended to harness additional soil resistance. The main disadvantage of the counterfort is that it is prone to tension failure at the point where it joins the main wall and its effect on increasing frictional grip within the backfill is uncertain.

Counterfort walls were a popular form of design from the 1700s to about 1860. A few counterfort walls were built in the late 1800s. A large number of these walls still survive. Masonry counterfort walls were quite different from the modern reinforced concrete counterfort walls: the former have only local piers added to the back of a wall whereas the latter have a base slab between the counterforts so that the weight of the backfill on the base increases the resistance to sliding and overturning.

Where a wall is constructed abutting a soil or rock formation which is naturally stable and will form a vertical face, the back face of the wall is often keyed into this material (Figure 1.30): see also 'skin walls' (Section 1.6). If the rock level varied across the site the section of the wall was often also varied to take advantage of rock where it occurred.

In a number of early cases, such as at Sheerness, the back face of the wall is curved, parallel to the front face (Figure 1.31). This design has also been found in London, Bristol, Gloucester and Leith. Notice the stepped heel in the Sheerness example. A modern version of the backward-leaning wall is shown in Figure 1.19.

Base

Apart from forming the interface between the foundation to the wall and the main structure of the wall itself, the base must provide sufficient resistance to horizontal sliding to prevent the bottom of the wall from kicking outwards. A number of methods have been used to try to

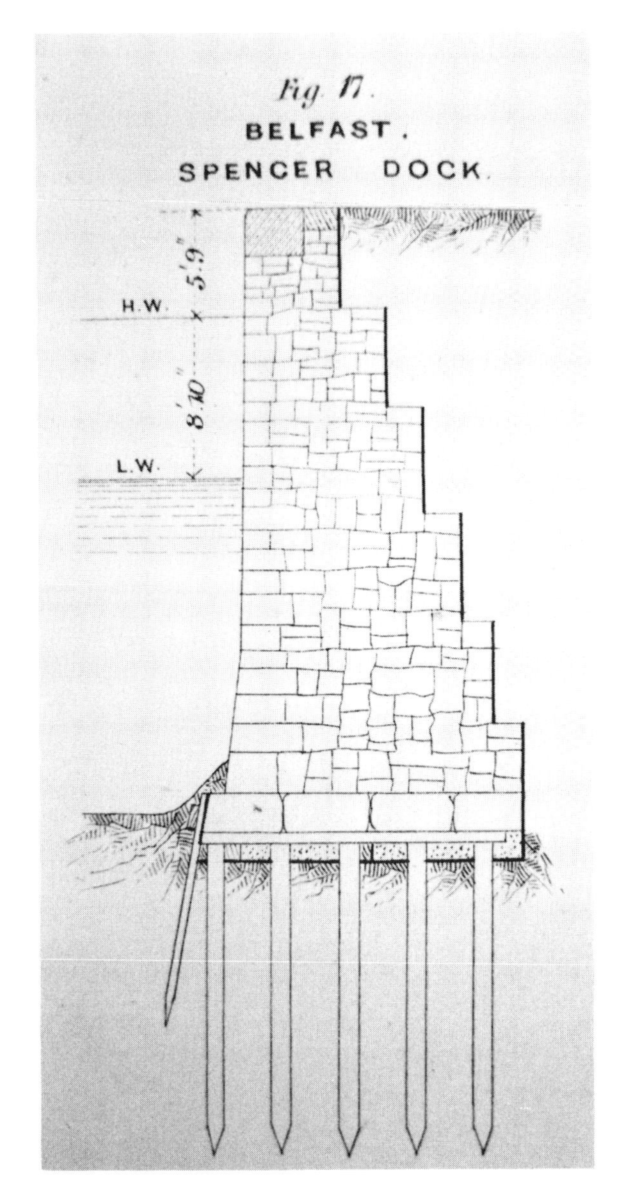

Figure 1.25 Spencer Dock, Belfast, 1872
Source: Vernon-Harcourt, L.F., *Harbours and Docks*, Vol. II, 1885, plate 14, figure 17, ICE.

increase this resistance to sliding. These generally involve modifying the shape by means of serrations (Figure 1.27), castellations (Figure 1.32), keying the wall into the ground (Figure 1.33) and sloping or stepping the base up to the front face (Figures 1.17, 1.34 and 1.35).

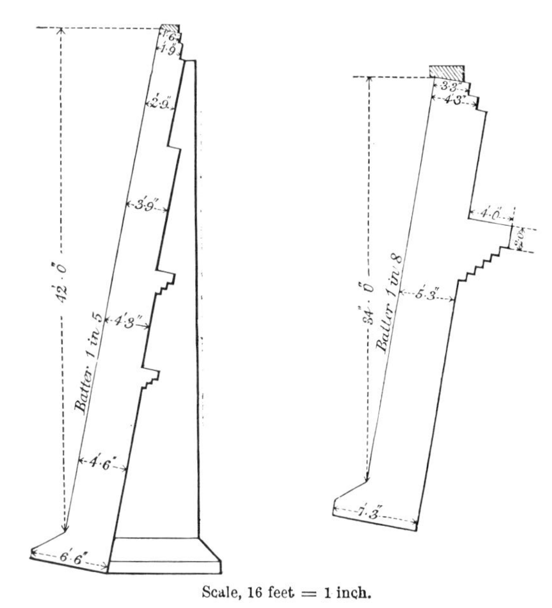

Figure 1.26 Retaining walls, showing oversails on the rear face, 1881
Source: Benjamin Baker, 'On the actual lateral pressure of earthworks', ICE Proc., Vol. LXV, 1881, Figures 48 and 49.

Sometimes intimate contact with the ground is obtained by casting concrete directly on to the ground or rock.

In some cases the method of forming the base of the structure gives added resistance, such as a wall which is founded on bagged concrete (Figure 1.36).

Toe

The toe of the wall increases its resistance to overturning (Figures 1.14 and 1.17). In some cases scour protection which is structurally separate from the wall has been provided (Figure 1.25). It is sometimes difficult to distinguish between a structural toe and scour protection. Scour protection is also often provided by metal or timber piles (Figure 1.37).

A toe which is expected to carry high loads may be strengthened by the addition of a row of toe piles (Figure 1.28) to take vertical loads, or raked piles to take some of the horizontal load (Figure 1.38).

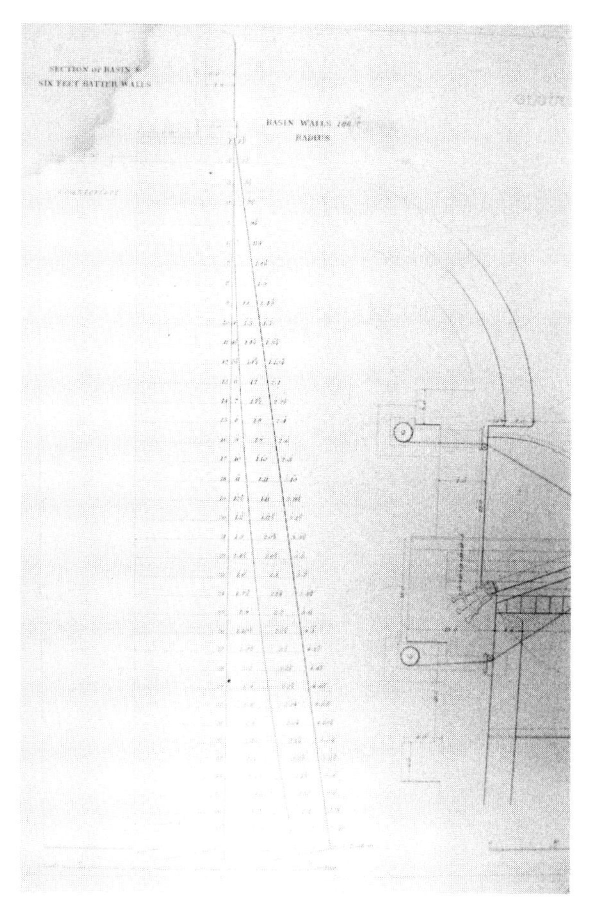

Figure 1.27 Section of wall at the old entrance, Sharpness, showing counterfort and serrated base, 1820
Source: Simms, *Public Works of Great Britain*, plate 94, ICE.

Figure 1.28 Glasson Dock, Morecombe Quay
Source: Stevenson, T., *Design and Construction of Harbours*, 1886, plate XIX, ICE.

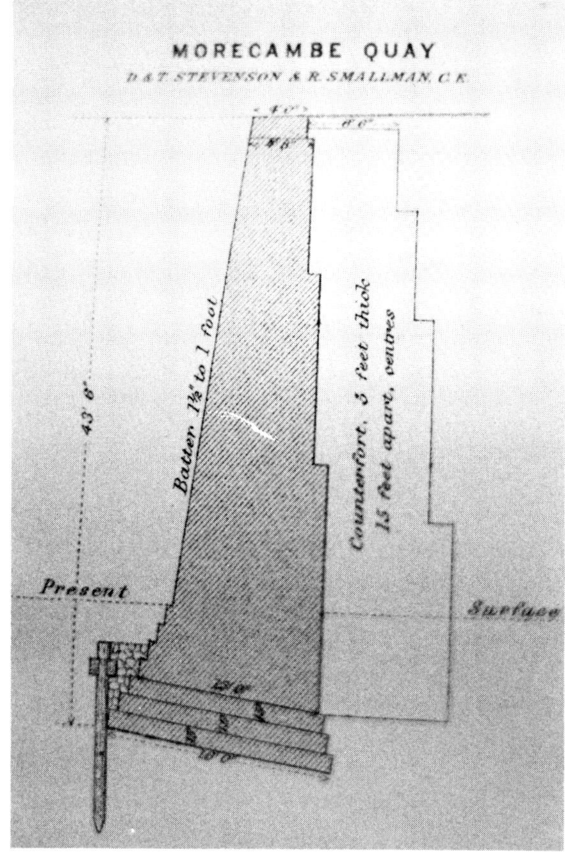

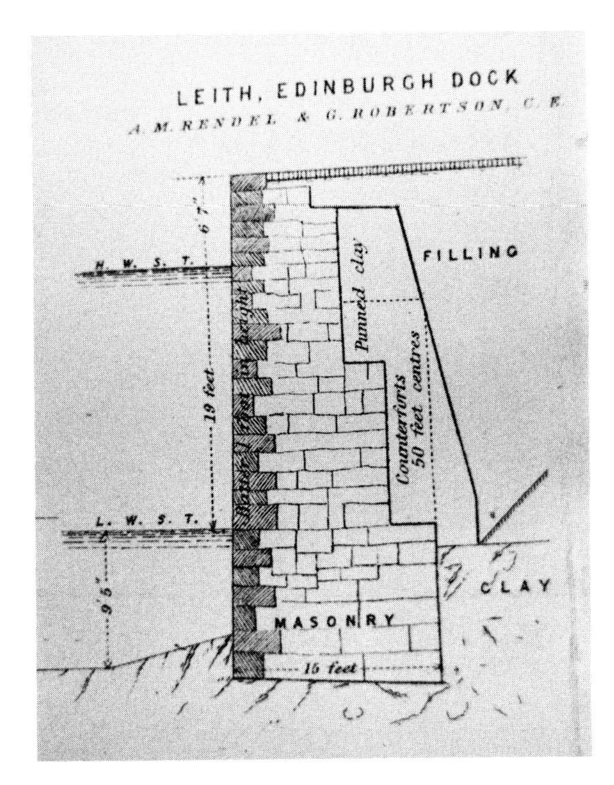

Figure 1.29 Edinburgh Dock, Leith
Source: Stevenson, T., *Design and Construction of Harbours*, 1886, plate XIX, ICE.

Figure 1.30 Cross-section of a Liverpool wall, 1898
Source: Du Plat Taylor, *Design, Construction and Maintenance of Docks, Harbours and Piers*, Eyre and Spottiswoode, 1949.

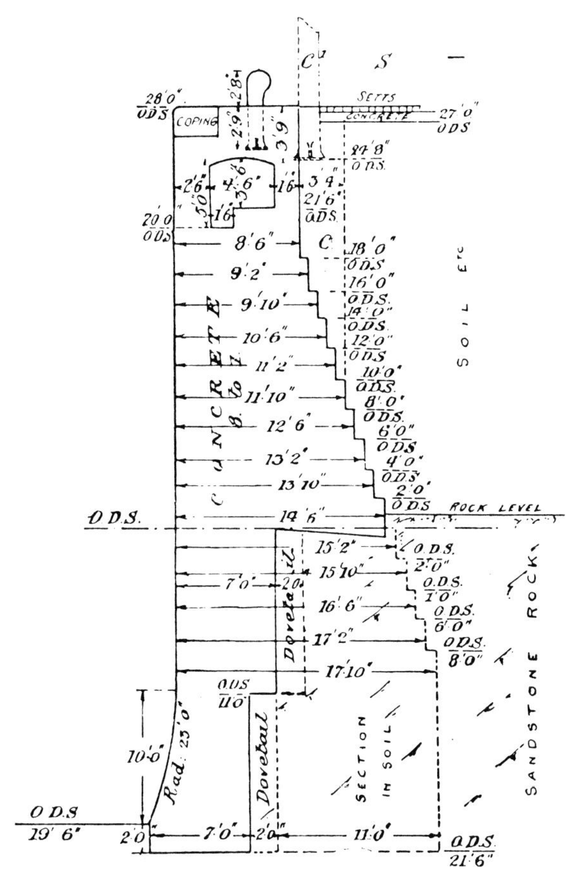

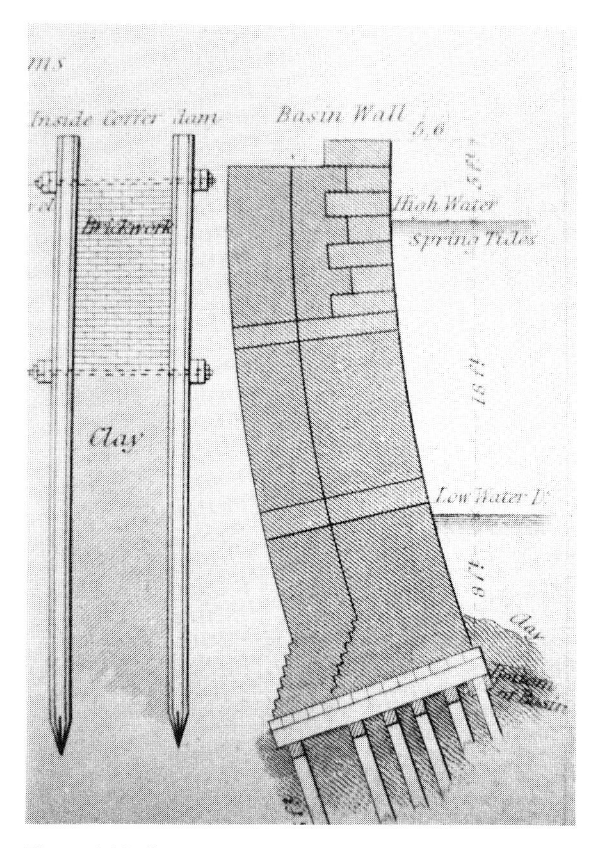

Figure 1.31 Curved dock wall with heel, Sheerness, 1813–27
Source: Rennie, J., *The theory, formation and construction of British and foreign harbours*, 1854.

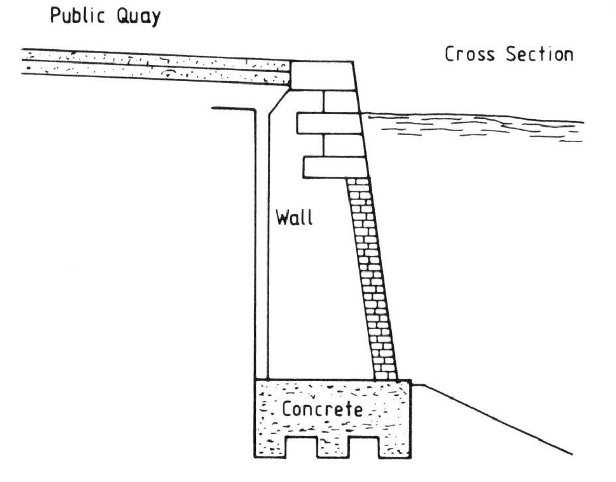

Figure 1.32 Baker's Quay, Gloucester Docks, 1829
Source: British Waterways Archives, Gloucester.

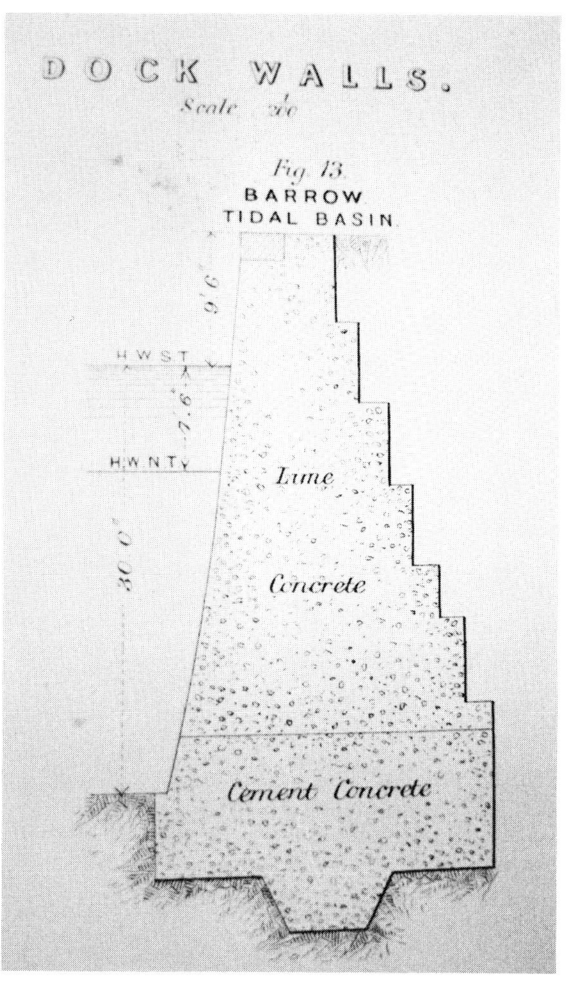

Figure 1.33 The tidal basin at Barrow, 1879
Source: Vernon-Harcourt, L.F., *Harbours and Docks*, Vol. II, plate 14, figure 13, ICE.

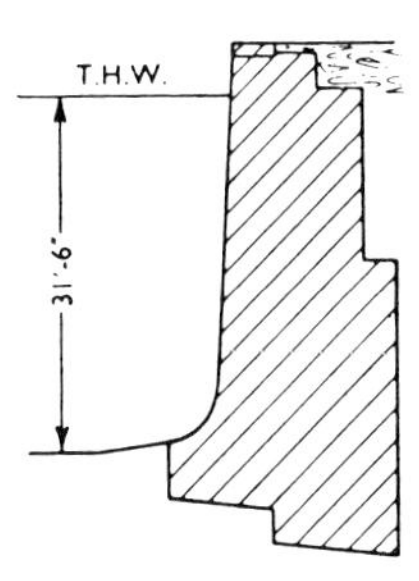

Figure 1.34 Surrey Commercial, Greenland Dock, London, 1898
Source: Greeves, I.S., *London Docks 1800–1980, A Civil Engineering History*, Thomas Telford, 1980.

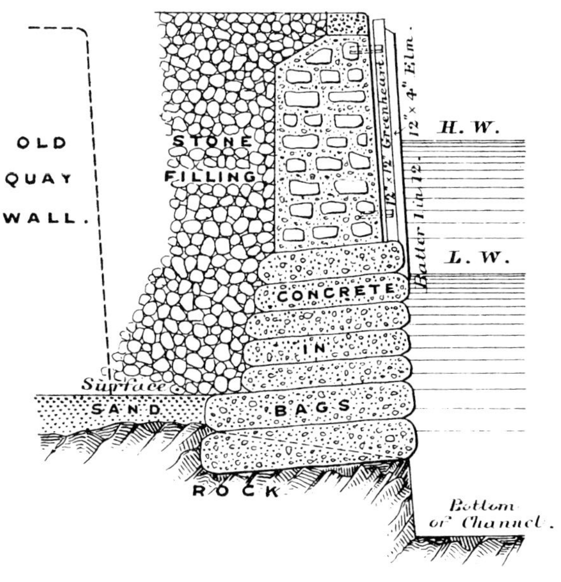

Figure 1.36 The end wall, Winton Pier, Ardrossan, 1892
Source: Robertson, R., ICE Proc., Vol. 120, 1895, plate 5, figure 9.

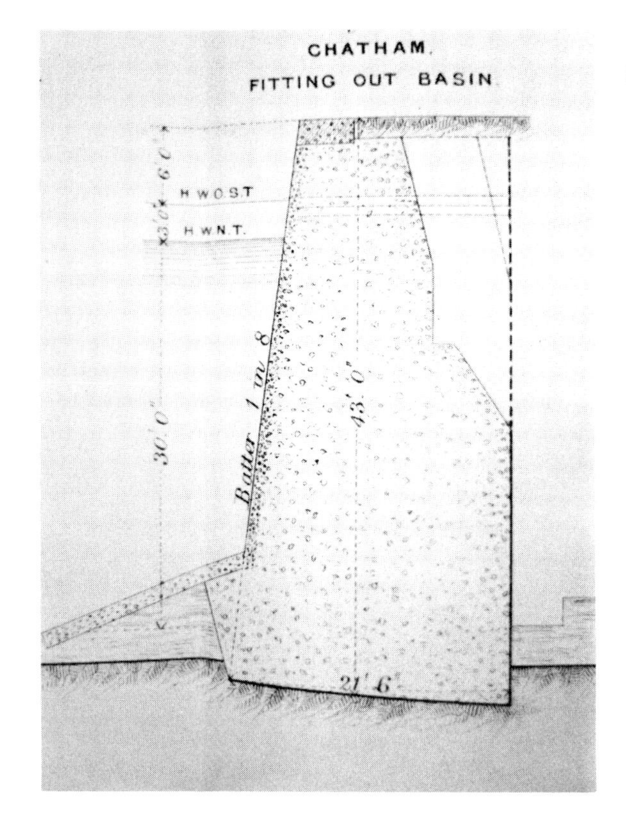

Figure 1.35 Fitting out basin, Chatham, circa 1880
Source: Vernon-Harcourt, L.F., *Harbours and Docks*, Vol. II, plate 14, figure 10.

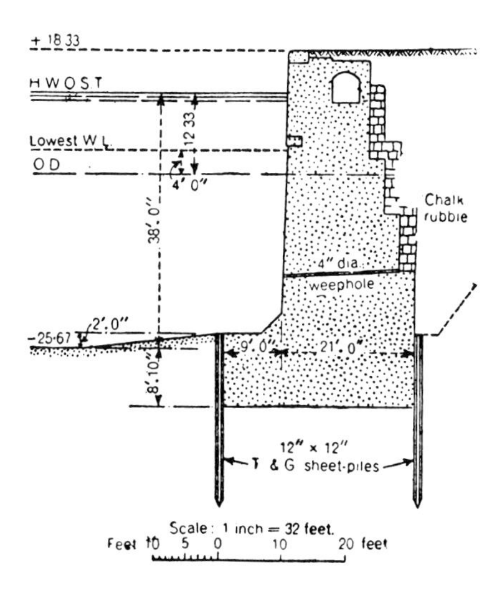

Figure 1.37 King George V Dock, Hull, built on running sand
Source: Cornick, H.F., *Dock and harbour engineering Volume I*, 1968, Charles Griffin.

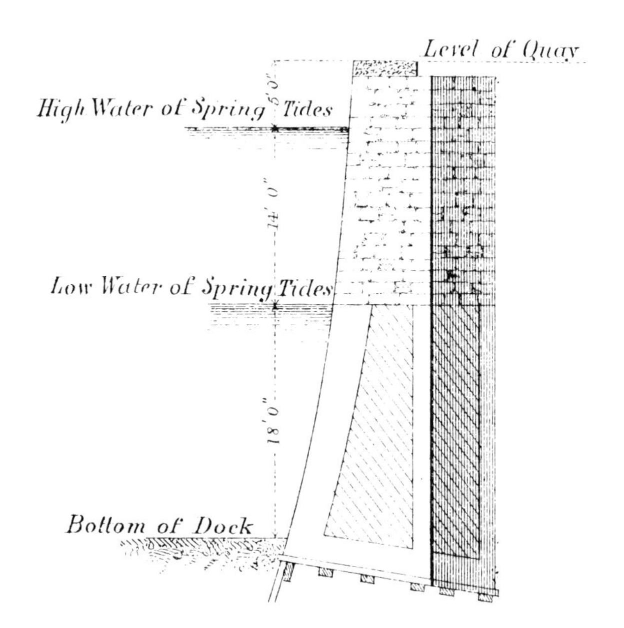

Figure 1.38 Southampton Dock wall, 1842, (later strengthened with land ties)
Source: Giles, A., *On the construction of Southampton Docks*, ICE Proc., Vol. 17, figure 4.

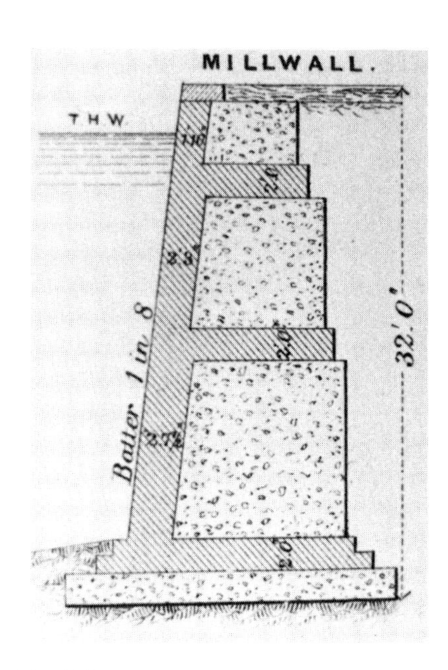

Figure 1.39 Millwall, London, 1866
Source: Vernon-Harcourt, L.F., *Harbours and Docks*, Vol. II, plate 14, figure 8.

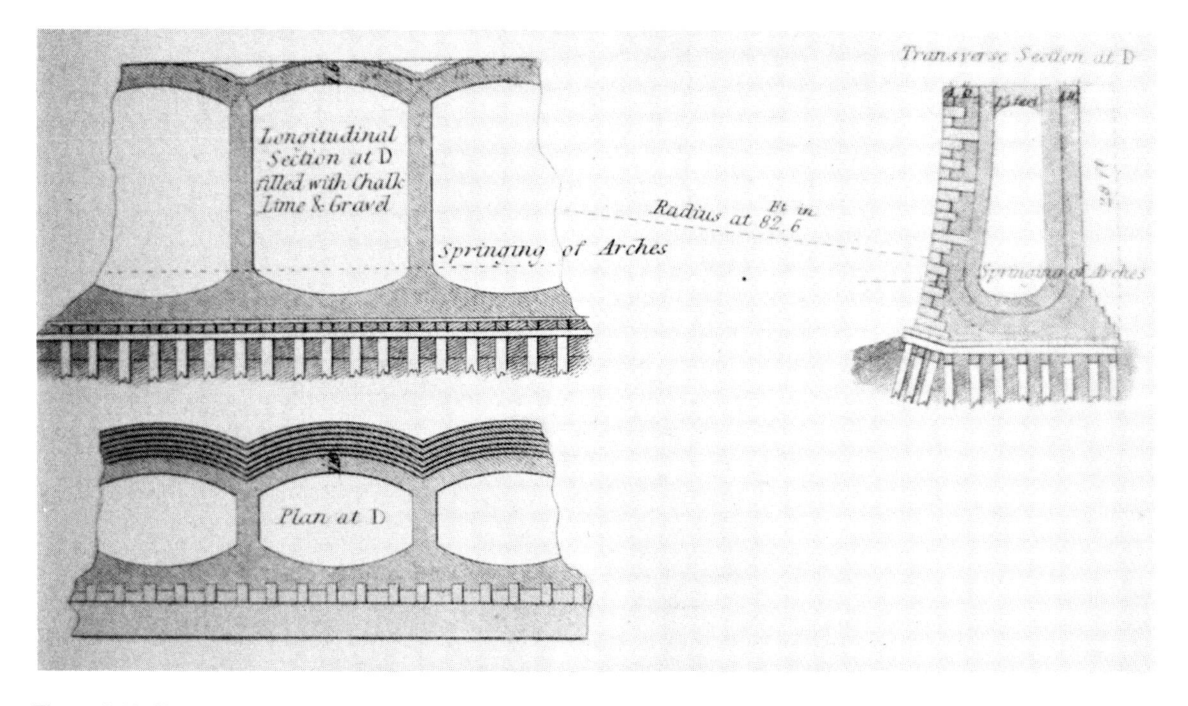

Figure 1.40 Sheerness Docks and river wall, 1813–27
Source: Rennie, plate 77 (see figure 1.31).

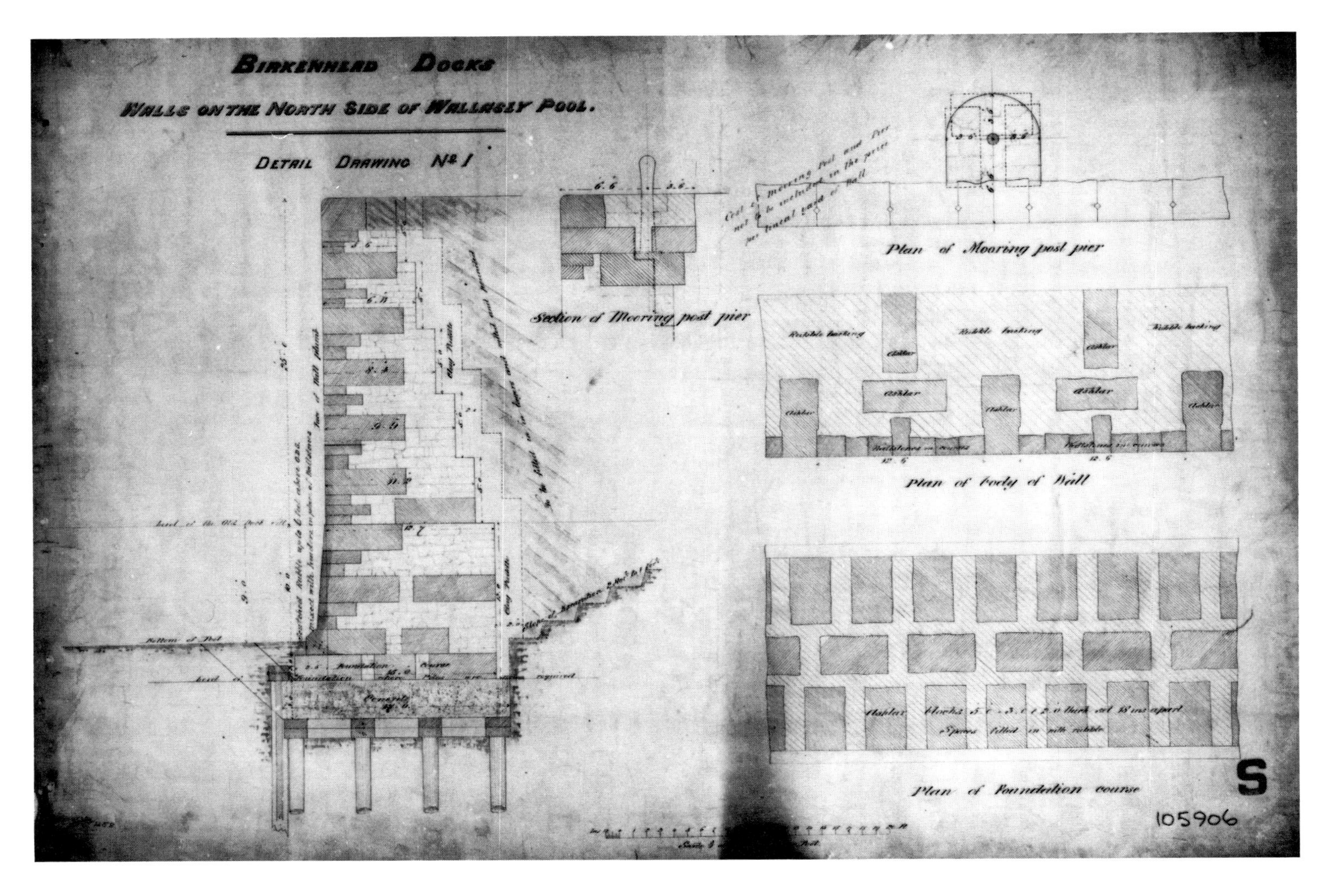

Figure 1.41 Dock wall, cast with internal ashlar blocks, Wallasey Pool, Birkenhead Docks, circa 1856
Source: Mersey Docks and Harbour Company.

Internal structure

Although the front face of a wall may be of masonry construction the main volume of material forming the core of the wall, and the back face, is often of mass concrete or some other material. In some cases the wall appears to have been built in 'lifts' with a masonry band inserted between each lift (Figure 1.39). In other cases the main shell of the wall is built of masonry and filling material is then added (Figure 1.16).

The arch form of construction (see above) avoids having to use a large volume of fill material, and there are other examples of voids being formed for the same reason (Figure 1.40). On occasions large stones are placed within the concrete fill, presumably to save on cement and possibly to reduce the heat of hydration, and there is at least one instance of ashlar blocks being used in this way (Figure 1.41).

The strength of the whole structure may be increased by tying the component parts together by keying (Figure 1.42) or by means of reinforcement. Old rails (Figure 1.43), cast iron bands (Figure 1.15), hoop iron 'bonds' (Figure 1.44) and the like were used for this purpose. In some cases the ties may have been used as part of the temporary works during construction and were then cast into the permanent works (Figures 1.45 and 1.46).

Evidence that metal work was used in wall construction from an early date is found in the 1795 specification for the Gloucester and Berkeley Canal[13] (see box).

Specification for metalwork used in the construction of the Gloucester & Berkeley canal (1795), Public Record Office, Kew

'All Cramps Chainbars and Landties or other Bars set into the stone work shall be let into the said stones (where necessary and ordered) in Groves so as to be flush with the surface of the course of stones and where it shall be necessary form parts thereof with lead the same shall be done by the said George Stroud Daniel Spencer and Thomas Cook, the lead being found by the said company.'

Backfilling

The material used to backfill behind the wall is important because it affects the loading on the wall. Materials with low angles of internal friction exert a greater pressure than those with high angles. In some cases the backfilling has not been designed and the original excavated material is replaced, but in the majority of cases the backfill is specified on the design drawings.

Generally a granular material is specified such as ballast (Figure 1.45), hand-packed broken rock or rubble (Figure 1.47), quarry refuse (Figure 1.24), or ashes (Figure 1.48). However, when a lock, dock or canal is designed to remain empty, or full, of water for a period of time it is usual to find that a layer of puddle clay has been placed against the back face of the wall (Figures 1.41 and 1.49), with a granular filling material behind the puddle clay. In one case, clay has been mixed with rock and has been used to prevent penetration of fine backfilling into rubble fill at the back of the wall (Figure 1.50). A particularly good example of designed backfilling is the Carron Dock at Grangemouth (Figure 1.51).

It should be noted that in many cases during excavation for a wall the original ground has been retained by means of temporary works behind the wall. These temporary works are often left in place and the backfill is placed over them. One effect of this is that, when the timber eventually rots, settlement may occur in the apron and this settlement is sometimes mistaken for other types of wall distress.

Other features

In addition to the features mentioned above there are a number of other characteristics of quays, docks and locks which have been noted.

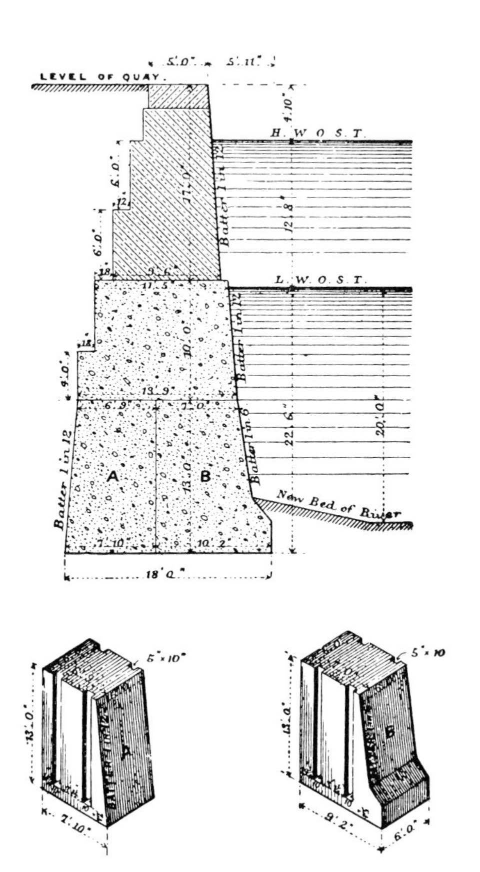

Figure 1.42 Wall made of precast blocks, Cork, 1877
Source: Cornick, H.F., *Dock and harbour engineering Volume I*, 1968, Charles Griffin, (see also Stoney, B.B. ICE Proc., 1887).

These are described below.

Bollards: Bollards are either mounted on their own detached foundation blocks or are integral with the wall structure. They may be built into the counterfort (Figure 1.52) or into the main part of the wall. In the latter case they may be tied back to mooring stones by wrought iron stays (Figure 1.53).

Ties: Ties have occasionally been used to improve the stability of walls. They are of wrought iron rod (Figures 1.48 and 1.54) or chain (Figure 1.46) construction and usually run back to an iron or timber pile driven at a slight rake behind the wall, or to a masonry or concrete block. As can be seen from the Southwold example (Figure 1.54), the visible part of the wall may not show any evidence of the existence of a tie rod. Sometimes walls on either side of a narrow strip, separating two basins, are tied together (see Figure 1.55).

Piles: Piles are used extensively in the foundations to support walls on soft soils. The older piles are of timber but cast iron was used in many instances and there are numerous examples of composite or semi-composite forms of construction using cast iron circular and sheet piles. The piles are used to support the base (Figure 1.49) of the wall at its normal level, or they may be used to support the entire wall at a point well above seabed level (Figures 1.56 and 1.57), in which case sheet piles will be used to retain the soil below. However, in a few cases this sheet piling is absent, the wall being suspended above a battered slope (Figure 1.58). The bearing piles are not always circular.

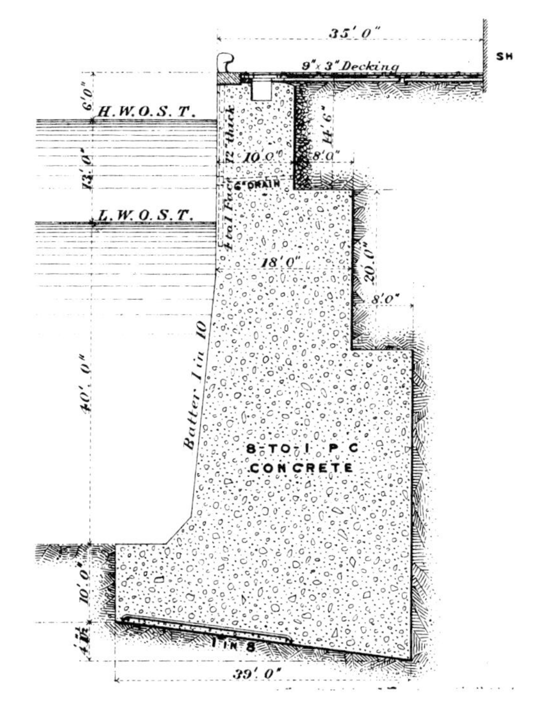

Figure 1.43 White Star dock, Southampton, 1907
Source: Wentworth Shields, ICE Proc, 195, plate 2, figure 3. (see 1.78)

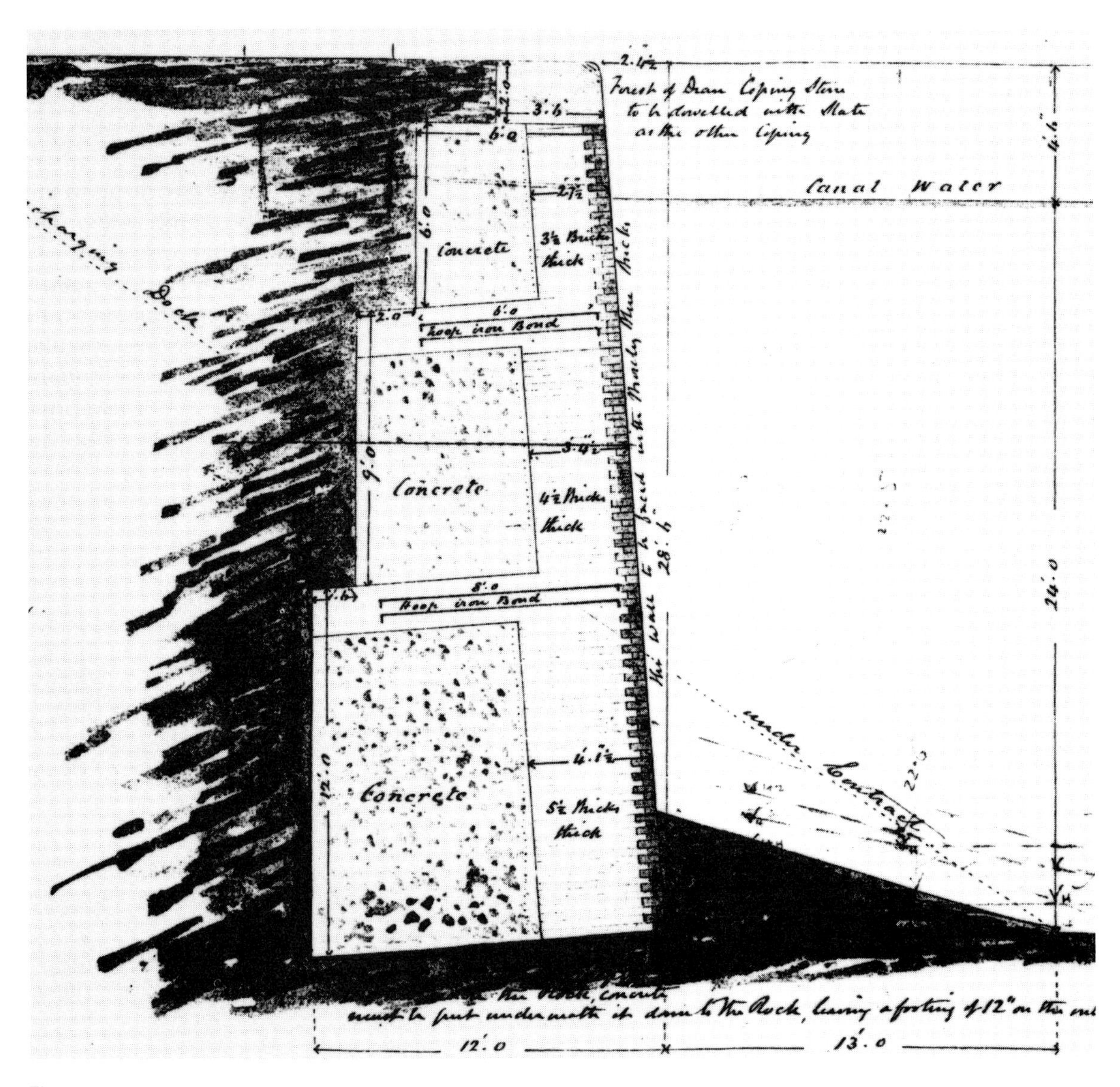

Figure 1.44 Dock wall in main basin at Sharpness, 1874
Source: British Waterways Archives, Gloucester.

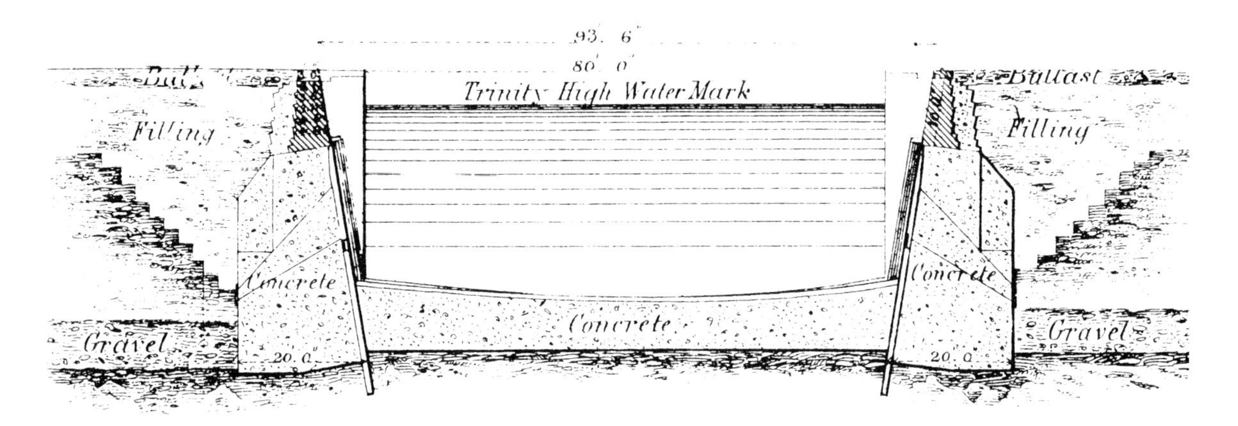

Figure 1.45 Victoria Docks, London, 1858
Source: ICE Proc., Vol. XVIII, 1858.

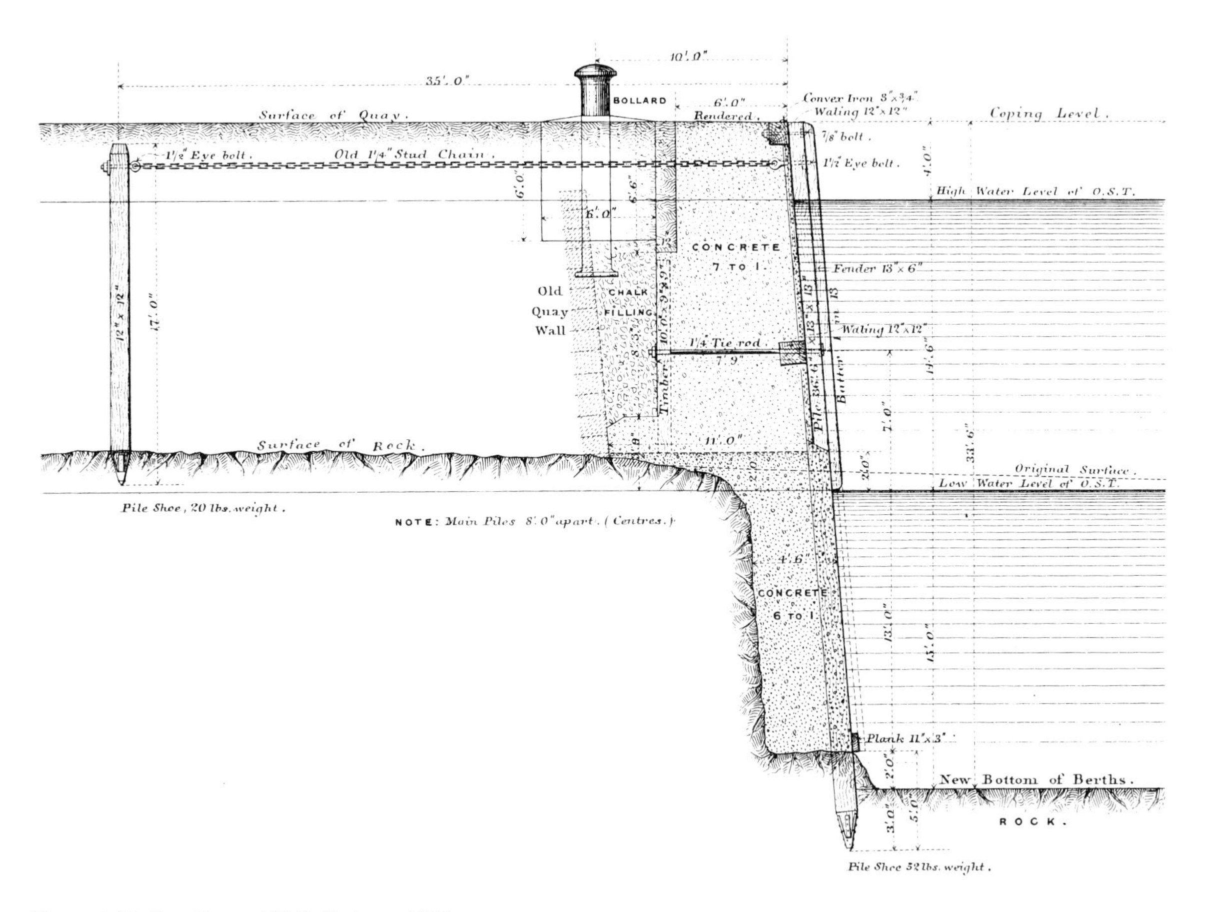

Figure 1.46 New Quay at Blyth Harbour, 1882
Source: Kidd, W. 'On the blasting and removal of rock under water and the construction of a deep water quay at Blyth Harbour.' ICE Proc, Vol. 81, 1885, plate 14, figure 7.

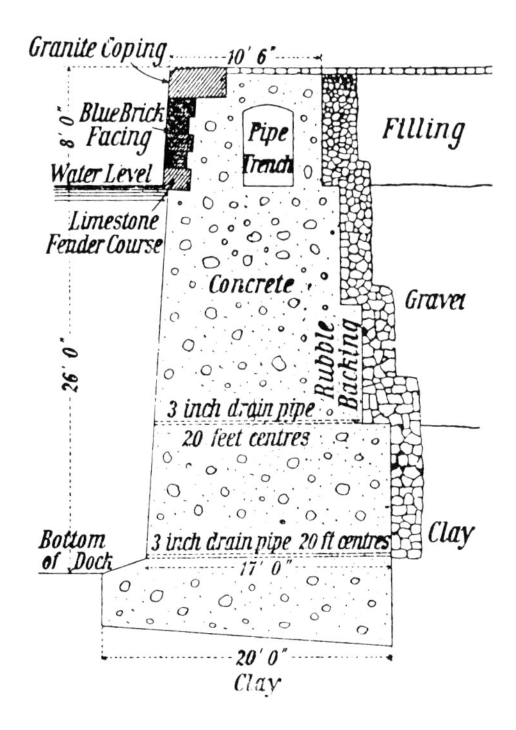

Figure 1.47 Dock wall at Manchester
Source: Cornick, H.F., *Dock and harbour engineering Volume I*, 1968, Charles Griffin.

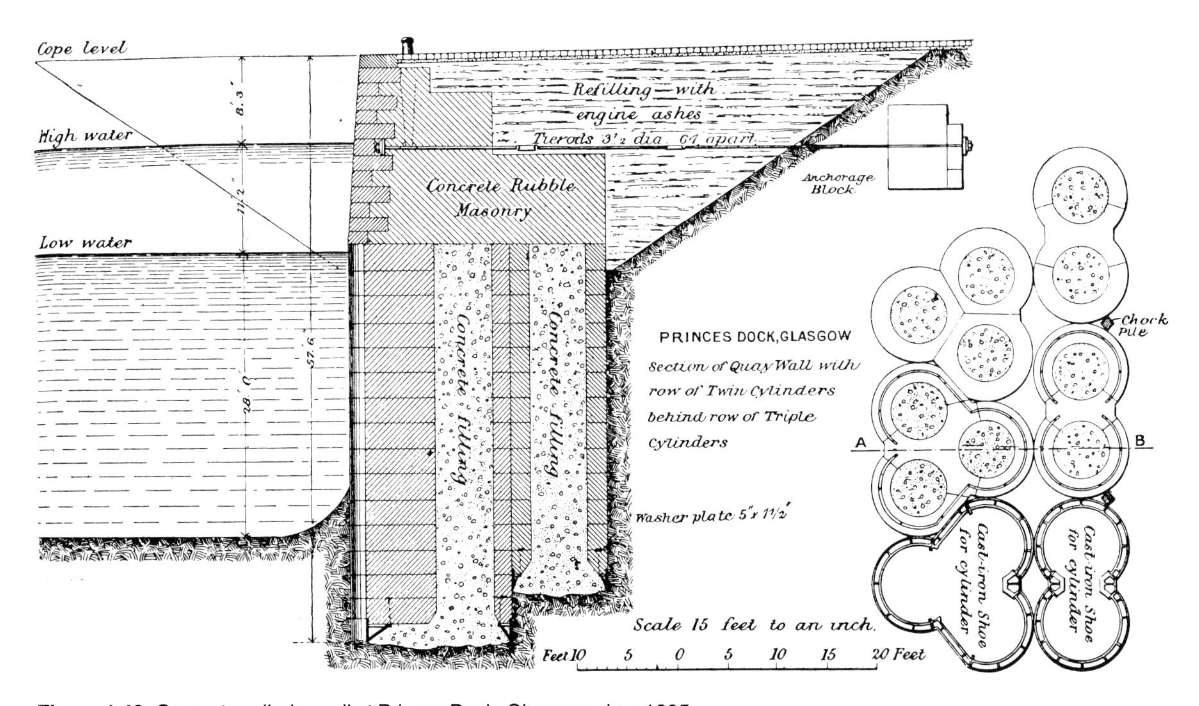

Figure 1.48 Concrete cylinder wall at Princes Dock, Glasgow, circa 1895
Source: Cornick, H.F., *Dock and harbour engineering Volume I*, 1968, Charles Griffin.

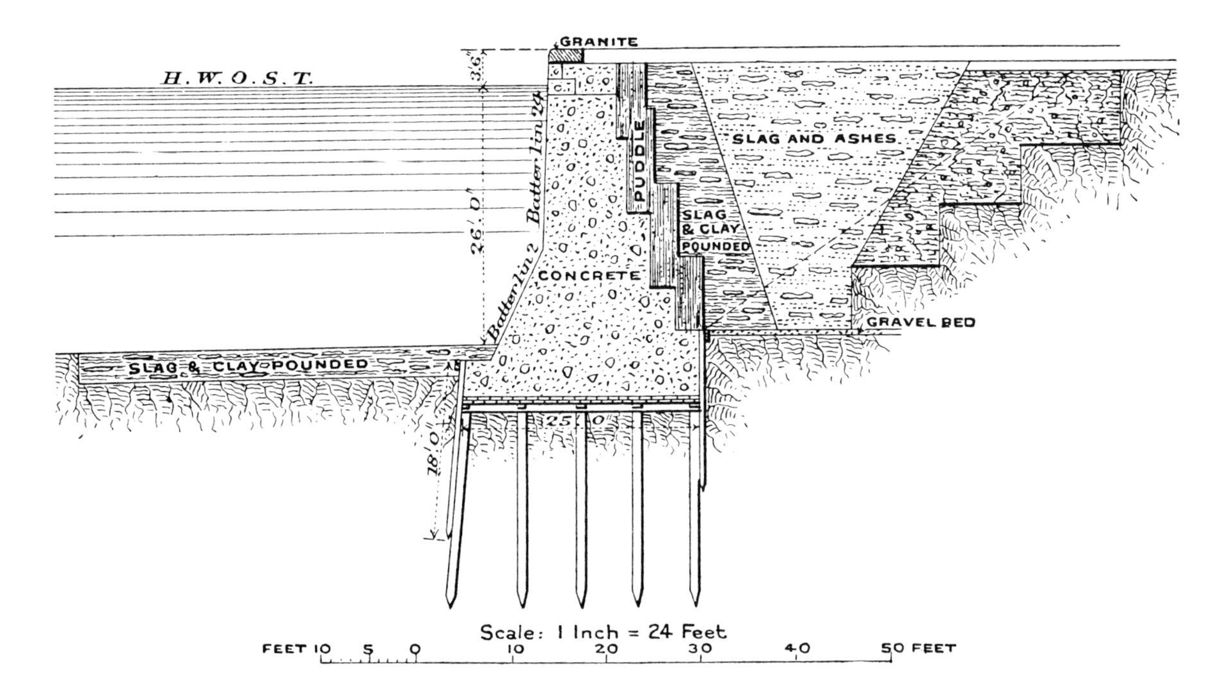

Figure 1.49 Designed backfill for the Carron Dock, Grangemouth, 1879
Source: Wentworth Shields, ICE Proc., Vol. 195, page 96, figure 18.

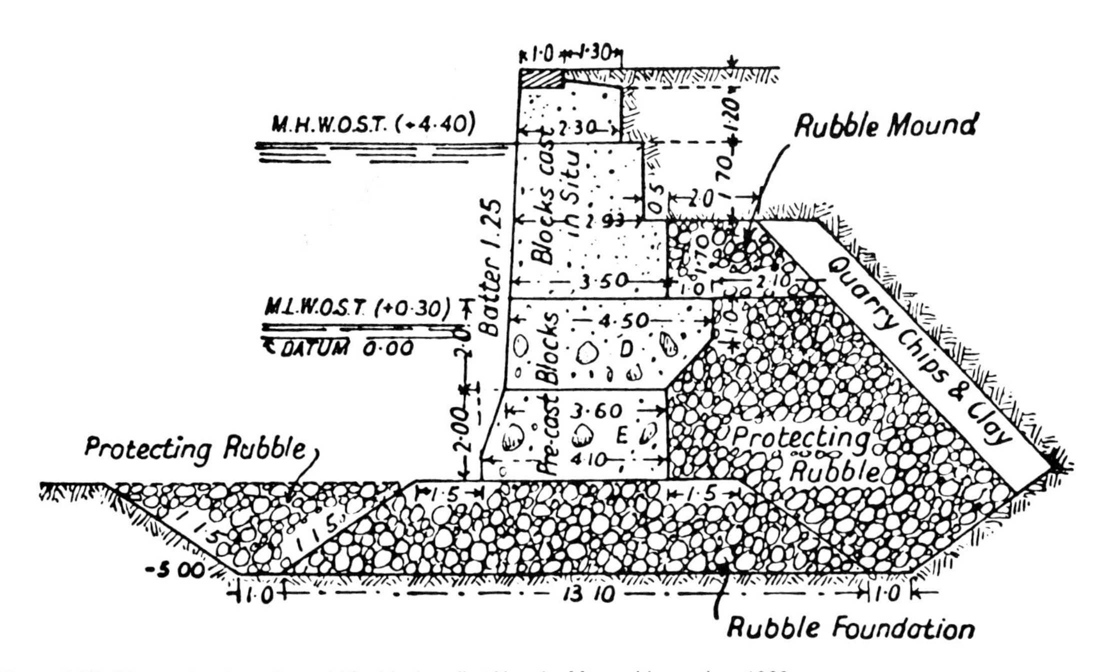

Figure 1.50 Clay and rock seal on rubble: block wall at Nacala, Mozambique, circa 1960
Source: Cornick, H.F., *Dock and harbour engineering Volume 1*, 1968, Charles Griffin, p.93.

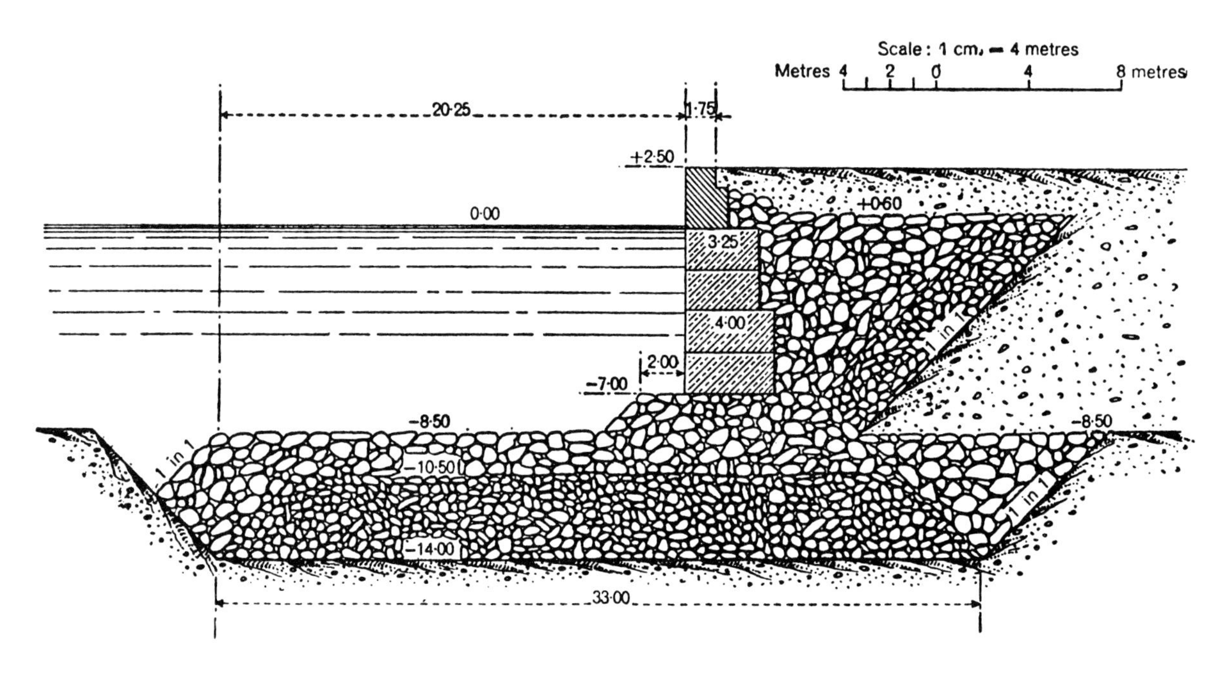

Figure 1.51 Quay wall at Trieste
Source: Cagli, H.C. *Italian Docks and Harbours*, ICE Proc., Vol. 1, 1935/36, Plate 4.

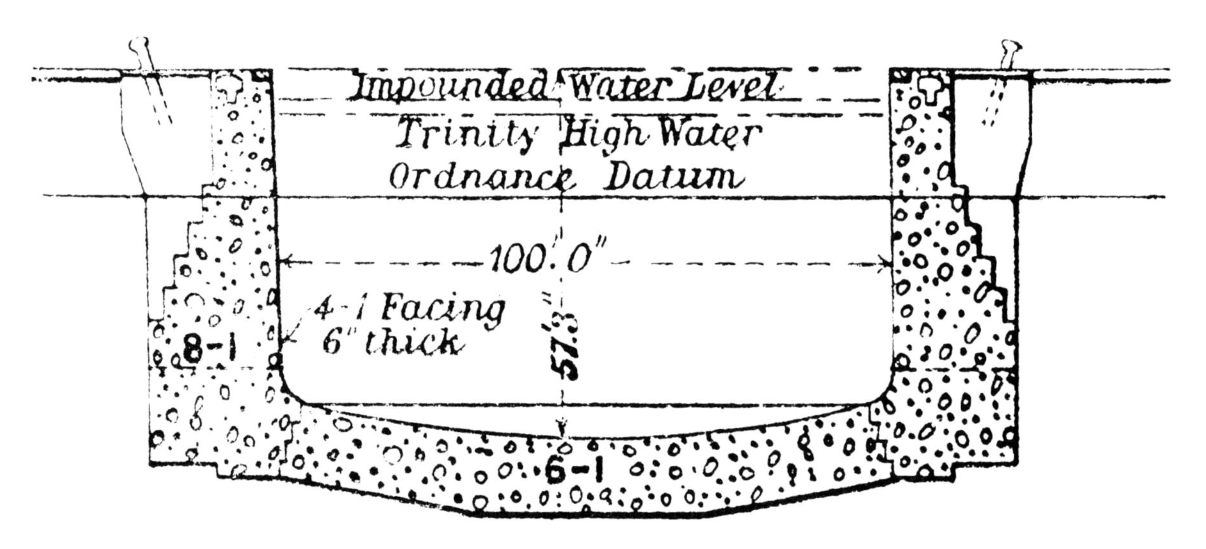

Figure 1.52 Transverse section through King George V Dock, London, 1921
Source: ICE Proc., Vol. 216, plate 7, figure 4.

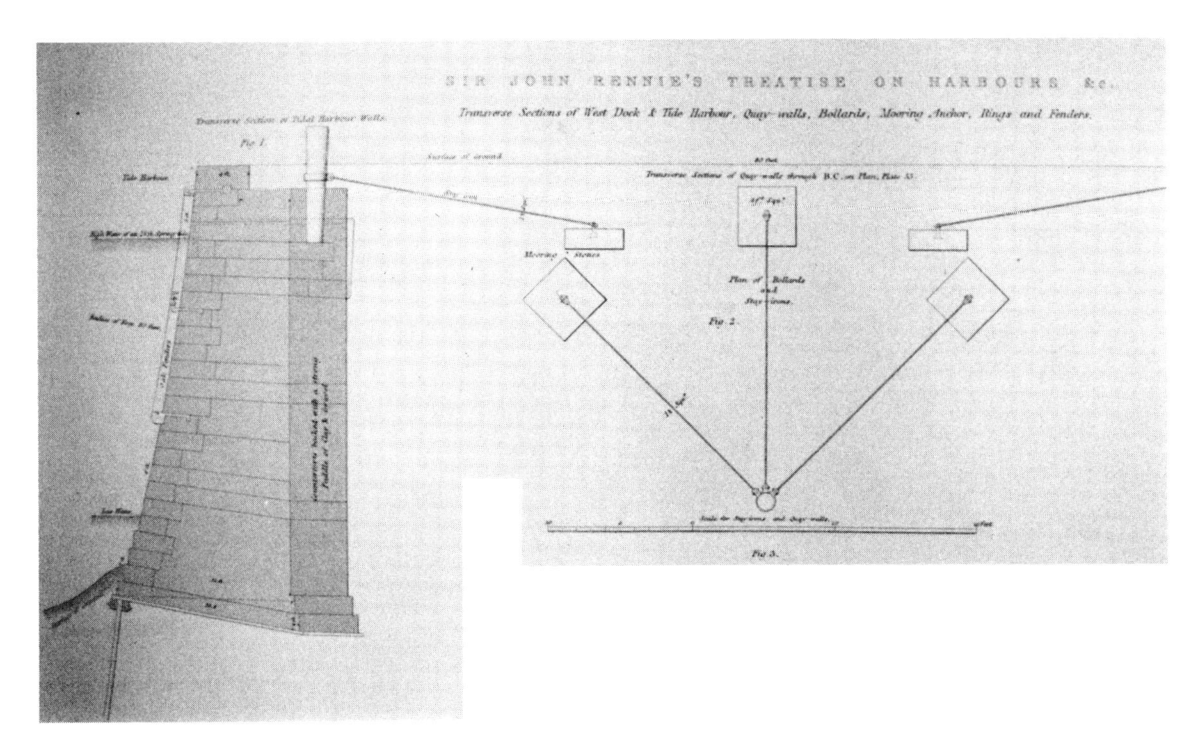

Figure 1.53 Section of the tidal harbour wall showing bollard and stay-irons, Hartlepool, 1834
Source: Rennie, plate 62, figure 1. (see Fig.1.31)

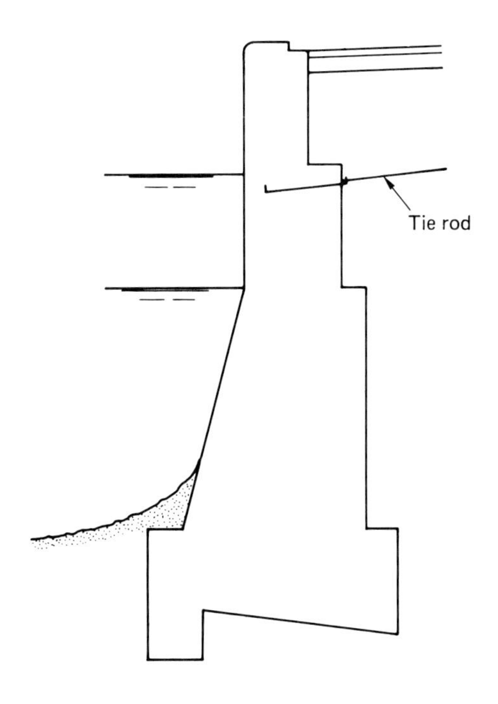

Figure 1.54 Cross-section of wall at the entrance to Southwold Harbour
Source: Waveney District Council.

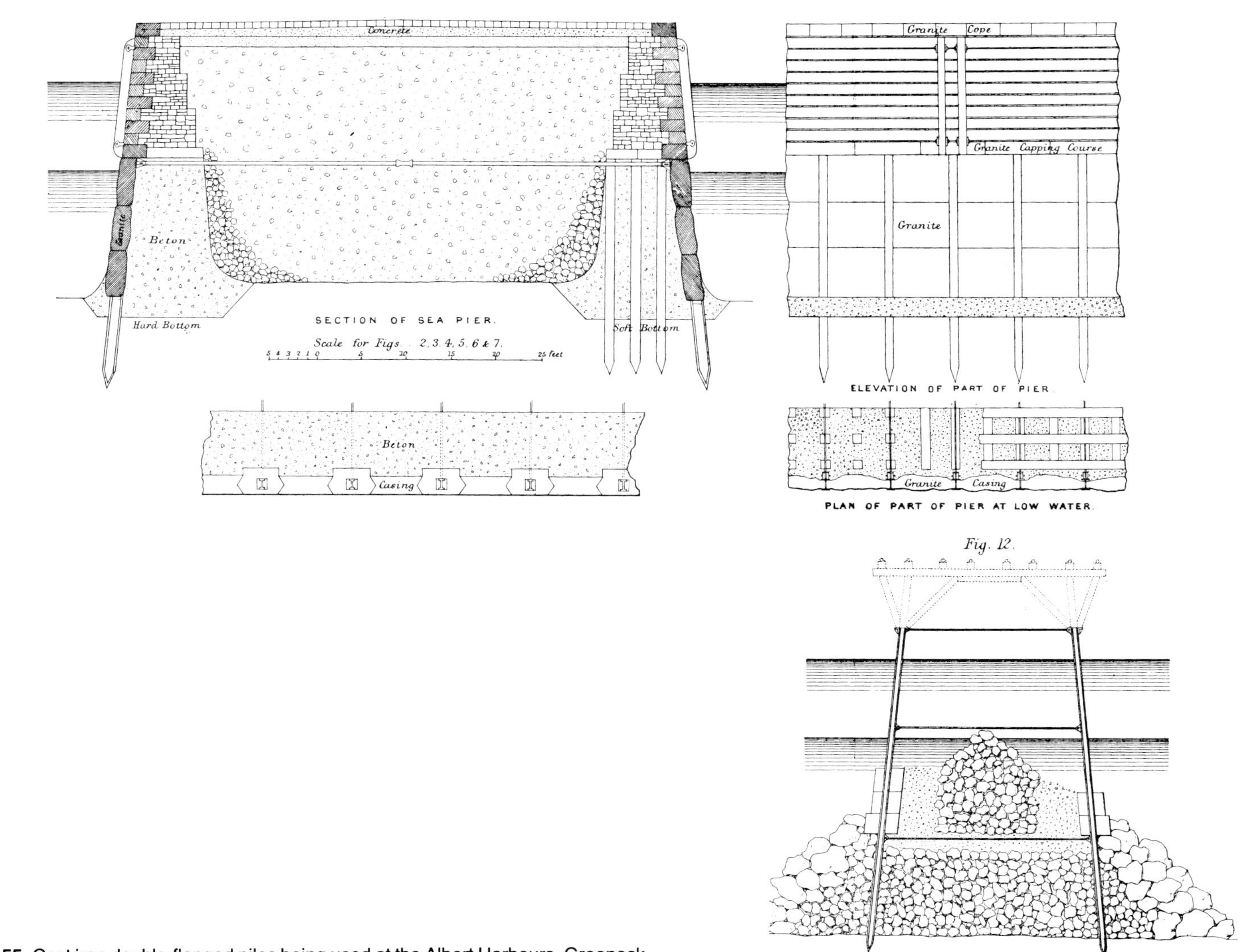

Figure 1.55 Cast iron double-flanged piles being used at the Albert Harbours, Greenock, circa 1860

Source: ICE Proc., Vol. XXII, 1862/63, plate 9, figures 3, 4, 6, 7 and 12.

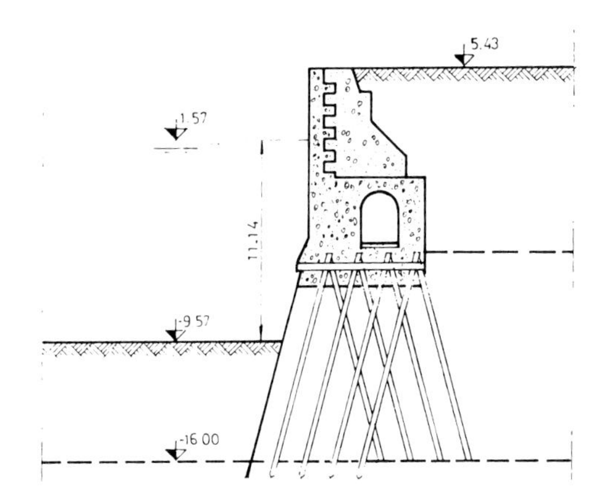

Figure 1.56 Section of a wall at Bremerhaven, 1897
Source: Final Report of the International Commission for the Study of Locks. Copyright by PIANC General Secretariat, Brussels, Belgium.

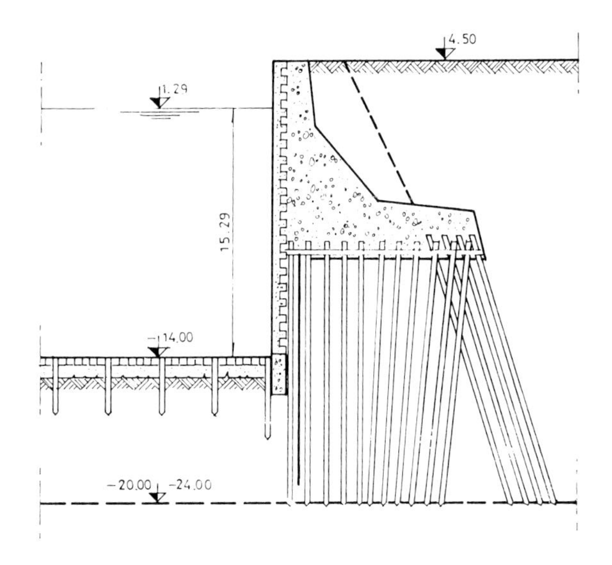

Figure 1.57 Section of a wall at Brunsbuttelkoog, 1914
Source: Final Report of the International Commission for the Study of Locks. Copyright by PIANC General Secretariat, Brussels, Belgium.

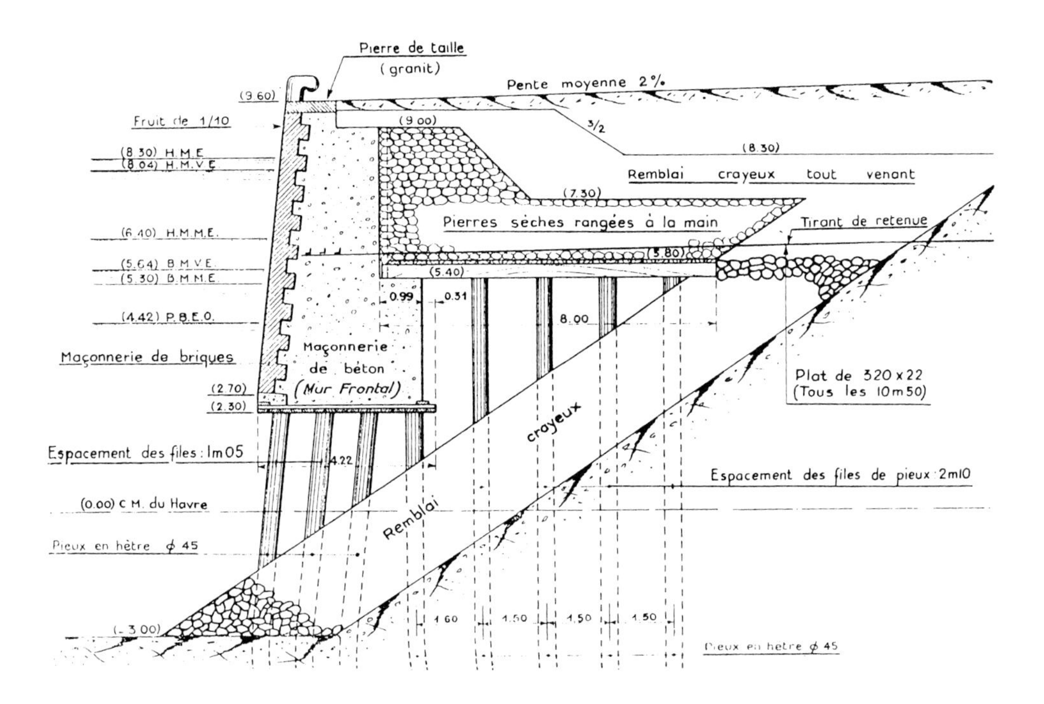

Figure 1.58 'Rouen' type quay wall
Source: Copyright by PIANC General Secretariat, Brussels, Belgium.

Examples of joist piles have been found as well as 'cast iron double flanged' piles (Figure 1.55).

Caissons: In a few cases walls have been constructed on caissons (Figures 1.59, 1.60, 1.61 and 1.48) and in one example on cast iron cylinders connected by cast iron sheet piles (Figure 1.62). Caissons may be constructed of brick, concrete blocks or *in situ* concrete.

Drainage: A number of walls have been designed with weepholes to reduce the water pressure behind the wall at periods of low water and to prevent a build-up of surface water (Figures 1.63, 1.37 and 1.47). In some instances these drains are fitted with tide flaps to reduce the flow of water in and out of the soil behind the wall.

Foundation rafts: Foundation rafts made in the form of grillages of timber sleepers (Figure 1.15) are often placed immediately below the base of the wall. These may be supported on timber piles or on tipped rubble.

Fenders and timber rubbing strips: These are often built into the wall at the time of construction (Figure 1.53) or may be used as permanent shuttering (see Figure 1.64).

1.3 Breakwaters

1.3.1 Function and design

The main components of a breakwater are shown in Figure 1.65. The primary function of a breakwater is to provide an area of calm water between it and the coast behind. There is thus no constraint on the design of the outer face other than that imposed by the nature of the incident waves, the currents and the seabed at that location. The inner face of the breakwater is similarly unconstrained but as it is commonly used as a berthing face against which vessels can moor, it is often nearly vertical.

The wave forces acting on the outer face of the breakwater impose severe localized forces on the individual components of the structure

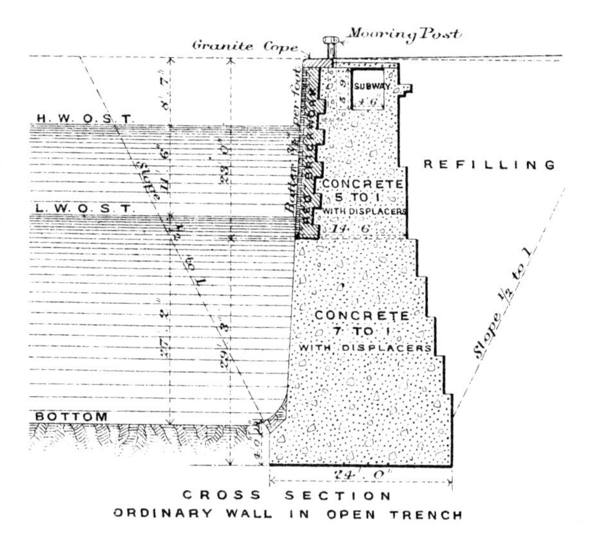

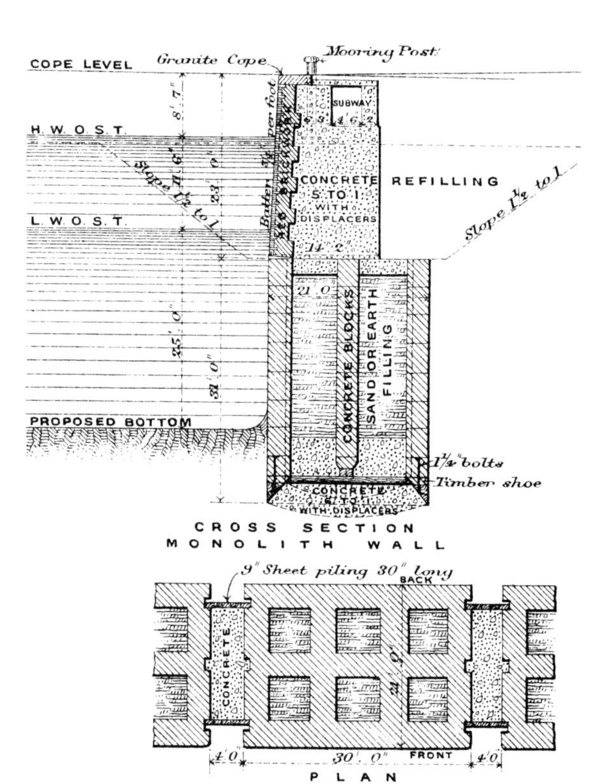

Figure 1.59 Caisson-type foundations at Rothesay Dock, Glasgow, 1901
Source: Mason, T., 'The improvement of the River Clyde and Harbour of Glasgow 1875–1914'. ICE Proc., Vol. 200, 1915, plate 1, figure 6.

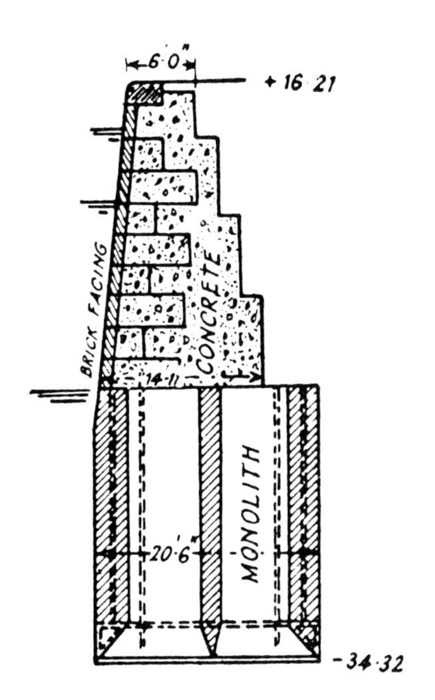

Figure 1.60 Monolithic foundations at St Andrews Dock, Hull, 1901
Source: Du Plat Taylor, page 82, figure 72.

as well as overturning and sliding forces on the structure itself.

Because of the exposed nature of the site on which a breakwater is constructed the method of construction has a considerable effect on the design; the use of large blocks, concrete bags and natural stones being much favoured, even in the internal portions of the structure. Certainly, the techniques of forming, transporting and placing large blocks (350 tons in Dublin) were quite advanced by the 1870s, as evidenced by the lively discussion on this subject in the Proceedings of the Institution of Civil Engineers at this time.[18] Much of the credit is given to the French, who used the technique in Algiers in about 1844.

Another popular method of construction was to erect a timber frame along the line of the breakwater and to form the structure around it, much of the timber being left *in situ* after the construction was completed (Figures 1.55 and 1.66).

1.3.2 Component characteristics

Outer face

The outer face may vary from being vertical to sloping outwards at a slope of 1:2 (Figures 1.67 and 1.69), or even flatter (Figure 1.70). The sloping faces usually become steeper towards the top of the wall and are often vertical at the top. Many of the outer faces are extended upwards beyond the main body of the wall to provide a protective wall on the outside edge of the breakwater which reflects waves and reduces overtopping by broken waves and spray (Figures 1.71 to 1.73). Some outer faces are stepped (Figure 1.74).

The blocks used in the outer face of breakwaters tend to be large, of high quality and securely jointed so that they can withstand the effect of large waves. Granite and limestone ashlar (Figure 1.73) were popular. Various special arrangements for placing blocks were devised, such as piano blockwork (Figure 1.75) and slicework (see below). In particularly severe conditions the blocks were tied back to the mass of the breakwater during construction.[19] An alternative method of dissipating wave energy is to cause the waves to break on the outer face of the breakwater, and this was no doubt the intention in the examples shown in Figures 1.68 and 1.70.

In a number of cases the whole of the breakwater, or a large proportion of it, was formed from mass concrete (Figure 1.76). In others the faces were formed by a timber frame which contained a rubble hearting (Figures 1.77 and 1.78).

Inner face

The inner face is often vertical or near-vertical. In other respects it is similar in construction to the outer face but may be composed of slightly smaller blocks.

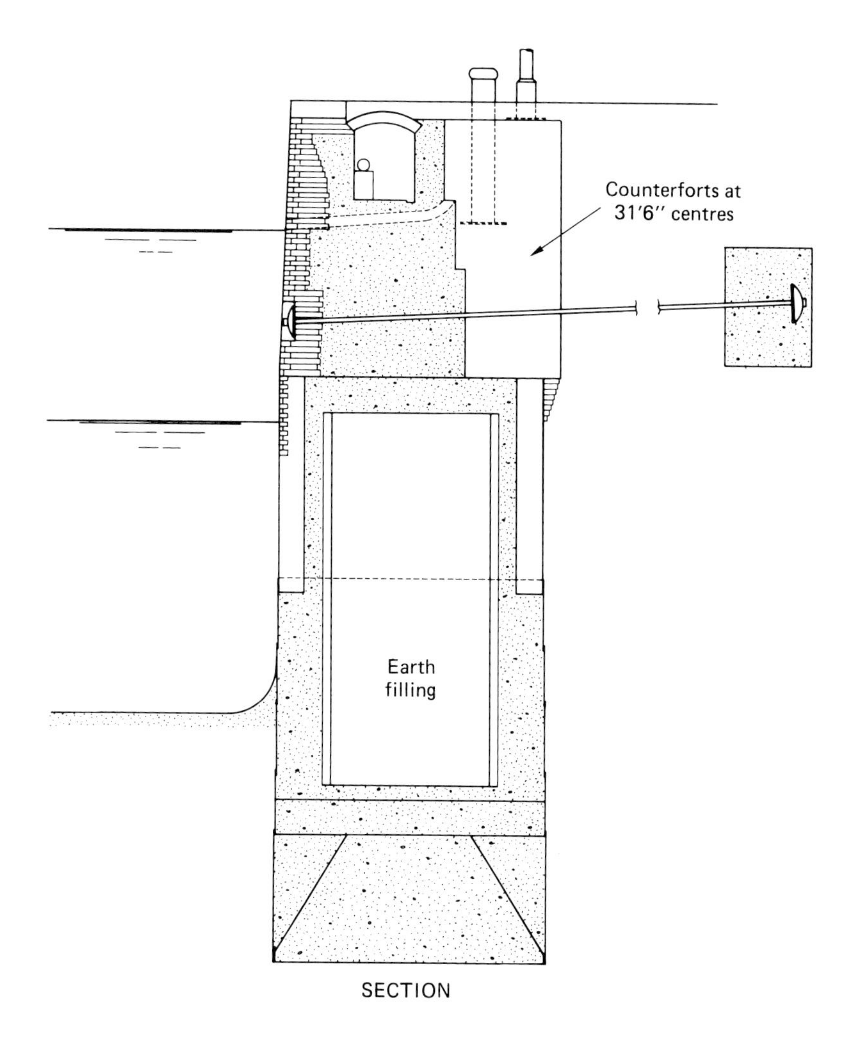

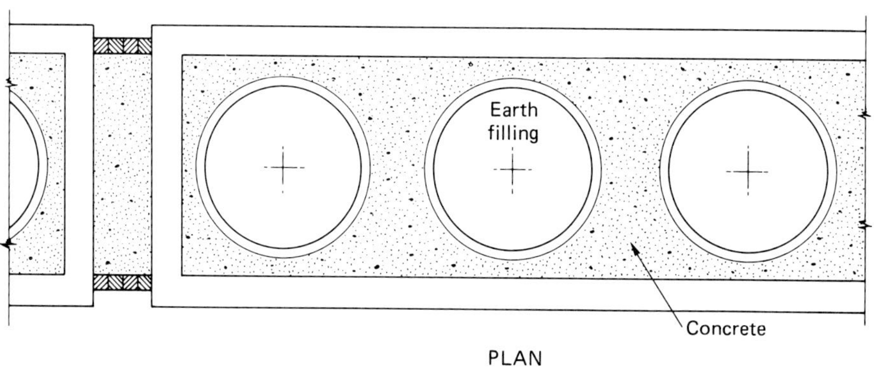

Figure 1.61 Broomielaw quay, Glasgow, 1900–2
Source: Clyde Port Authority and Glasgow District Council.

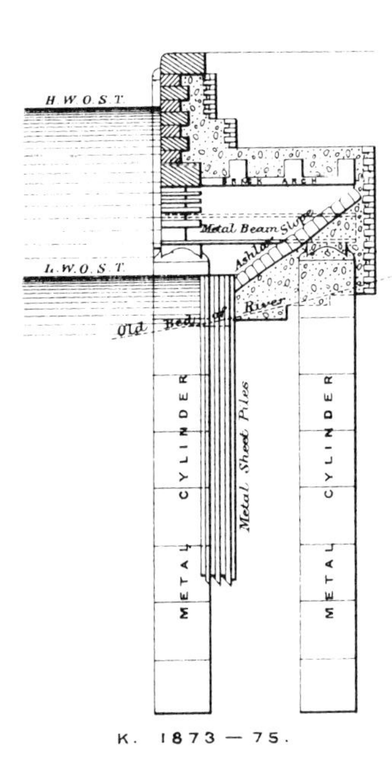

Figure 1.62 Quay wall at Newcastle on cast iron cylinders, 1873
Source: Scott, A., 'Deep water quays Newcastle-upon-Tyne'. ICE Proc., Vol. 119, 1894, plate 8, figure 7.

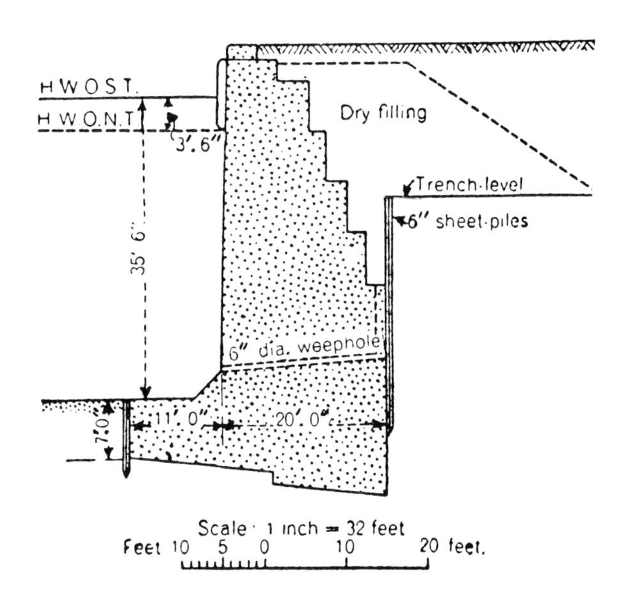

Figure 1.63 Dock wall at Immingham
Source: Cornick, H.F., *Dock and harbour engineering Volume I*, 1968, Charles Griffin.

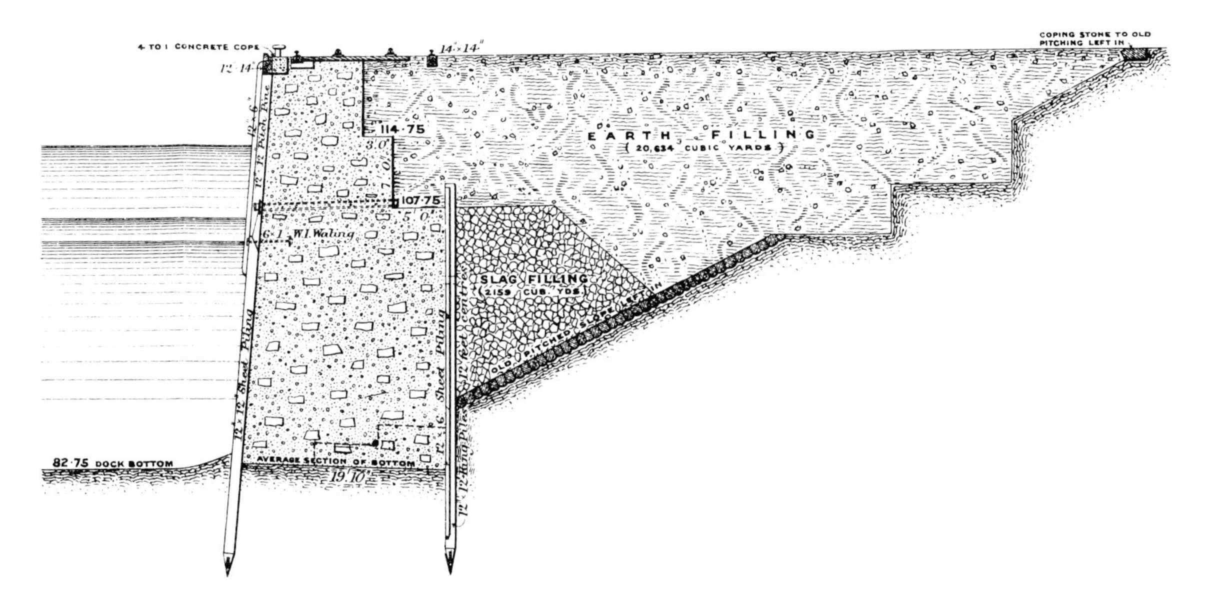

Figure 1.64 Concrete quay on south side of basin at Barrow, 1901
Source: Savile, L.H., 'Lowering of the sill of the Ramsden Lock, Barrow in Furness'. ICE Proc., Vol. CLVIII 1903/4, Part IV, plate 4, figure 2.

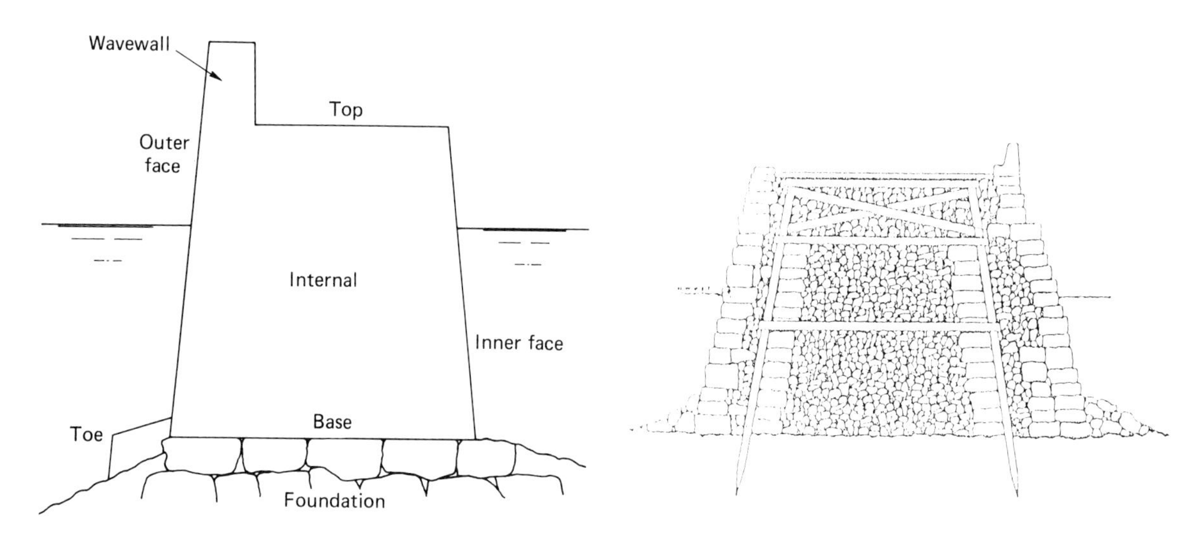

Figure 1.65 The main components of a breakwater
Source: Livesey Henderson.

Figure 1.66 Section of the North Pier, West Hartlepool, 1847–58
Source: Dock and Harbour Authority Dec. 1926, page 53.

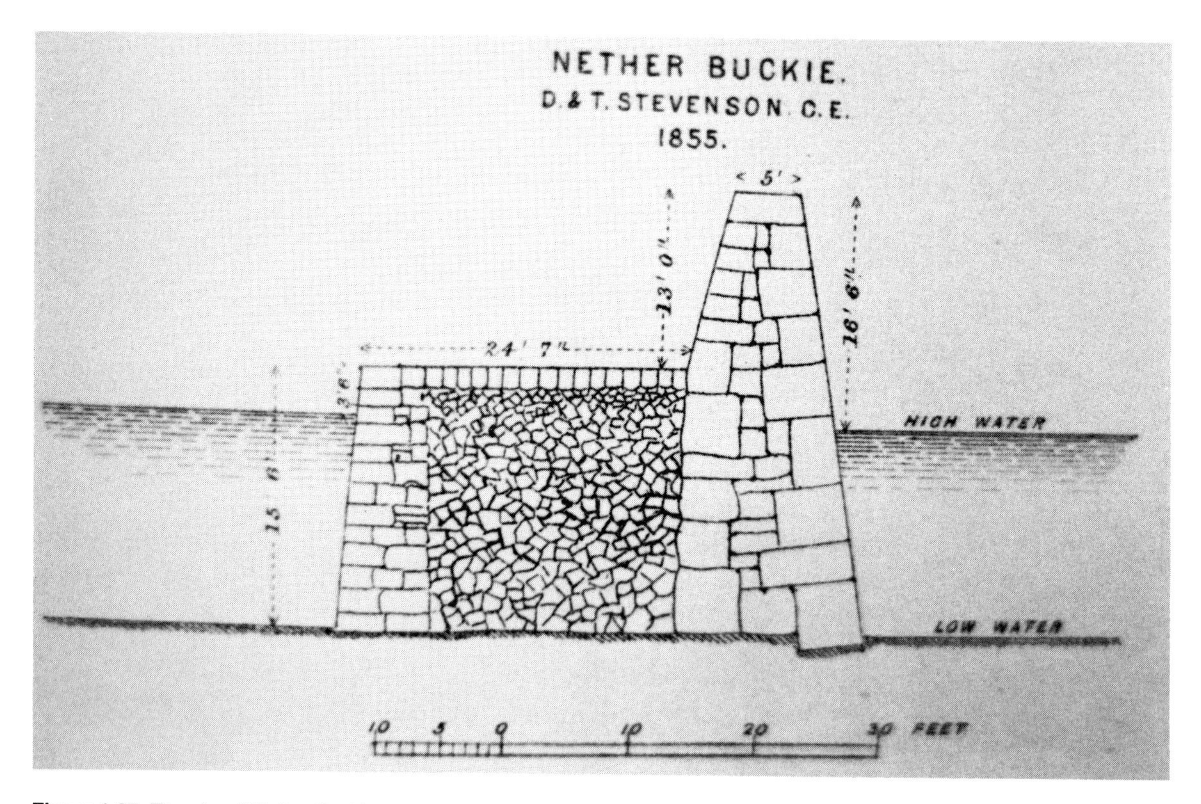

Figure 1.67 The pier at Nether Buckie, 1855
Source: Stevenson, plate IV. (see Fig.1.28)

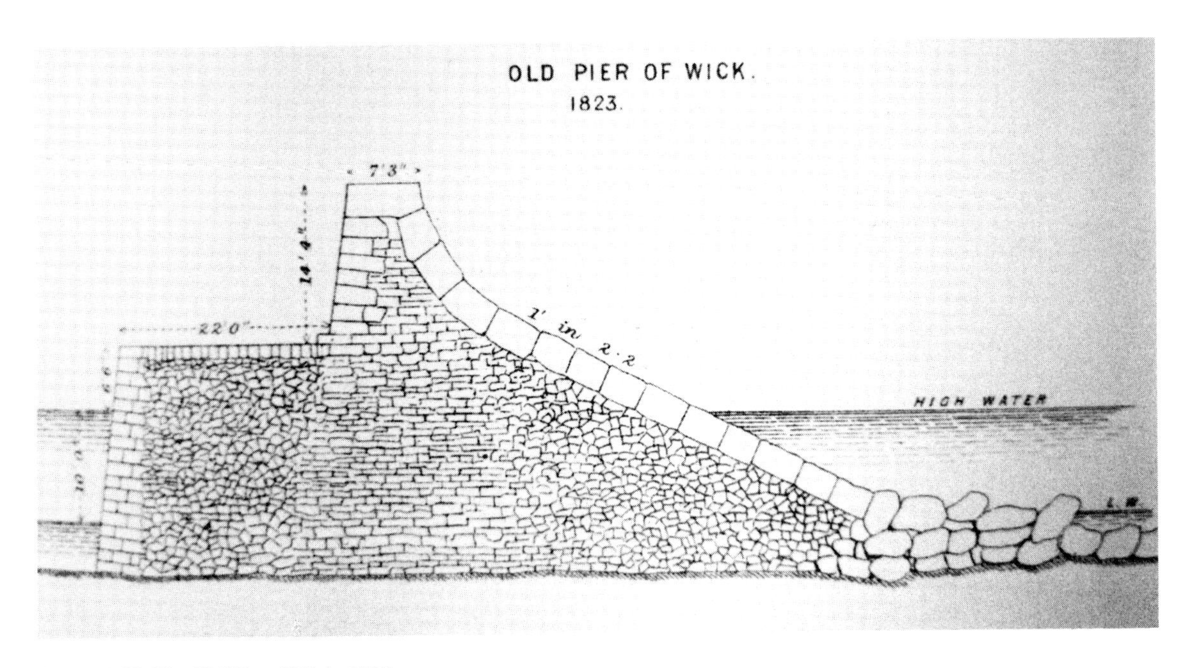

Figure 1.68 The Old Pier at Wick, 1823
Source: Stevenson, plate IV. (see Fig.1.28)

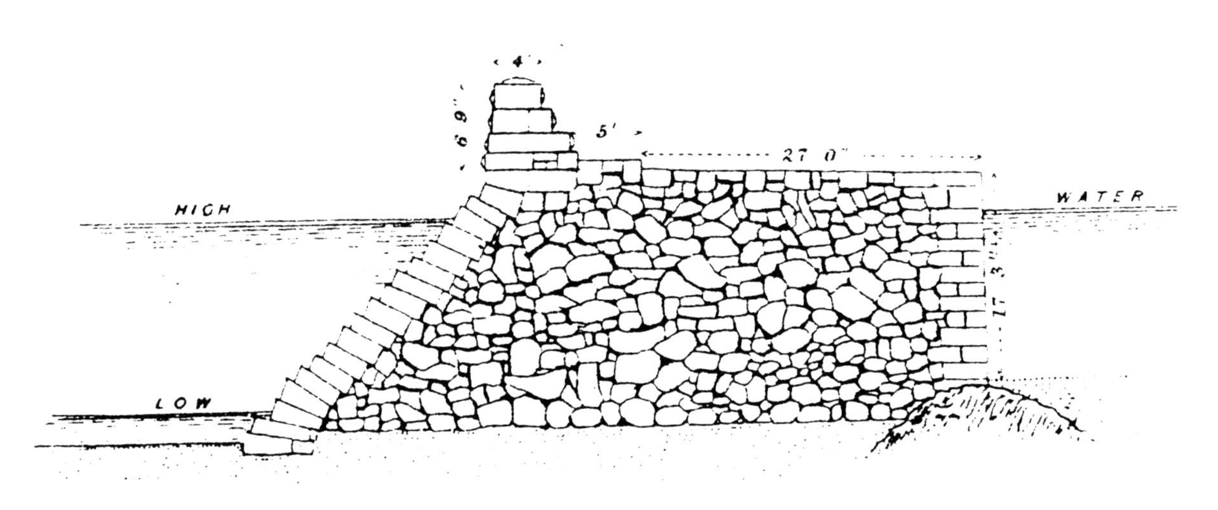

Figure 1.69 The pier at Hynish, Argyllshire, 1843
Source: Stevenson, plate IV. (see Fig.1.28)

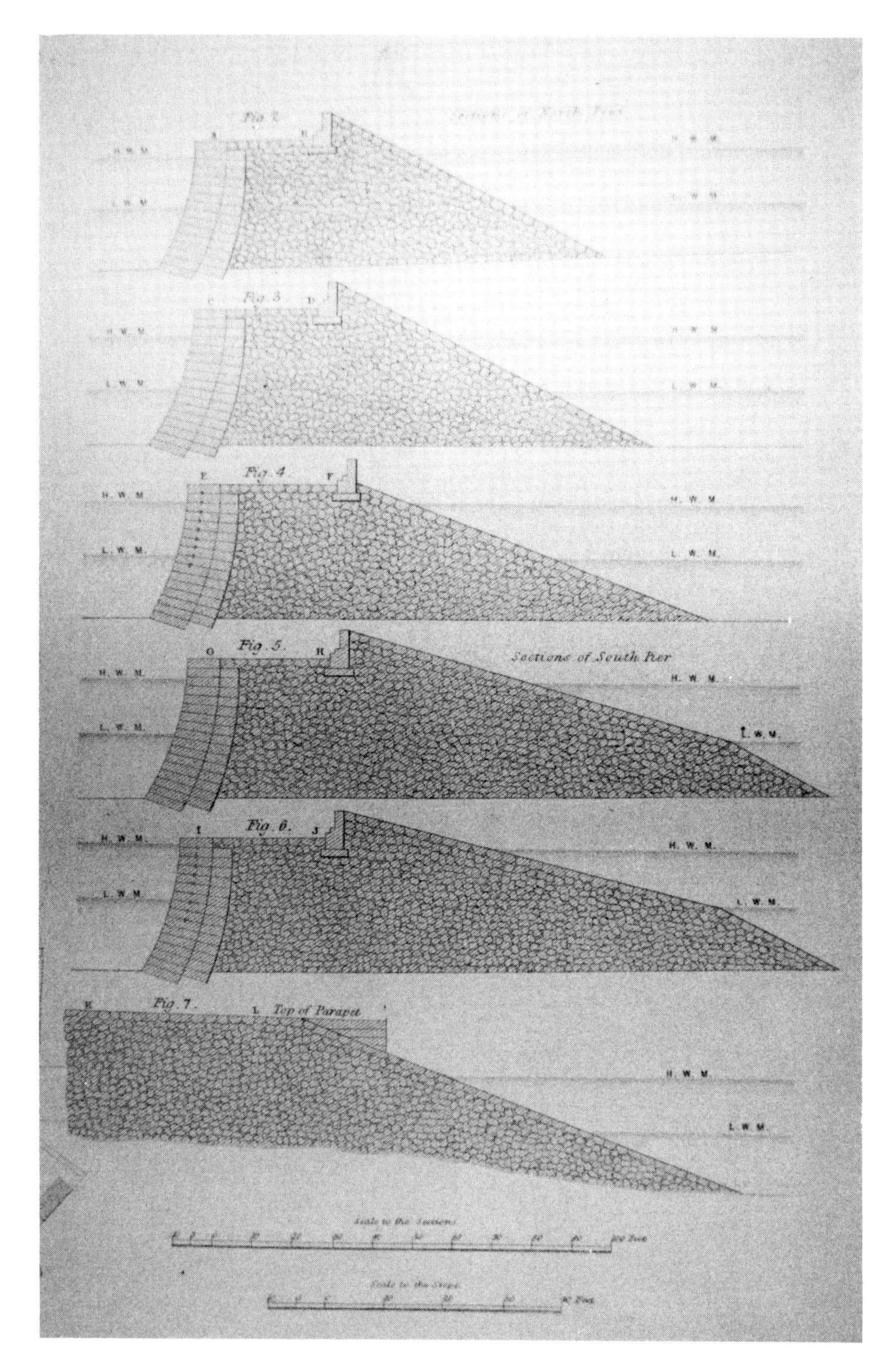

Figure 1.70 Breakwater at Donaghadee Harbour, 1821
Source: Rennie, 1854, plate 83, figures 2 and 6. (see Fig.1.31)

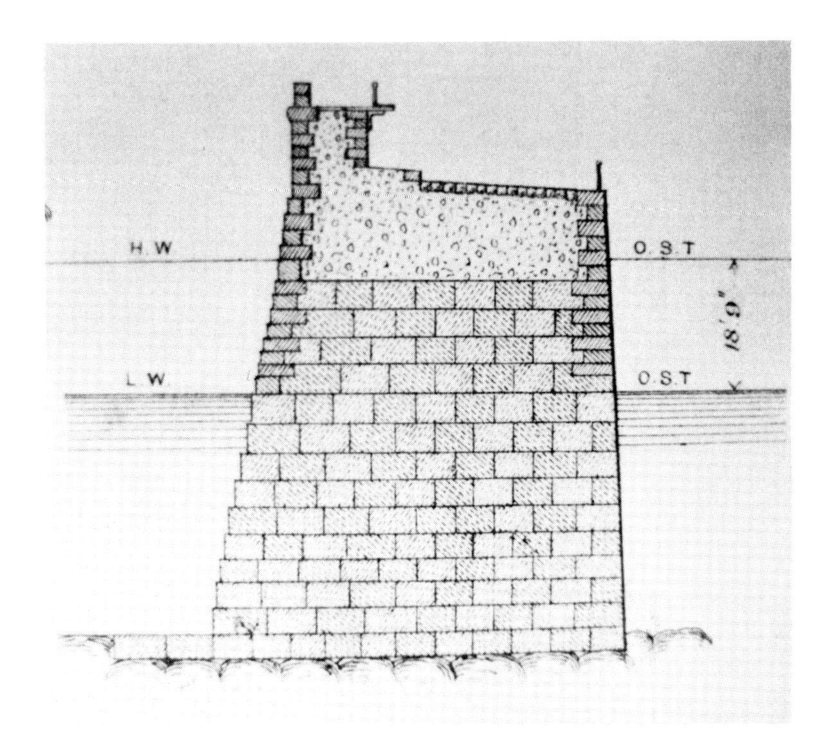

Figure 1.71 Dover breakwater, 1866
Source: Vernon-Harcourt, L.F., *Harbours and Docks*, Vol. II.

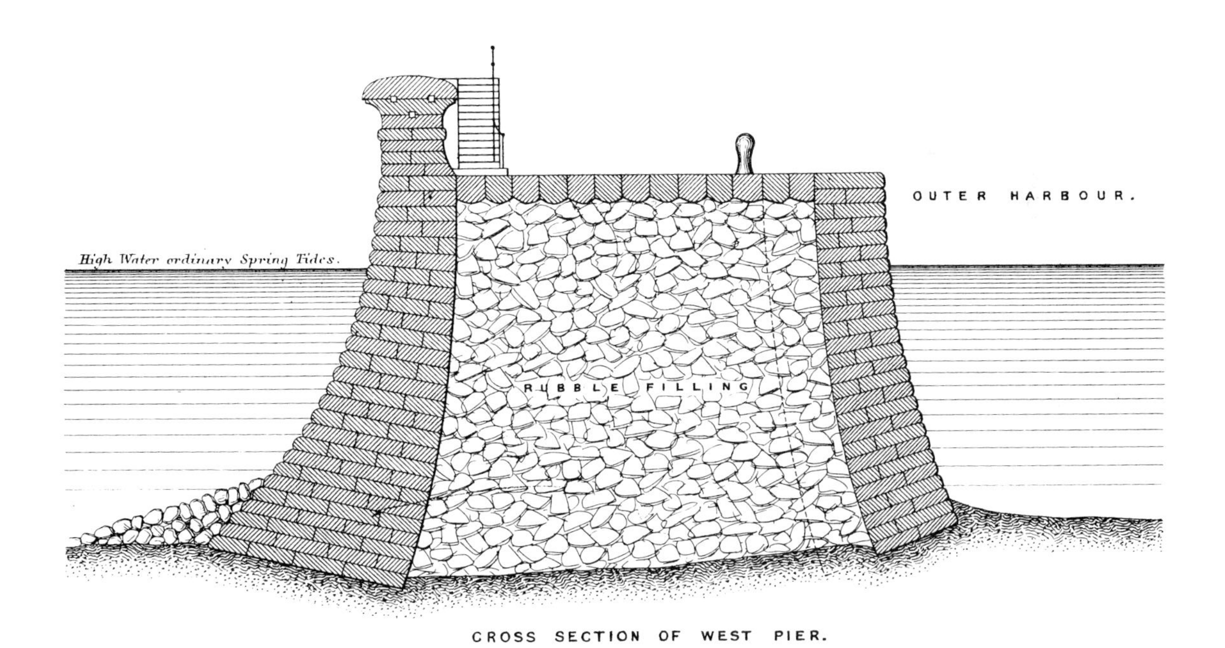

Figure 1.72 The West Pier at Whitehaven, 1831
Source: Williams, J.E., 'Whitehaven harbour and dock works'. ICE Proc., Vol. 55, 1878, plate 5, figure 1.

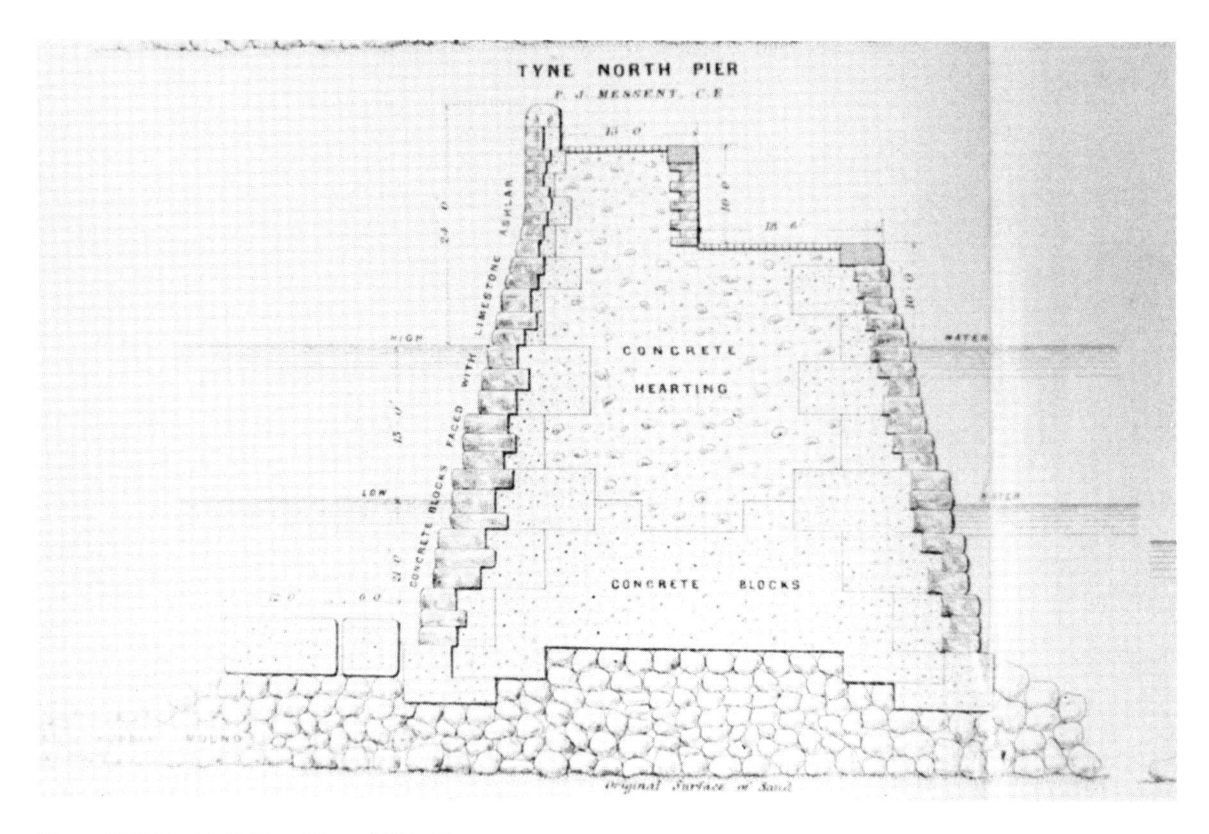

Figure 1.73 The North Pier at Tyne, 1855–95
Source: Stevenson, plate VIII. (see Figure 1.28)

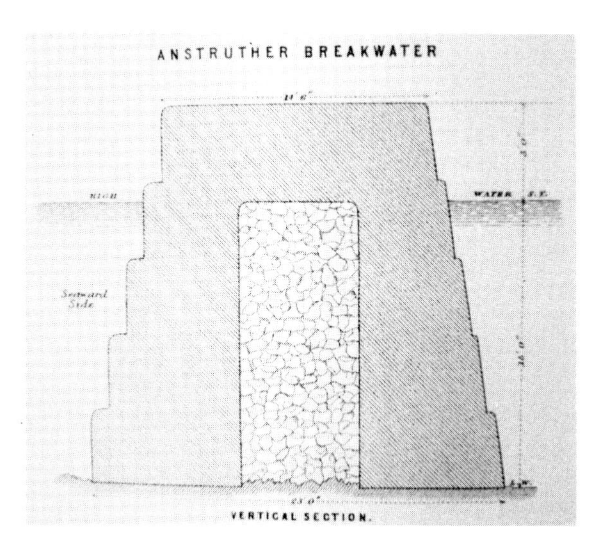

Figure 1.74 The breakwater at Anstruther
Source: Stevenson, plate XXI. (see Figure 1.28)

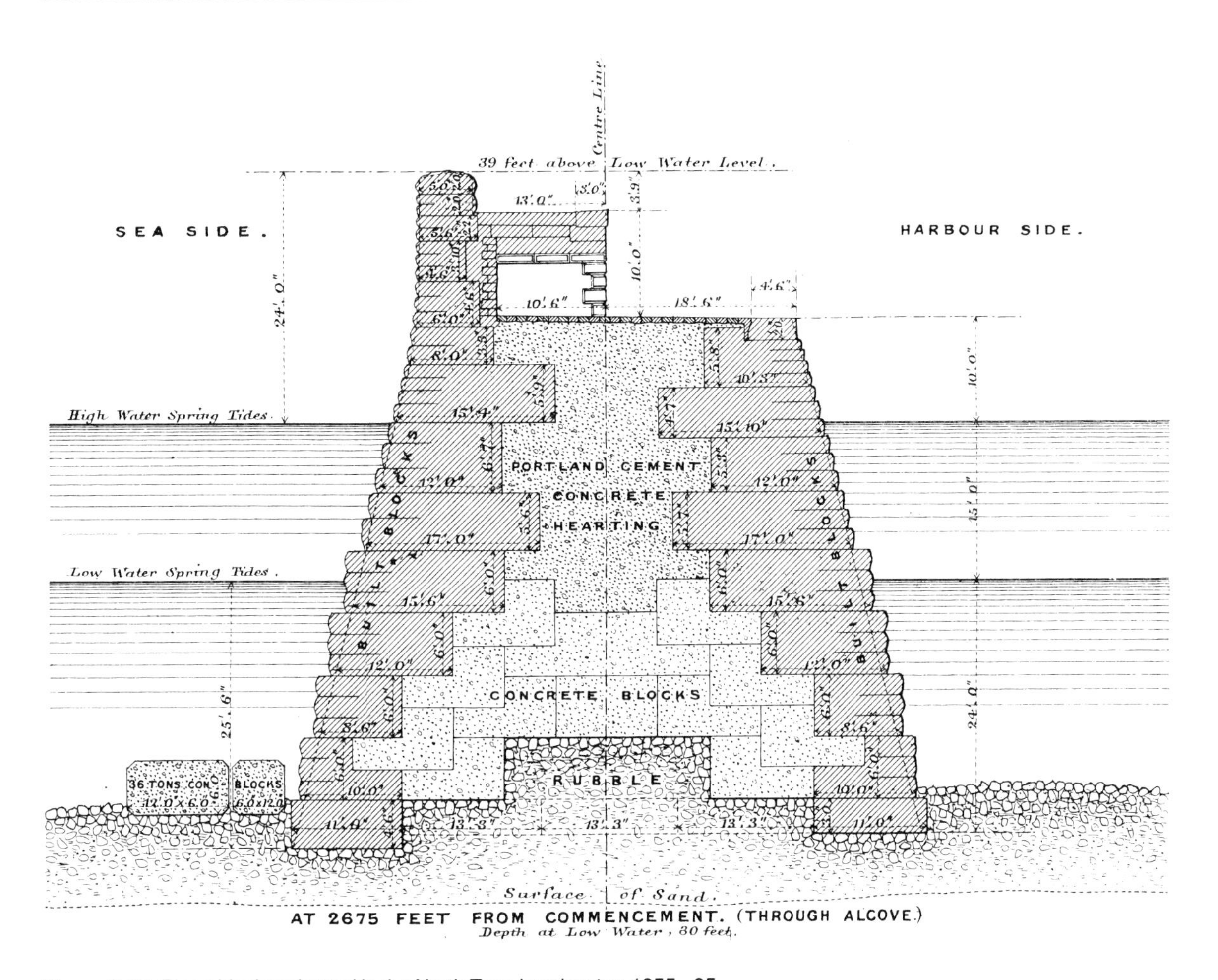

Figure 1.75 Piano blockwork used in the North Tyne breakwater, 1855–95
Source: Messant, P.J., 'Discussion on concrete work for harbours'. ICE Proc., Vol. 87, 1886, plate 8.

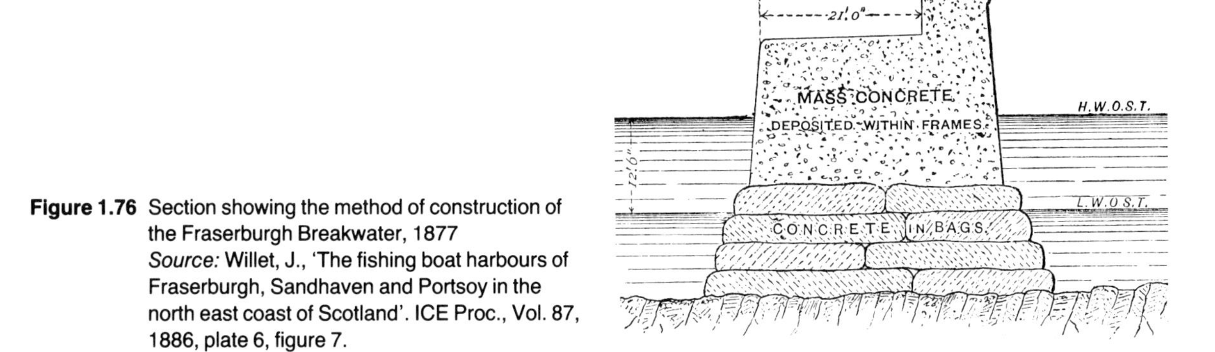

Figure 1.76 Section showing the method of construction of the Fraserburgh Breakwater, 1877
Source: Willet, J., 'The fishing boat harbours of Fraserburgh, Sandhaven and Portsoy in the north east coast of Scotland'. ICE Proc., Vol. 87, 1886, plate 6, figure 7.

Figure 1.77 The ancient manner of constructing the Cob in Lyme Regis, sixteenth century
Source: Smiles, S., *Lives of the Engineers*, Vol. II, John Murray, London, 1874.

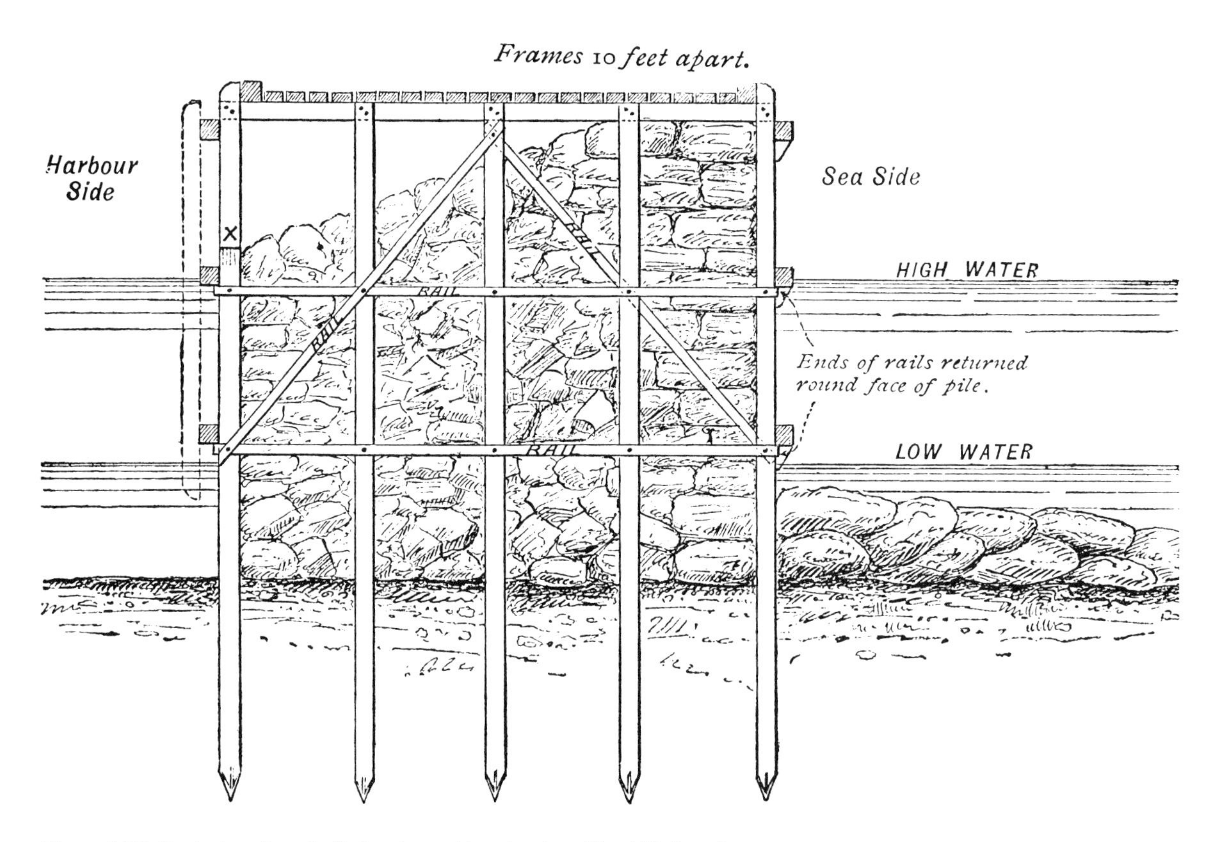

Figure 1.78 Typical section of a timber-framed breakwater with rubble hearting
Source: Shield, W., *Principles and practice of harbour construction*, Green and Co. 1895. Figure 28.

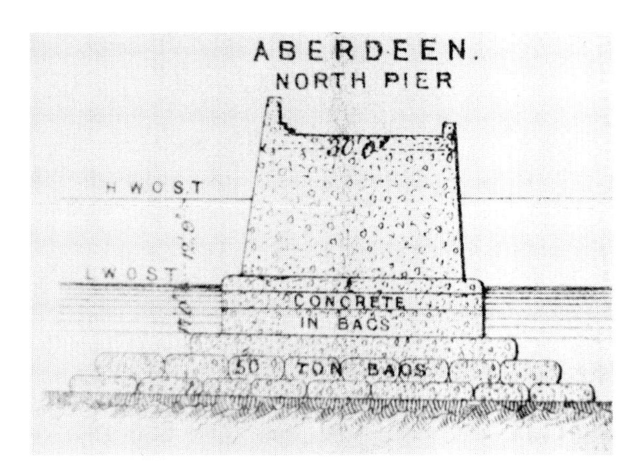

Figure 1.79 The North Pier, Aberdeen, 1877
Source: Vernon-Harcourt, plate 6, figure 16. (see 1.39)

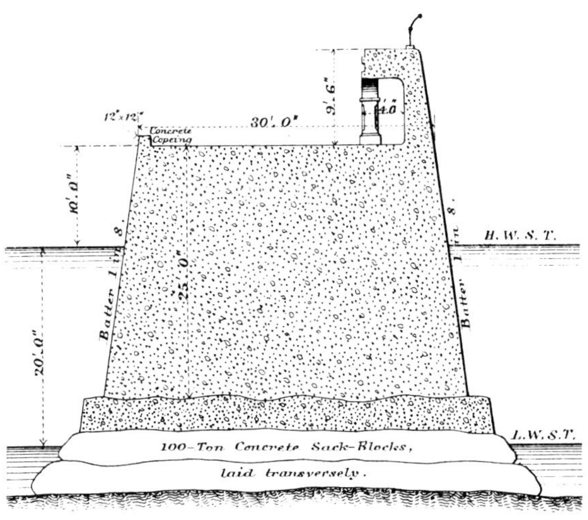

Figure 1.80 Breakwater at Newhaven, 1880
Source: Carey, A.E., 'Harbour improvements at Newhaven, Sussex'. ICE Proc., Vol. 87, 1886, plate 3, figure 7.

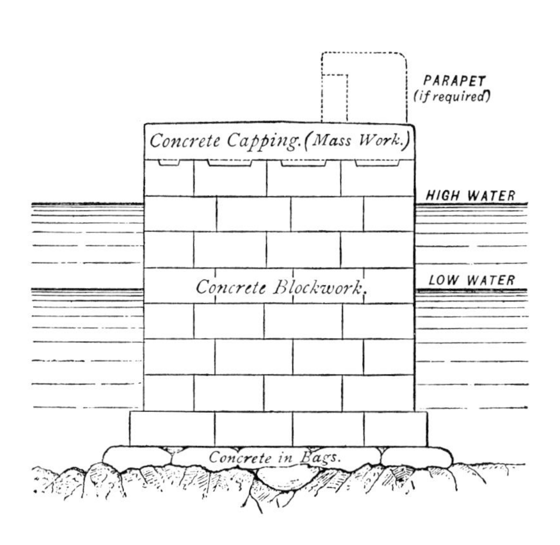

Figure 1.81 Breakwater formed of concrete blocks, with foundation bagwork keyed into rock
Source: Shield, W., *Principles and practice of harbour construction*, Longman Green and Co., 1895.

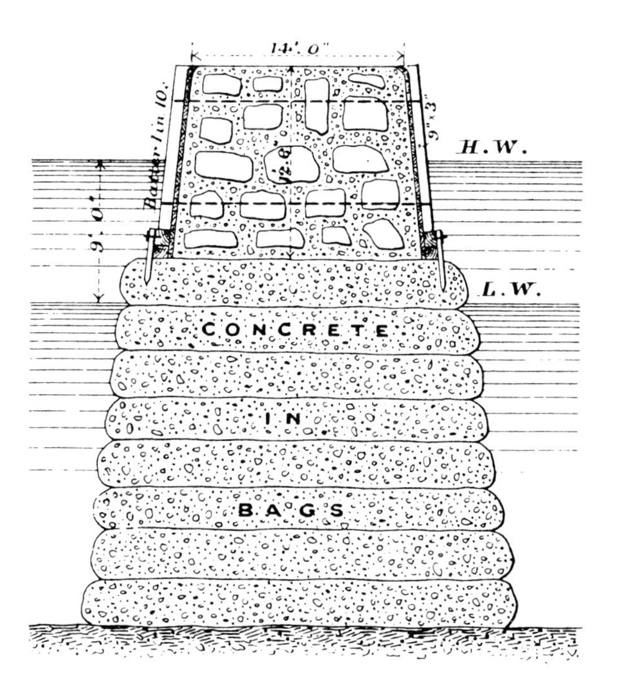

Figure 1.82 The outer portion of the breakwater at Ardrossan, 1892
Source: Robertson, R., 'Ardrossan Harbour extension,' ICE Proc., Vol. 120, 1895, plate 5, figure 10.

Base

The base of the breakwater was generally difficult to construct, because of the need to work under water in exposed sea conditions. A few bases are composed of large bags filled with concrete (Figures 1.76, 1.79 and 1.80). In some examples the bagwork has been keyed into the rock beneath to give resistance to horizontal sliding (Figure 1.81). In others the bags are massive and become a major part of the breakwater surmounted by a relatively small masonry structure (Figure 1.82).

Where it is possible to remove the overlying material, concrete may be cast directly on to a serrated rock formation.

In sandy locations rubble may be used to form a wide platform on which to construct the breakwater (Figure 1.73). This technique has been extended to the point at which the rubble mound is so massive, and reaches such a height, that many of the waves probably break before they reach the masonry construction (Figures 1.83 and 1.84).

Toe

The toe of a breakwater does not really perform the same function as the toe of a retaining wall. The breakwater toe usually acts as an anti-scour device and a wave spoiler. Toes are usually formed from either rubble (Figure 1.72), or concrete or stone blocks (Figures 1.85 and 1.75).

Top

The top surface of a breakwater performs a function which is not so pronounced in other types of waterfront wall. It is used to prevent water entering the internal zones of the structure, thereby preventing washout of hearting and the imposition of water pressure outwards on the outer skin of the structure. The surface may be constructed to a fall to aid the removal of water (Figure 1.71) or be sealed with granite pitching (Figures 1.86 and 1.87), or macadam (Figure 1.88).

Internal structure

Breakwaters may be constructed of solid masonry or concrete (Figure 1.89) or of masonry/concrete walls with a stone or boulder hearting (Figures 1.90 and 1.91). Some walls have only thin facings to the loose rubble hearting (see Figure 1.88). In some examples the blocks are held more rigidly in place by means of 'bag joggles'(Figure 1.85), granite fillets or rail cramps (Figure 1.92), the latter being used to tie in those blocks at the corners which would otherwise be prone to movement.

In some cases the blocks are laid on end at a slight angle to form 'slicework' which is an easier method of construction when using cranes. An early example of a form of slicework (1767) but restricted to the outer skin of the breakwater, is shown in Figure 1.91. A modern version of this technique is shown in Figure 1.93.

1.4 Sea walls for coastal defence

1.4.1 Function and design

Vertical sea walls are usually found at the tops of beaches where their main function is to prevent the sea encroaching further inland. Although this type of hard defence is now generally not favoured, due to the erosion caused by turbulence at the toe, there are still many vertical sea walls in existence.

The front face of the wall must resist forces imposed by the attacking waves and is often also designed to reflect the wave crests or breaking waves away from the top of the wall. The back face of the wall is usually hidden because the structure normally acts as a retaining wall or as a protective covering to a cliff face. There is thus no constraint on the shape of the hidden face.

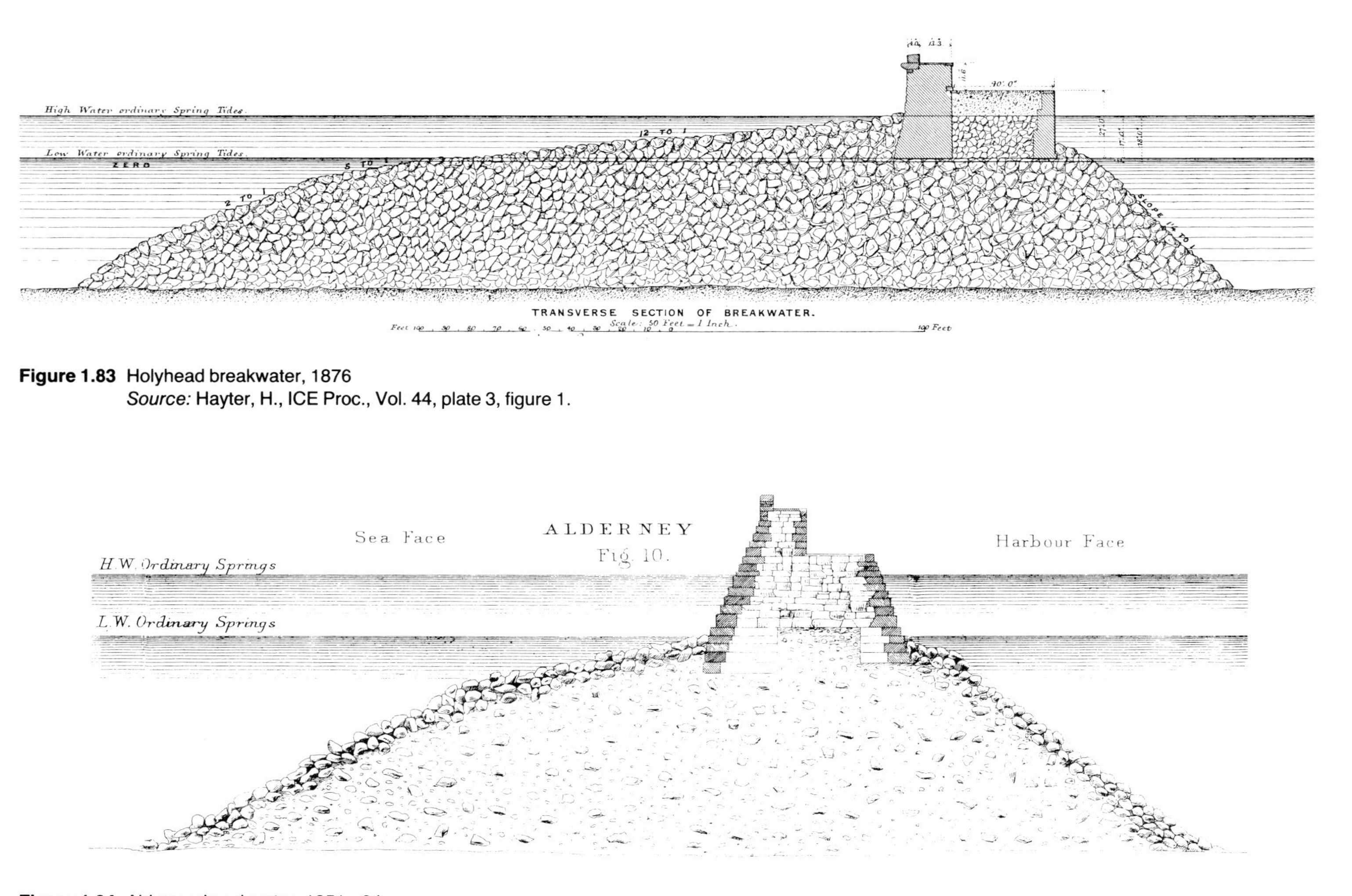

Figure 1.83 Holyhead breakwater, 1876
Source: Hayter, H., ICE Proc., Vol. 44, plate 3, figure 1.

Figure 1.84 Alderney breakwater, 1851–64
Source: ICE Proc., Vol. 22, plate 9A, figure 10.

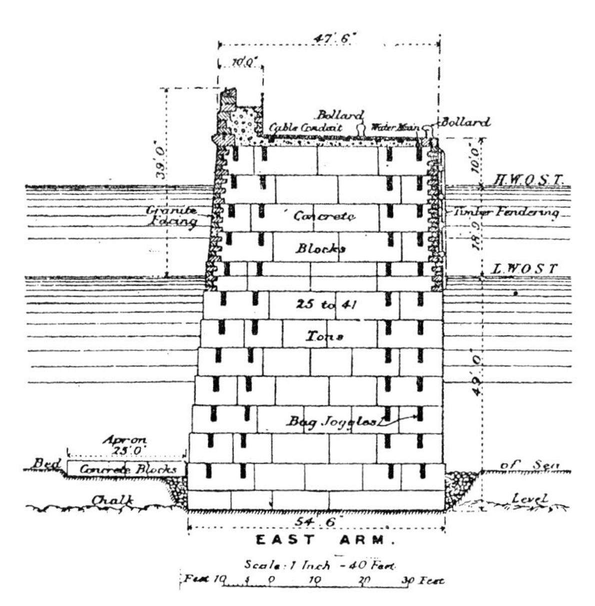

Figure 1.85 Section of the breakwater at Dover, 1898–1909, showing the disposition of bag joggles
Source: Brysson Cunningham, *A treatise on the principles and practice of harbour engineering*, figure 220.

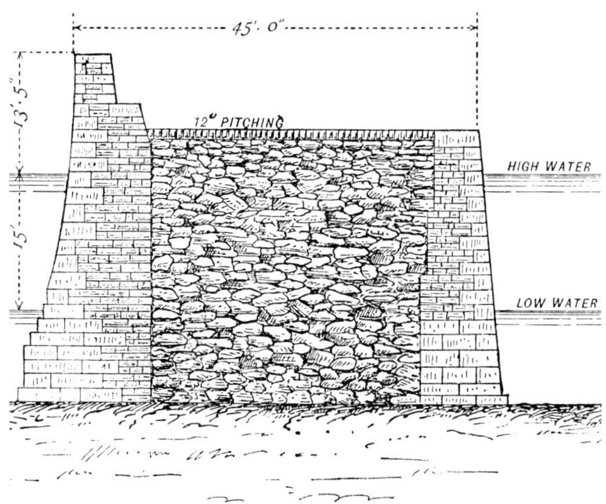

Figure 1.86 Kilrush Pier, Shannon, Eire, 1843
Source: Shield, W., *Principles and practice of harbour construction,* Longman Green and Co., 1895.

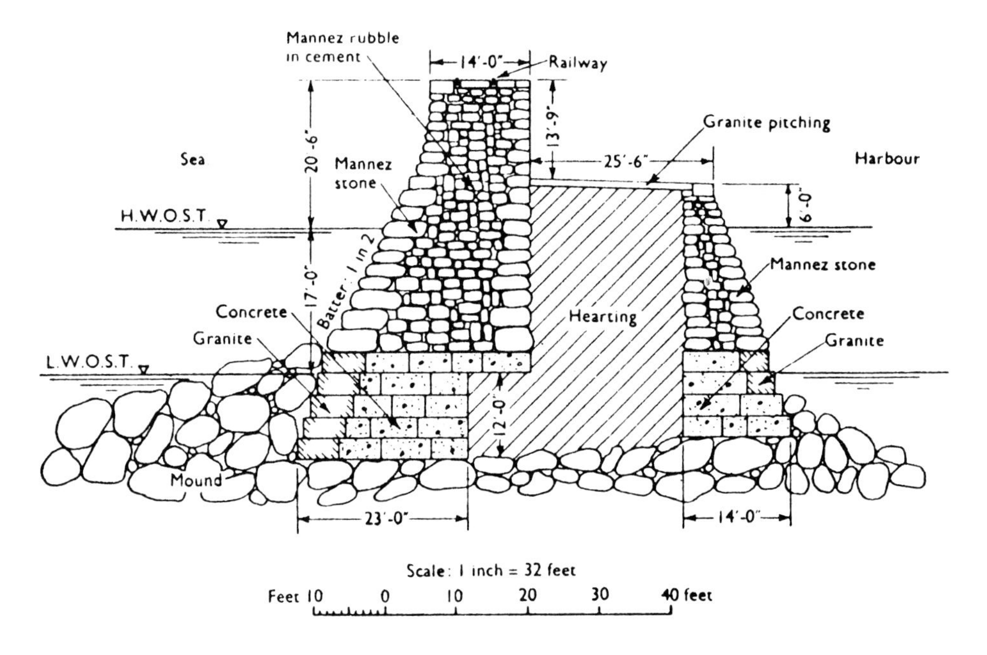

Figure 1.87 Section through Alderney breakwater, 1851–64
Source: King, J.L. and Bishop, R.W., 'Maintenance of some rubble breakwaters' ICE Proc., 1950/51, Vol. 9.

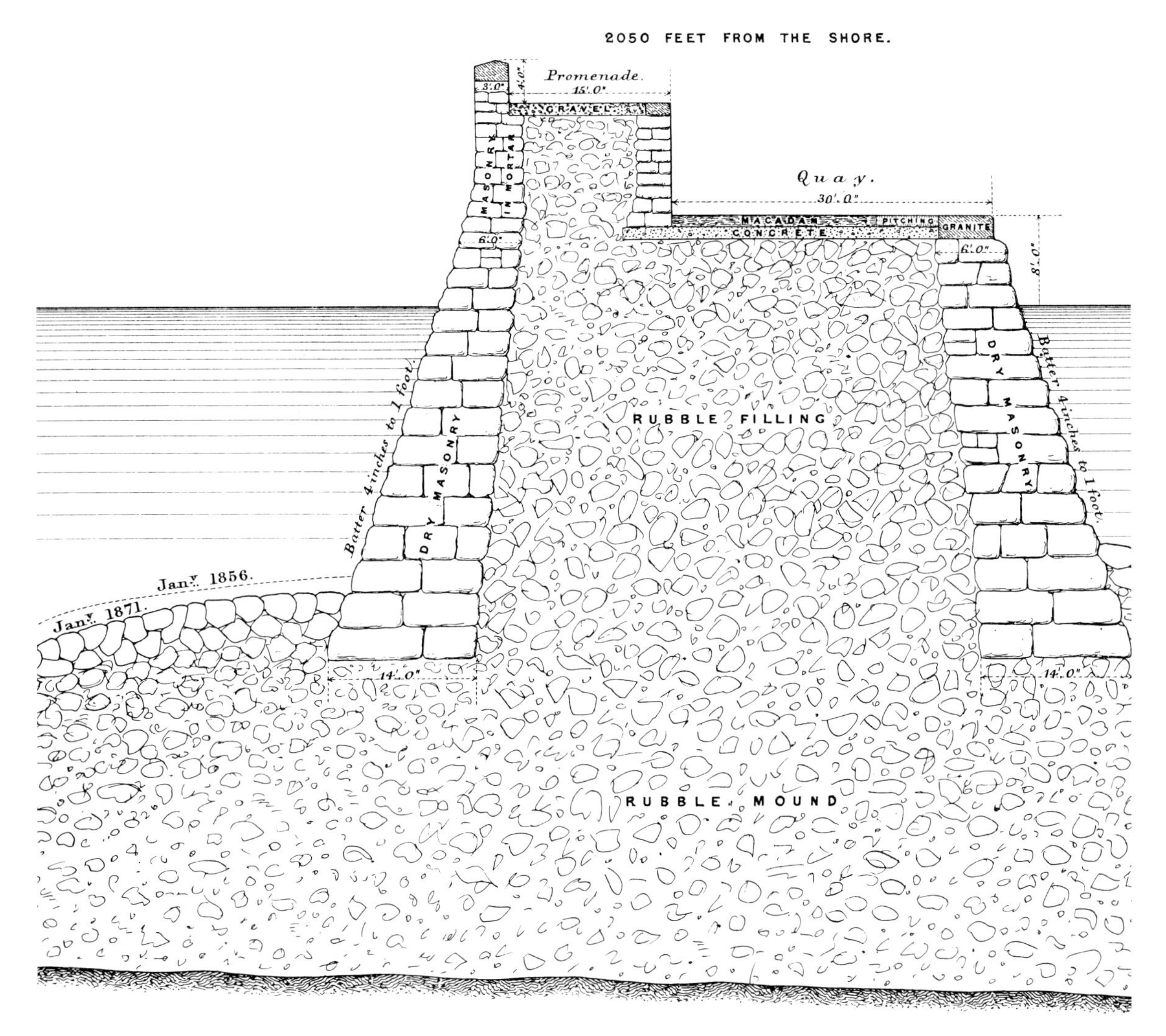

Figure 1.88 Section of breakwater at St Catherine's Harbour, Jersey, 1856
Source: Vernon-Harcourt, ICE Proc., Vol. 37, 1873/74. Part 1, plate 6.

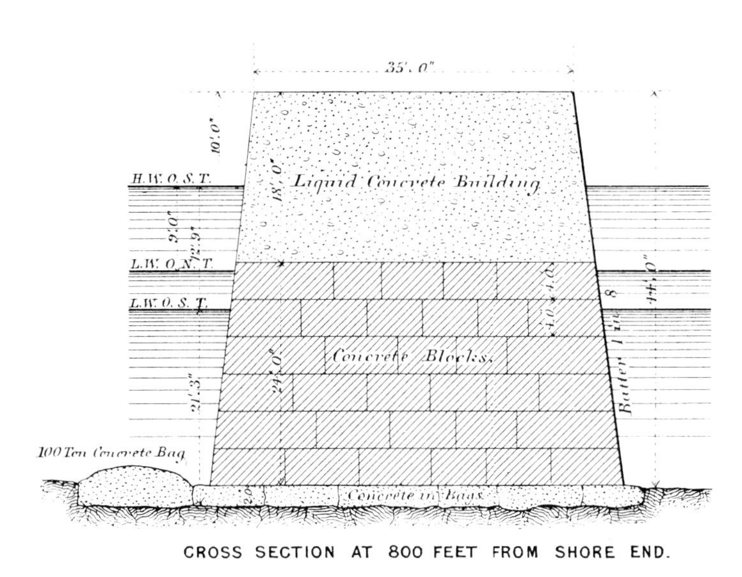

Figure 1.89 The south breakwater at Aberdeen, 1873
Source: Cay, W.D., 'New south breakwater at Aberdeen'. ICE Proc., Vol. 39, 1874, plate 10, figure 21.

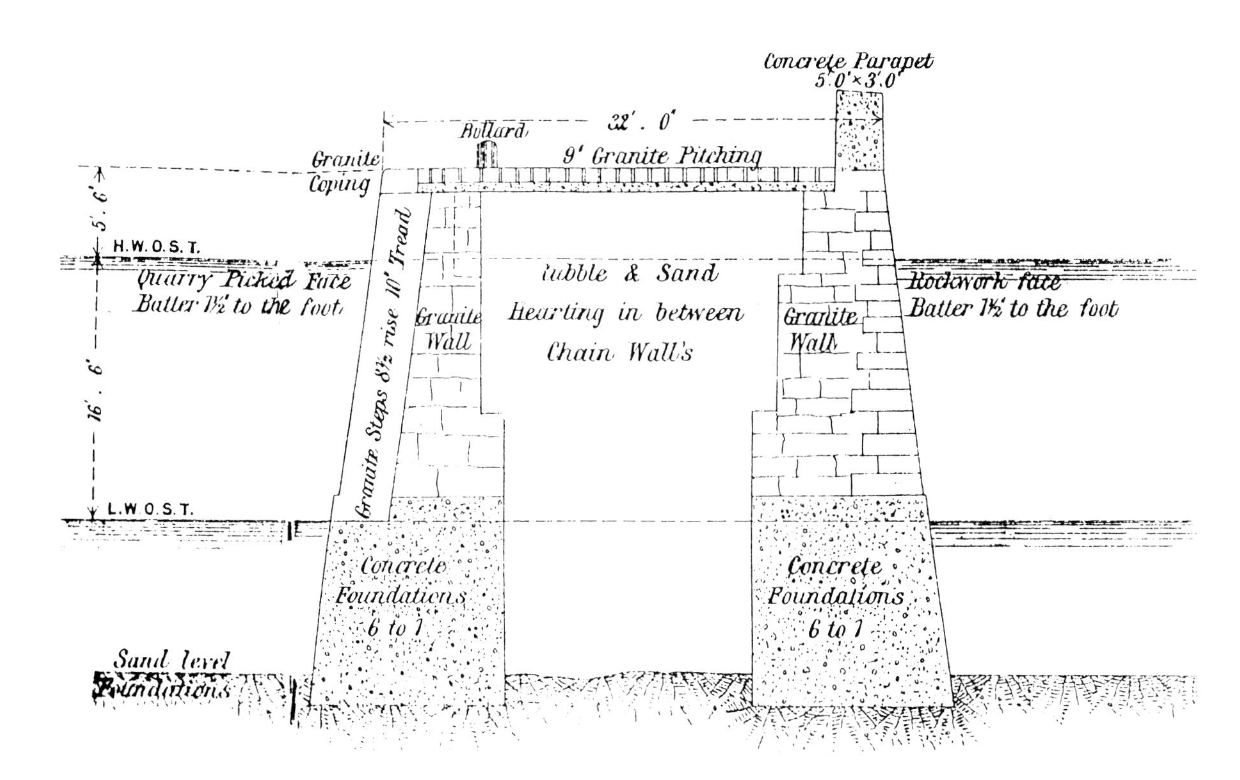

Figure 1.90 St Mary's Quay, Isles of Scilly
Source: Beckett Rankine.

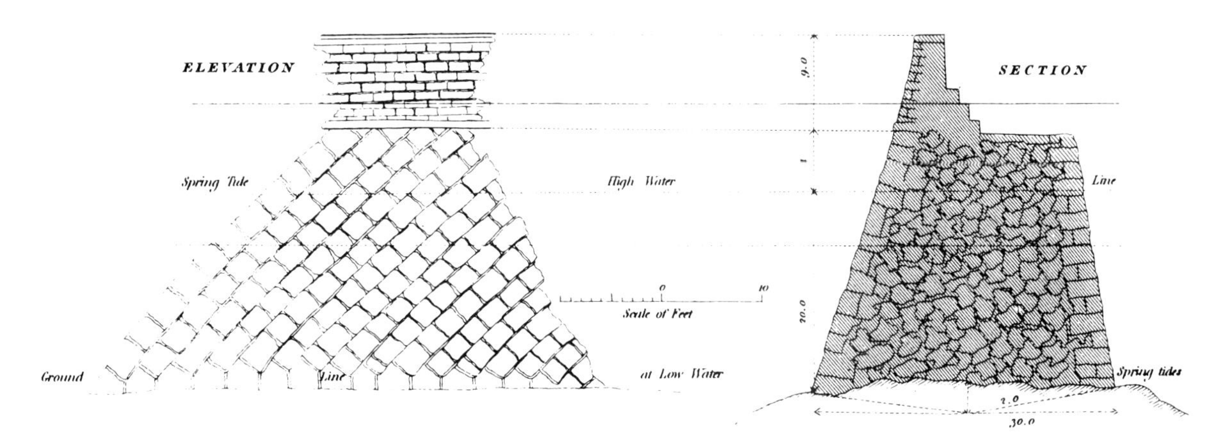

Figure 1.91 Section and elevation of the North Pier at Eyemouth, 1767
Source: John Smeaton FRS (Ed. Skempton, A.W.), Thomas Telford, 1981.

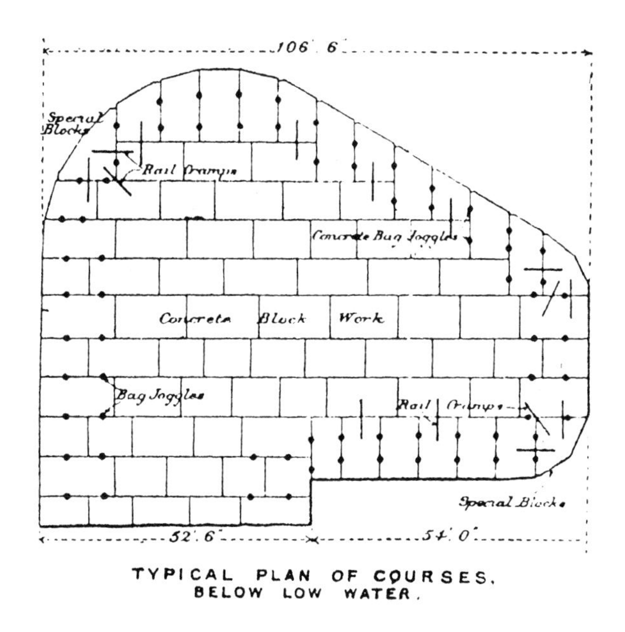

Figure 1.92 Typical sectional plan of the East Arm Pier Head at Dover, 1898–1909, showing position of rail cramps
Source: Brysson Cunningham. *A treatise on the principles and practice of harbour engineering.*

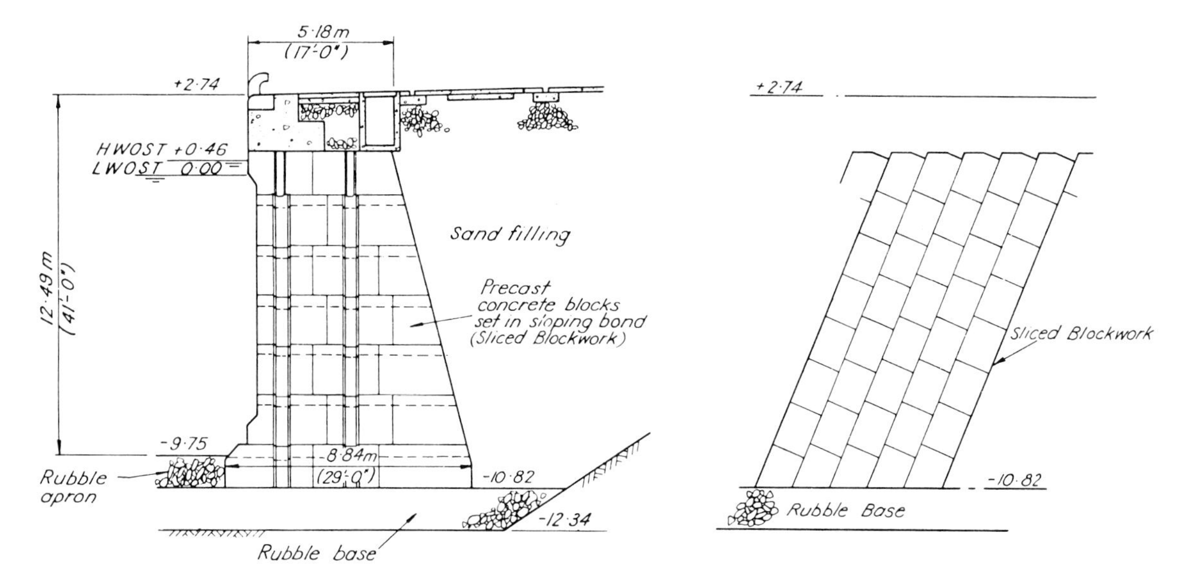

Figure 1.93 Sliced blockwork in Lagos Harbour, 1963–6
Source: National Ports Council. *Research project on port structures.* Bertlin and Partners, 1969.

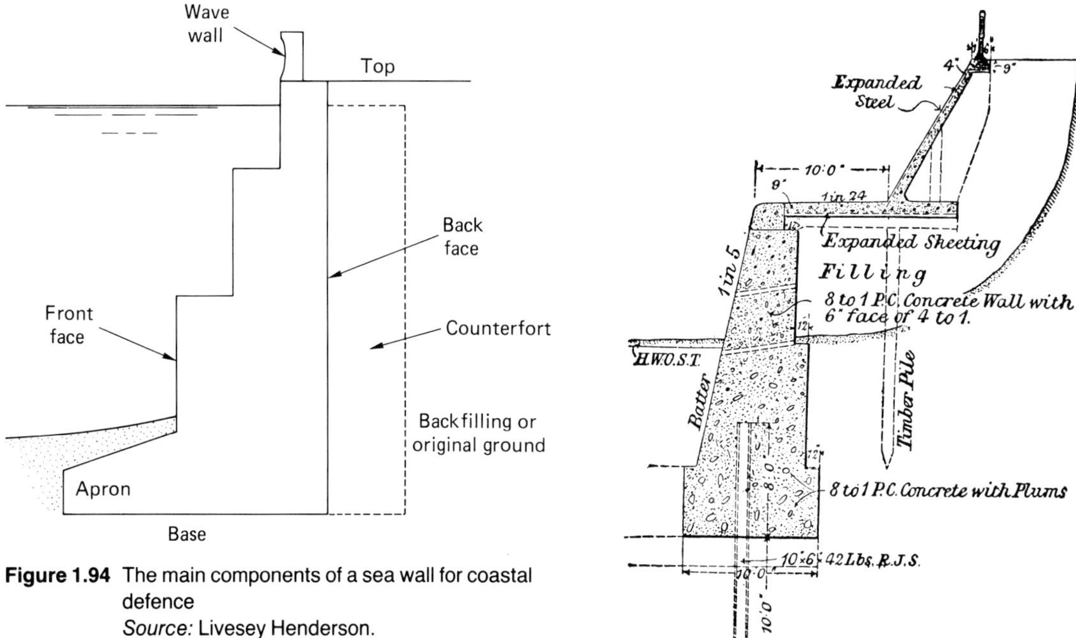

Figure 1.94 The main components of a sea wall for coastal defence
Source: Livesey Henderson.

Figure 1.95 The sea wall at Hornsea, 1907
Source: Ernest Matthews, *Coast Erosion and Protection.* Charles Griffin and Co. London, 1918. Figure 70.

Where the sea wall is a retaining structure, the top surface of the wall and the surface of the backfill need to be able to cope with the considerable volumes of water and spray which are thrown on to them in storm conditions. The toe apron of the wall is subjected to the scouring action of waves breaking against and above it.

The main components of a sea wall for coastal defence are shown in Figure 1.94. More details of sea walls are given in CIRIA Technical Note 125.[20]

1.4.2 Component characteristics

Outer face

The outer face of the wall may be truly vertical, battered slightly backwards (Figure 1.95) or curved (Figure 1.96(a)). Frequently this curve is continued upwards to the cope of the wall, forming a concave cross-section which has the effect of throwing the waves back towards the sea (Figures 1.97 and 1.98). Some walls have a stepped front (Figures 1.99 and 1.100).

Front faces are usually constructed from mass concrete, concrete blockwork or granite masonry. Concrete blockwork is sometimes faced with granite or basalt (Figure 1.97). In some cases the walls have been refaced with concrete on top of the original facing.

Back face

The back face is often vertical. In some cases a stepped effect as in dock walls is used (Figures 1.101 and 1.96(b)). Where the wall backs on to a cliff the back face may be integral with the cliff face (Figure 1.96(c) and see also 'skin walls', (Section 1.6)). Even in relatively modern walls counterforts may be used (Figures 1.97 and 1.100).

Base

The bases of sea walls are usually constructed on rock or clay substrata. Where firm material does not exist the base is often extended seawards to form a platform (see below).

Apron

Sea wall aprons are usually formed from mass concrete and protected at the seaward side by timber or steel sheet piles (Figure 1.98). Frequently the apron has been undercut by the sea and a new apron has been constructed below and in front of the original structure. Sometimes this occurs more than once and a series of new aprons is installed (Figure 1.102).

Internal structure

The internal structure of a vertical sea wall is usually simple, of mass concrete or blockwork throughout. Granular or rubble infilling is infrequent except in cases where the sea defences are widened to allow access to the beach or for a slipway (Figure 1.101).

Drainage

The water which falls on the landward side of the sea wall in times of storm or very high tides must be drained off in some convenient manner. Surface water is sometimes allowed to escape to the sea through scupper holes running through the wall to the front face at promenade or road level. Water which has percolated through the backfilling material may be drained off through a low-level drain running parallel to the wall immediately behind the back face (Figure 1.96(b)).

1.5 Retaining walls and flood defences

1.5.1 Function and design

Retaining walls and flood defences are generally found inland along the sides of rivers. They retain the river banks and prevent them from being eroded as well as acting as a flood defence at times of very high water levels. Canal walls are also included in this category. They do not usually have to be designed to cope with severe wave attack (although lakeside and canal walls may be) nor are they usually subjected to surcharge loadings from above (except when they form the edge of a road).

These walls often need to be modified because of a change in the hydraulic characteristics of the river, or to improve their flood defence function. It is therefore common to find that they have been raised by the addition of a higher coping, or that a new wall has been built in front of the original one. New walls are often integral with the old one. On the River Thames, new walls have been added twice in some locations. The main components of a retaining or flood defence wall are shown in Figure 1.103.

1.5.2 Component characteristics

In many respects these walls are constructed like dock walls, with near-vertical or battered front faces and stepped-back faces. Counterforts are also used to increase friction with the backfilling material (Figure 1.104).

River wall bases are frequently extended forwards to form an apron which provides scour protection. This apron may be formed of concrete or masonry or, quite frequently, a rubble mattress (Figure 1.105). It is also common to find that extra scour protection has been added to the wall some time after its original construction date.

Some rivers and canals have small masonry walls at water surface level to provide resistance against wave attack (Figure 1.107). This may consist of large coping stones placed above a timber or sheet-piled wall (Figure 1.106). In other instances there may be timber relieving platforms (Figure 1.108).

1.6 Skin walls

Skin walls were used when it was considered that the natural rock or hard clay material behind the wall was capable of standing to a

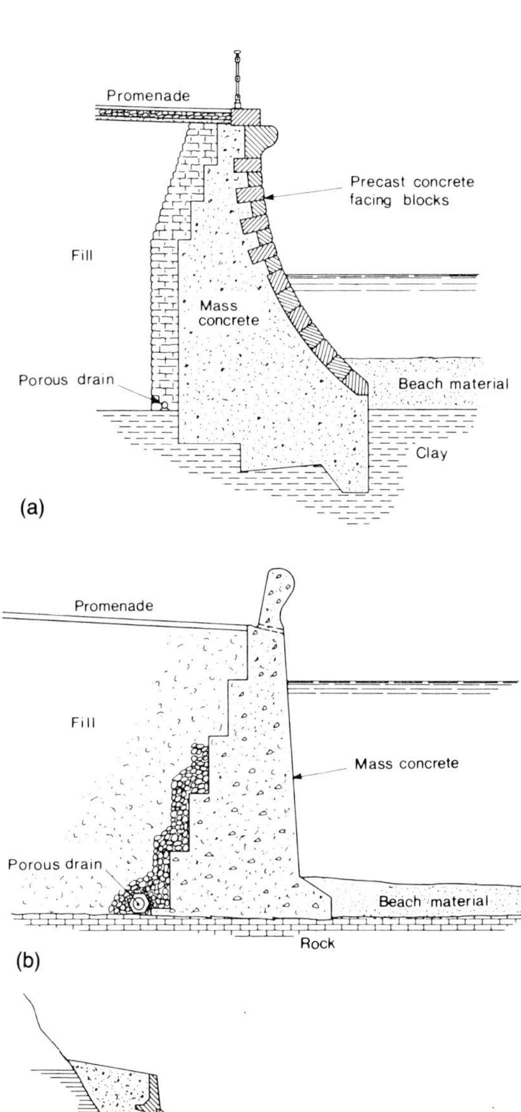

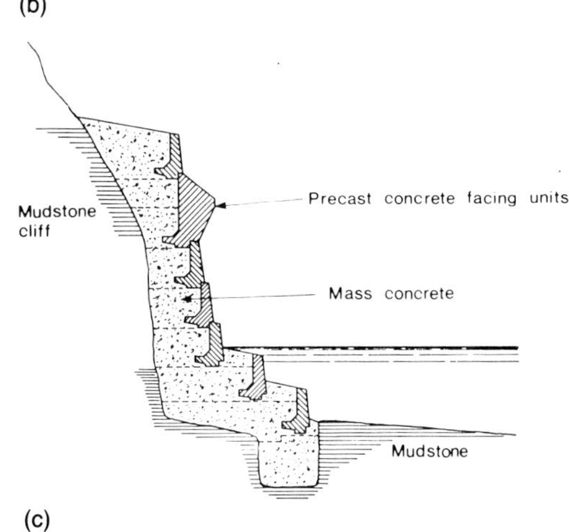

Figure 1.96 Typical vertical sea wall sections
Source: CIRIA TN 125.

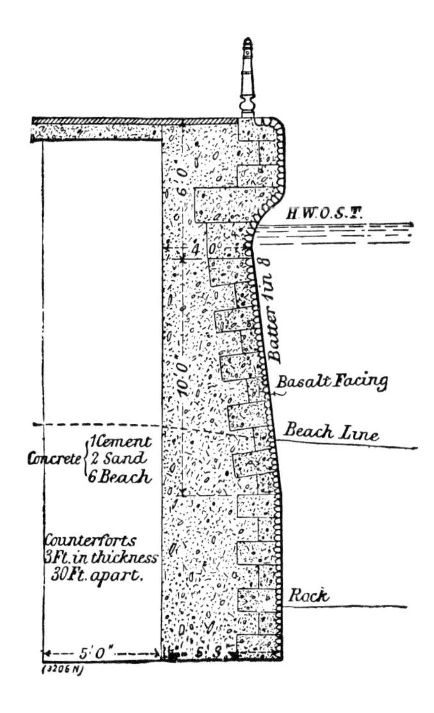

Figure 1.97 The sea wall at Caroline Place, Hastings, circa 1910
Source: Ernest Matthews, *Coast Erosion and Protection.* Charles Griffin and Co., London, 1918.

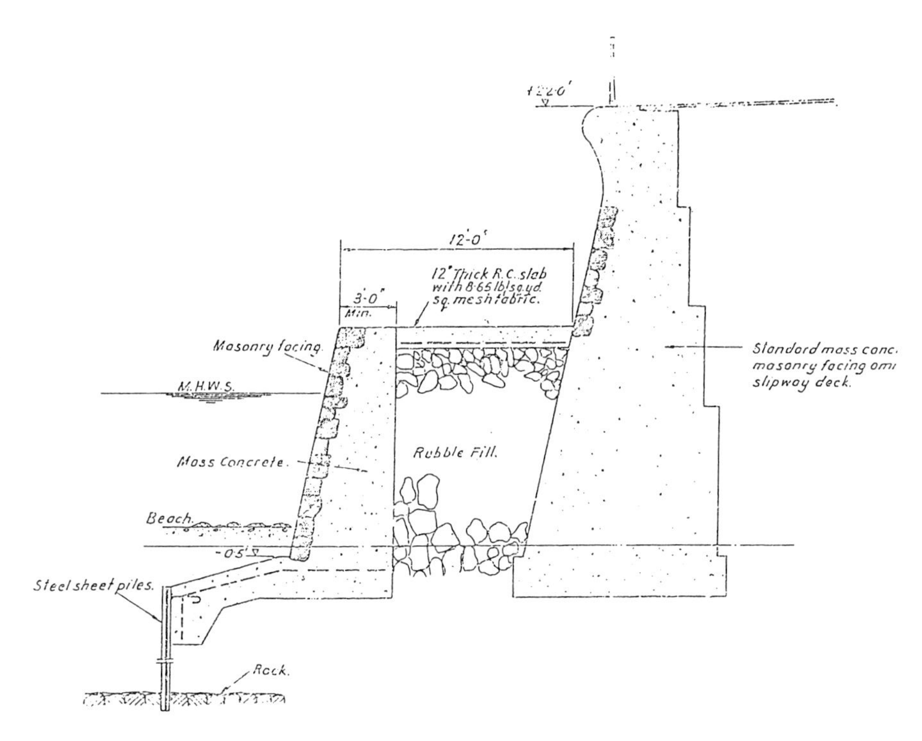

Figure 1.98 The sea defences at Penzance Harbour
Source: Penwith District Council.

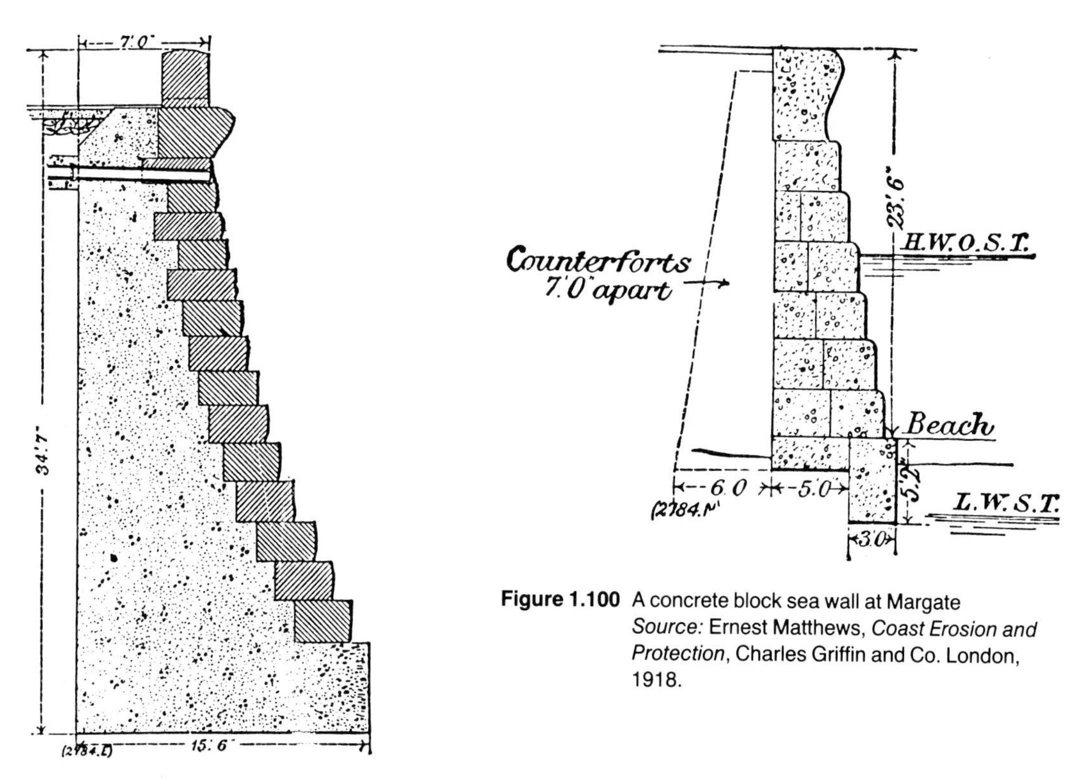

Figure 1.100 A concrete block sea wall at Margate
Source: Ernest Matthews, *Coast Erosion and Protection*, Charles Griffin and Co. London, 1918.

Figure 1.99 The Royal Prince's Parade sea wall, Bridlington, 1905
Source: Ernest Matthews, *Coast Erosion and Protection*, Charles Griffin and Co., London, 1918.

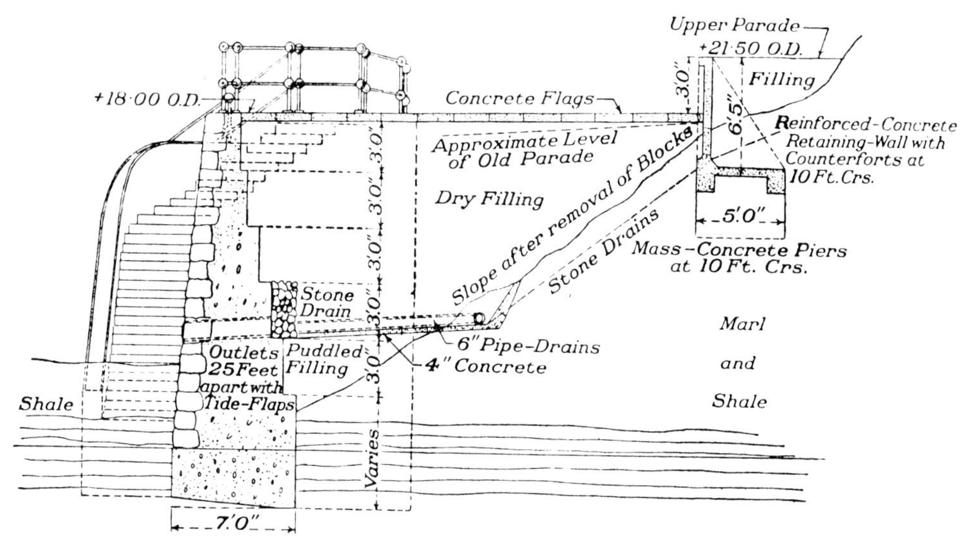

Figure 1.101 A section of the sea defences at Lyme Regis
Source: Clark, F.H. (1935) ICE Proc. SEP No. 177.

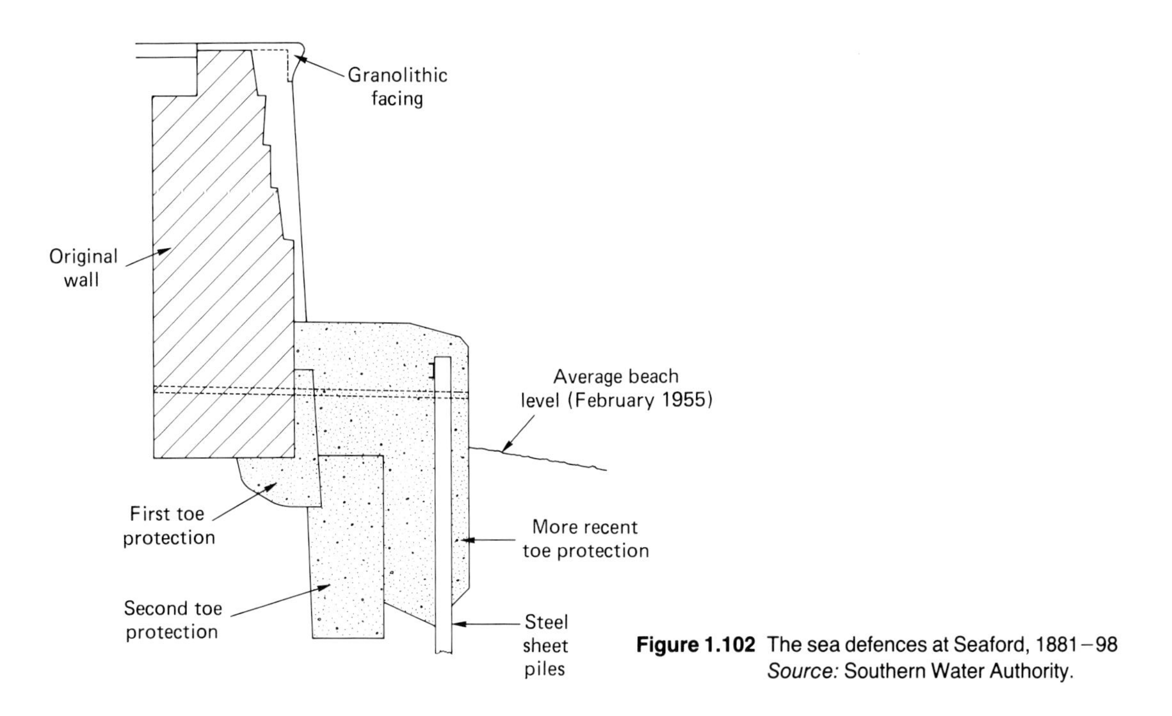

Figure 1.102 The sea defences at Seaford, 1881–98
Source: Southern Water Authority.

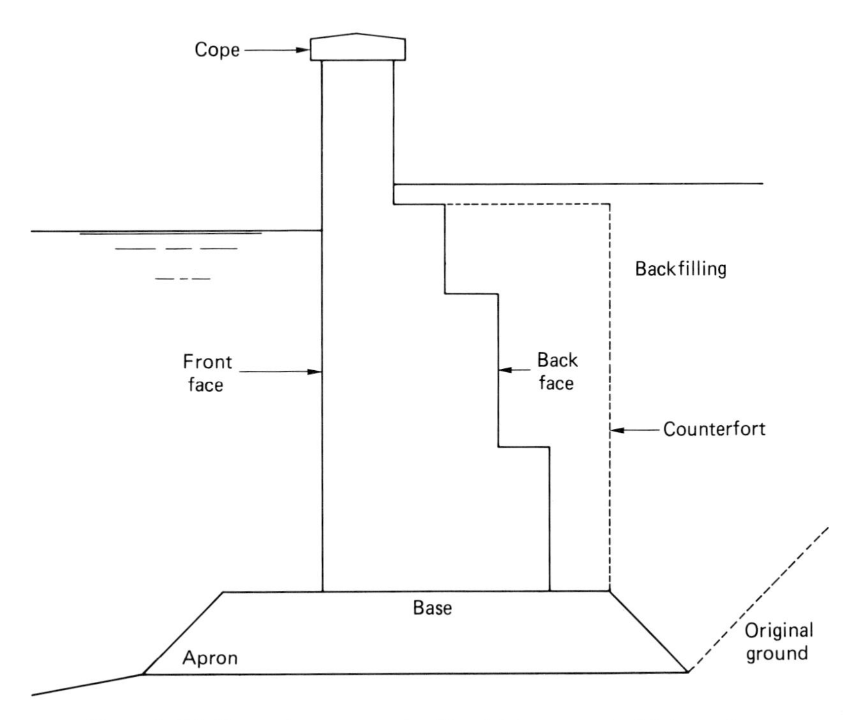

Figure 1.103 The main components of flood defences and retaining walls
Source: Livesey Henderson.

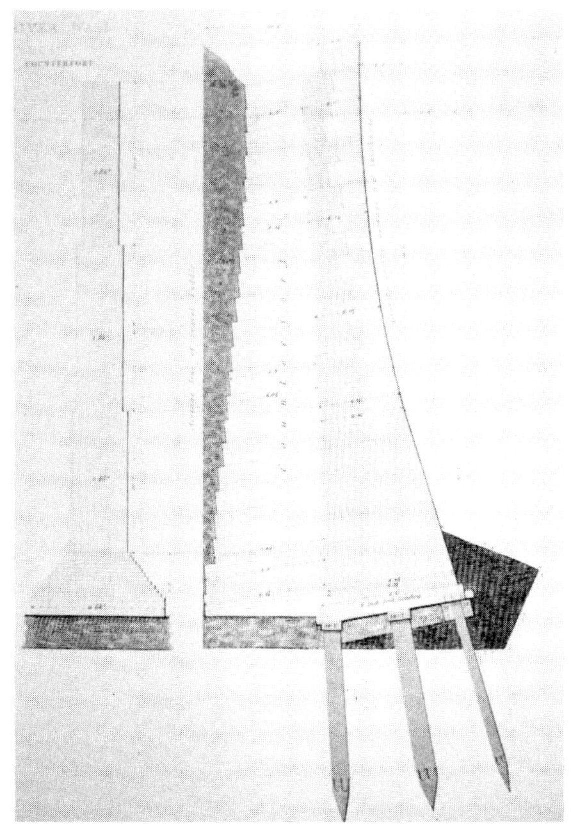

Figure 1.104 The Thames river wall at Fishmongers Hall, 1837
Source: Simms, *Public Works of Great Britain*, plate 96.

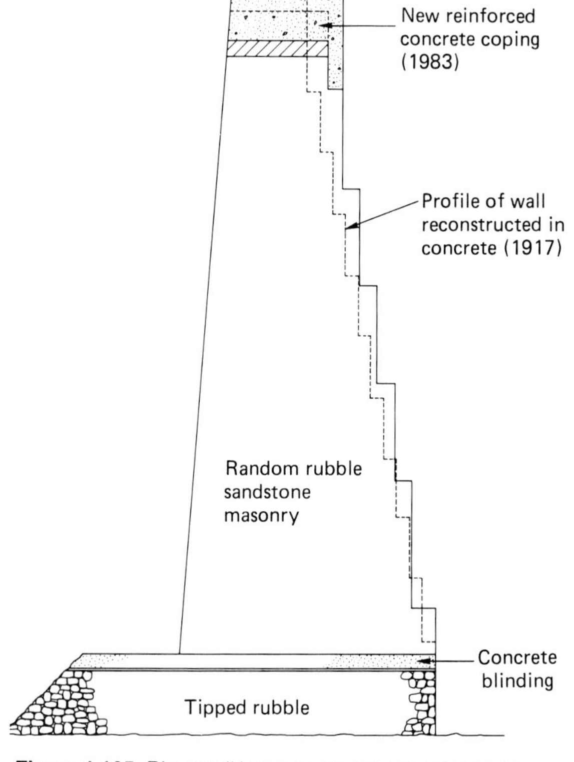

Figure 1.105 River wall between Manchester Dock and Chester Basin entrances, Liverpool
Source: Ronald Leach & Associates.

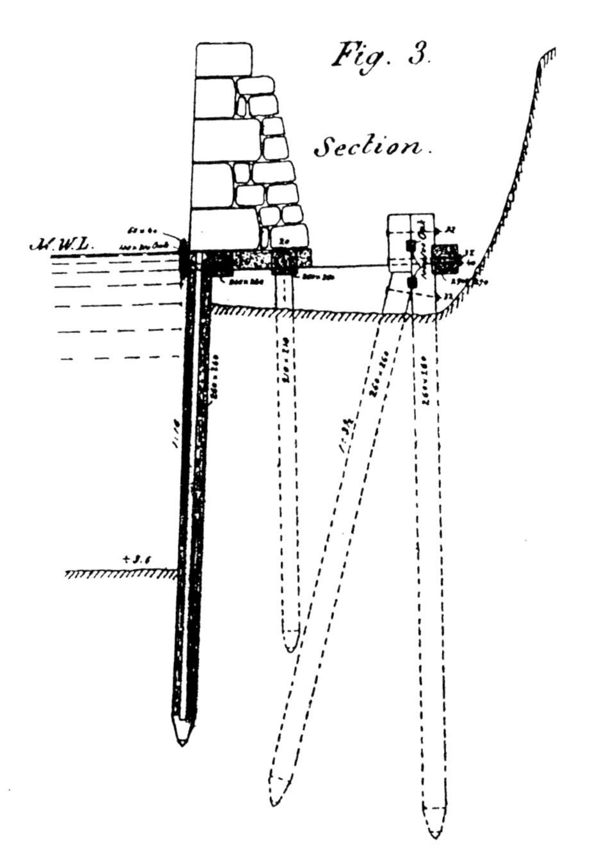

Figure 1.106 River wall on piles
Source: PIANC Congress 1912.

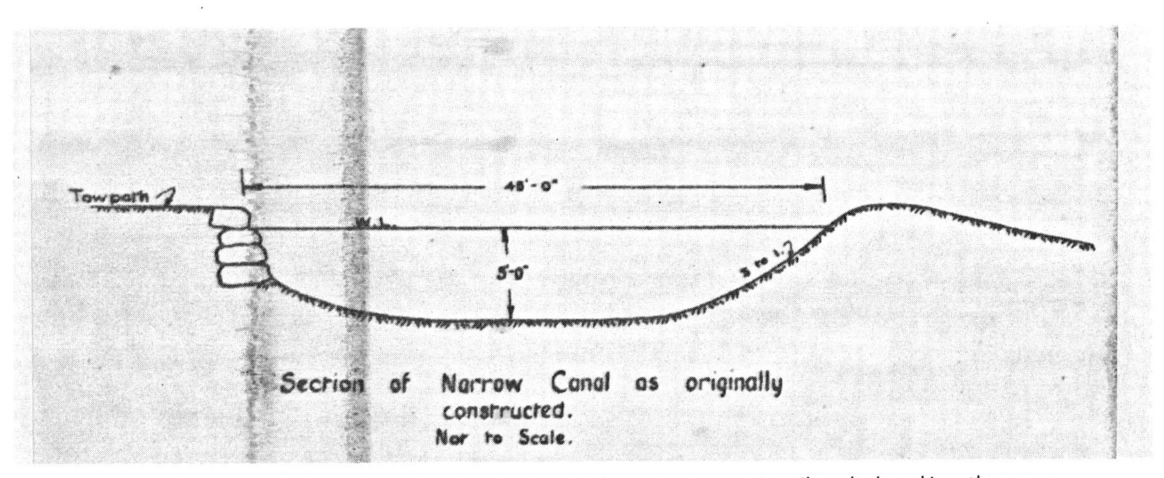

Figure 1.107 Section of Grand Union Canal wall, showing the masonry upper portion, designed to reduce scour by the wash from passing vessels
Source: British Waterways, South East Region, Grand Union Canal, Drg 2084.

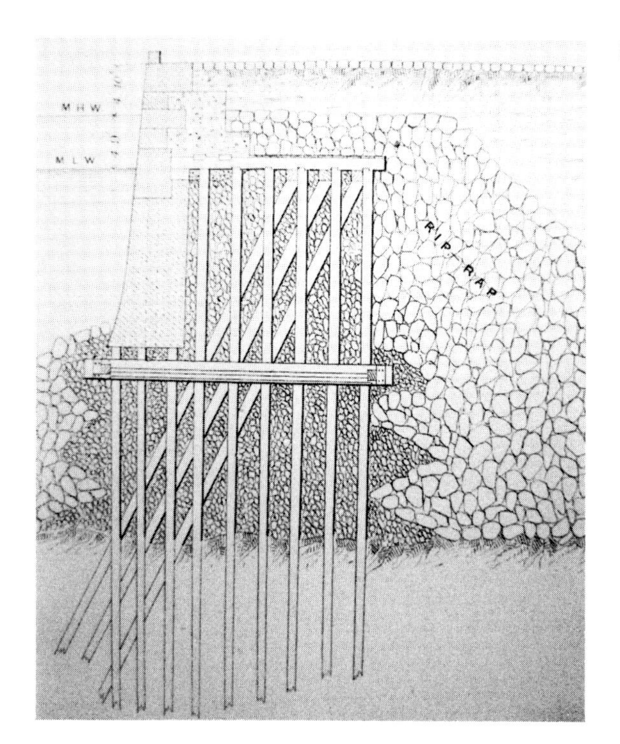

Figure 1.108 The North River, New York, circa 1883
Source: Vernon-Harcourt, plate 14, figure 18. (see 1.39)

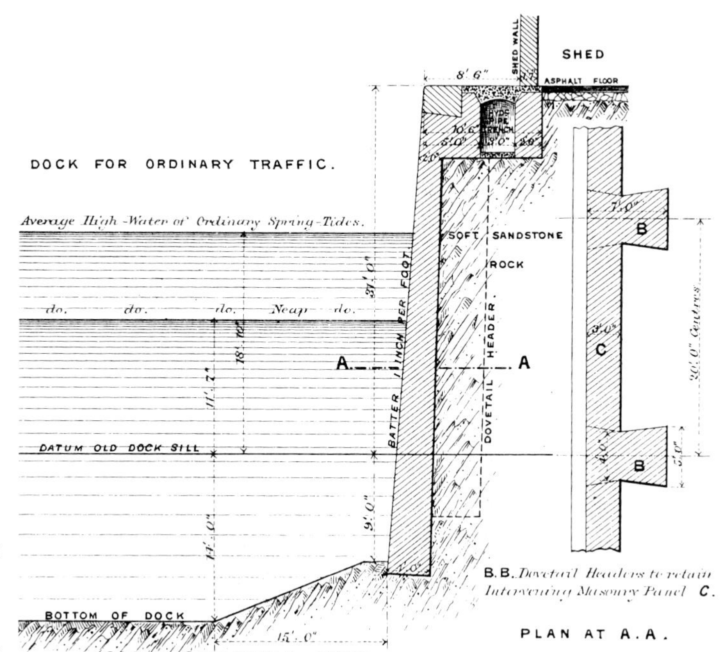

Figure 1.109 Section and plan of the Herculaneum Dock, Liverpool, 1873
Source: Lyster, G.F., 'Recent dock extensions at Liverpool with a general description of the Mersey Dock Estate, the port of Liverpool and the River Mersey'. ICE Proc., Vol. 100, 1890, plate 3.

vertical profile. In these circumstances it was only necessary to cover this material with a thin layer of concrete or masonry to protect it from the erosive action of the water, or perhaps to enhance the finish of the front face.

Walls of this type have been found in Liverpool (Figure 1.109), where the whole wall is a thin skin, and in Seaham (Figure 1.110), where the bottom half of the wall is of this type and the top half is more conventional. Kinipple[11] describes this form of construction which was used in the Garvel Graving Dock in Greenock. Outer skin walls can be found as sea walls on the Isle of Thanet, where they cover chalk outcrops, at Glentishen, where a masonry facing has been placed over shale, and at Robin Hood's Bay, near Scarborough.

An important aspect of a skin wall is the means by which it is attached to the ground behind. In the case of the Herculaneum Dock in Liverpool this was by the use of keys into the soft rock behind (Figure 1.109).

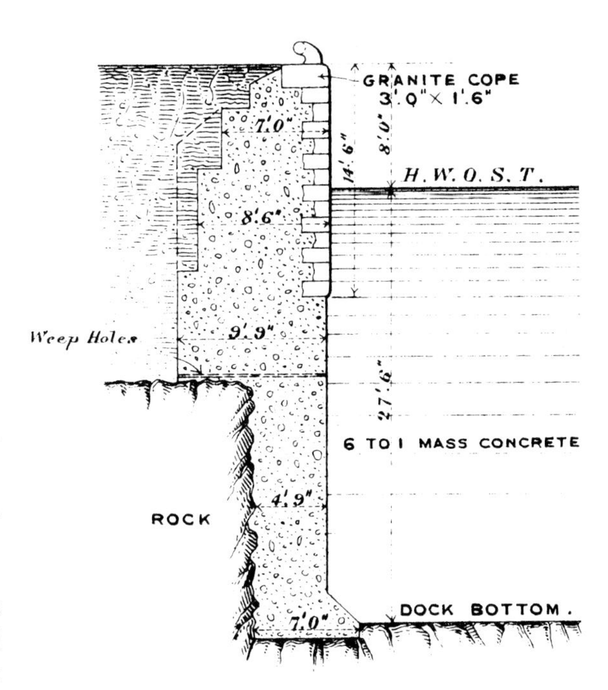

Figure 1.110 The new dock wall at Seaham, circa 1900
Source: Gask, P.T., 'The construction of the Seaham dock works.' ICE Proc., Vol. 165, plate 6, figure 6.

1.7 Bridge piers and abutments

Early bridge piers, like dock walls, were founded on piled foundations (unless hard material was discovered in the river bed), but the pier structures were more akin to the early breakwaters, being formed of rubble masonry inside a skin of ashlar (see Figure 1.111). One problem with the early piers was that they were prone to degradation from river currents and debris, and for this reason 'starlings' were built around the bottoms of the piers to provide some protection[21] (see Figure 1.112). As the starlings increased in size they began to restrict flow through the bridge openings and made the problem worse.

Designs for solid ashlar piers, which would alleviate this problem, were already in existence. An example by the Swiss engineer Labelye[22] is shown in Figure 1.113. An elegant combination of ashlar and rubble, for the Royal Border Bridge, is illustrated in Figure 1.114.

Bridge abutments were constructed in much the same way (see Figure 1.115). However, in the case of canal bridges the abutment was often modified to provide sufficient space for the towpath (see Figure 1.116). Canal bridge abutments are sometimes propped apart at the toe by an inverted arch, similar to those found in some docks.

1.8 Historical failures and repairs

A number of old waterfront walls failed soon after, or even during, construction. Others were modified to accommodate changes in service conditions. For example, many docks were deepened to accept deeper-draught vessels. Repairs and modifications are often hidden by the

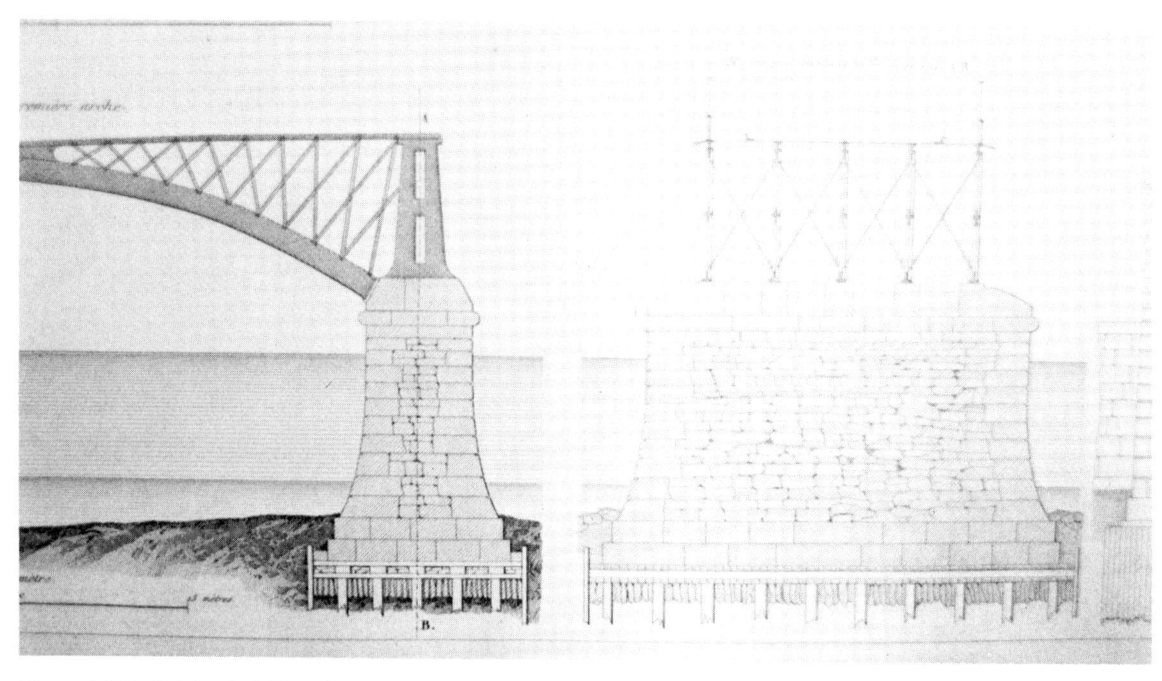

Figure 1.111 Details of a bridge pier
Source: Sganzin, plate 49, figure 261. (see Figure 1.8)

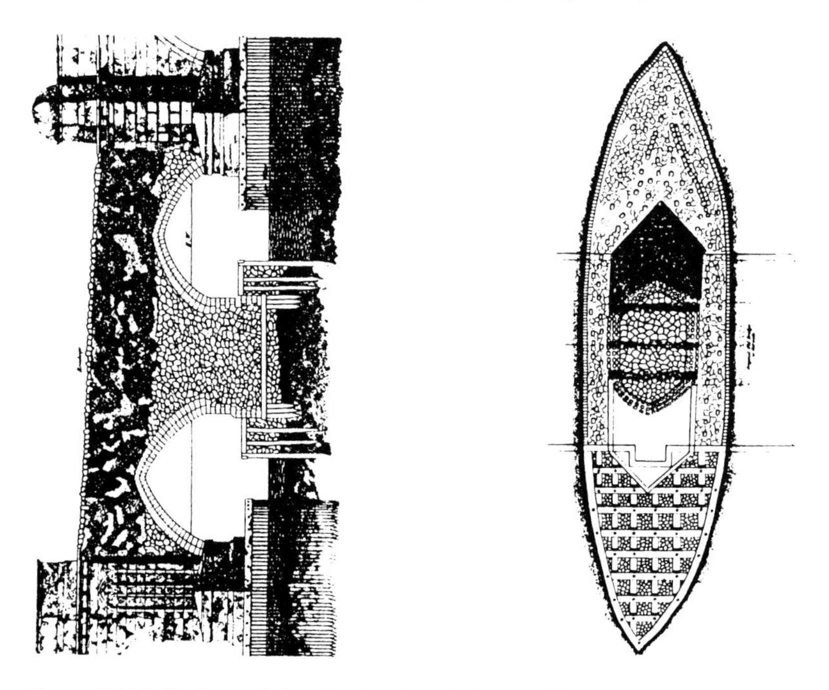

Figure 1.112 Section and plan of a pier of the old London Bridge as it appeared in 1826
Source: Home, G., *Old London Bridge*, John Lane, The Bodley Head, 1931.

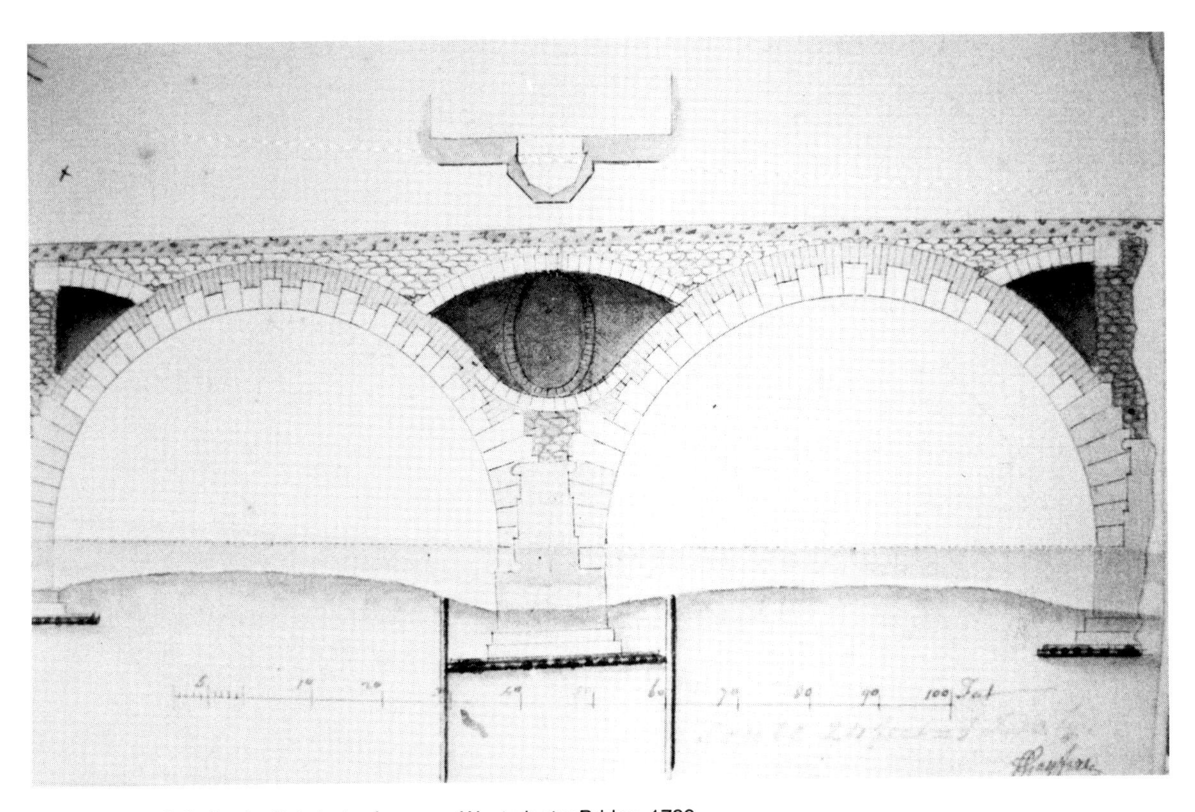

Figure 1.113 A design by C. Labelye for a new Westminster Bridge, 1739
Source: Labelye, C., *Short account for building a bridge at Westminster*, 1739.

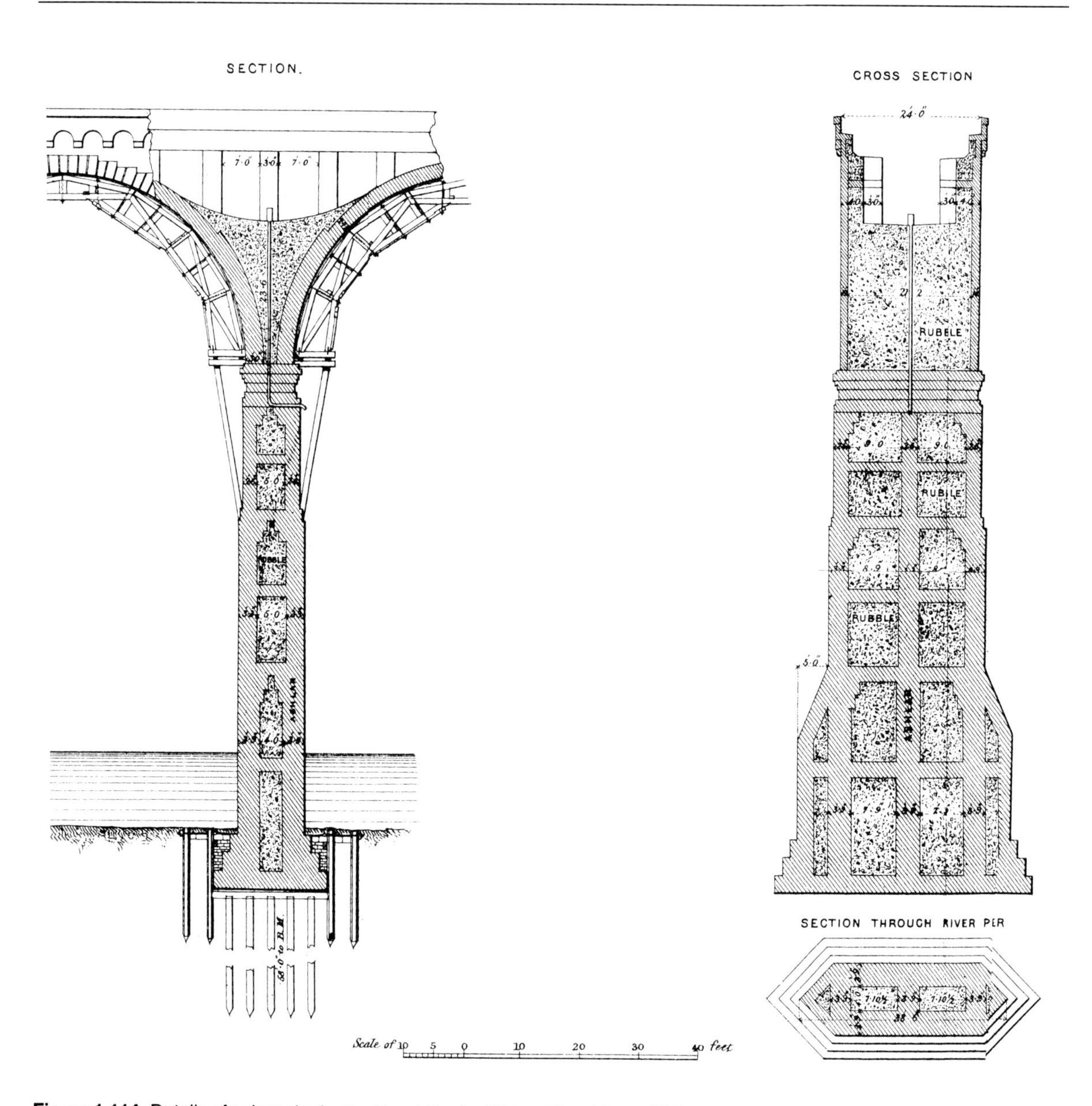

Figure 1.114 Details of a river pier for the Royal Border Bridge, River Tees, 1850
Source: Bruce, G.B. ICE Proc., Vol. 10, plate 3.

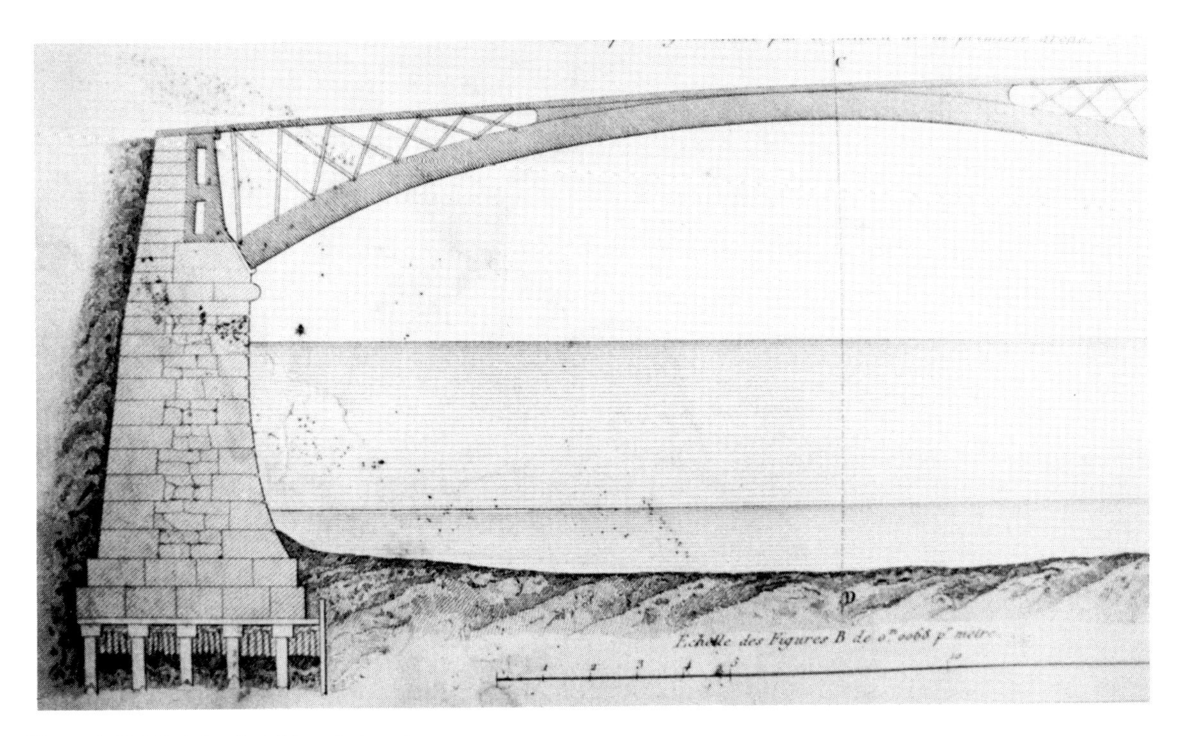

Figure 1.115 Details of a bridge abutment
Source: Sganzin, plate 49, figure 261. (see Figure 1.8)

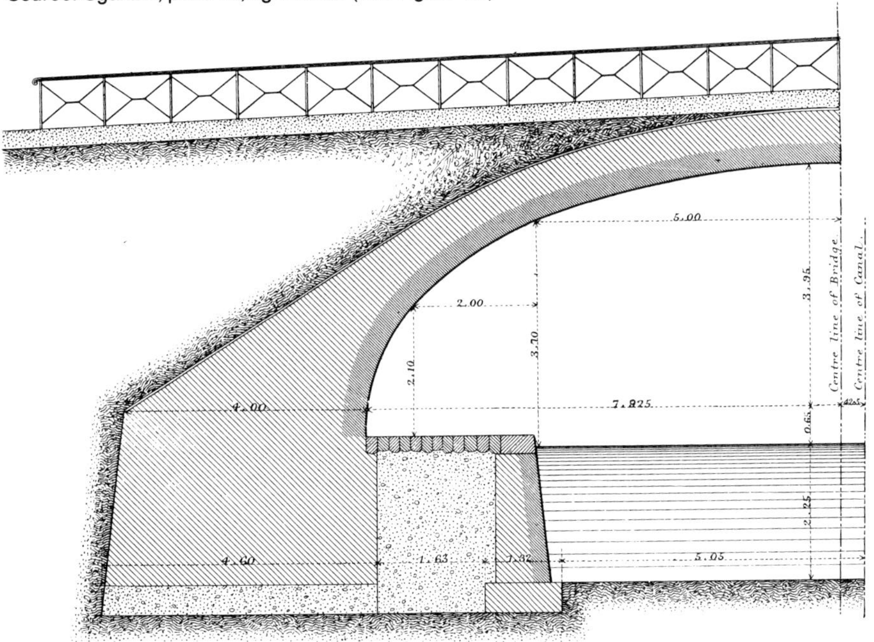

Figure 1.116 Typical detail at a canal bridge abutment, Eastern Canal of France, 1882
Source: Dawson, W.B., 'The Eastern Canal of France', ICE Proc., Vol. LV, 1878/79, Part 1, plate 8, figure 3.

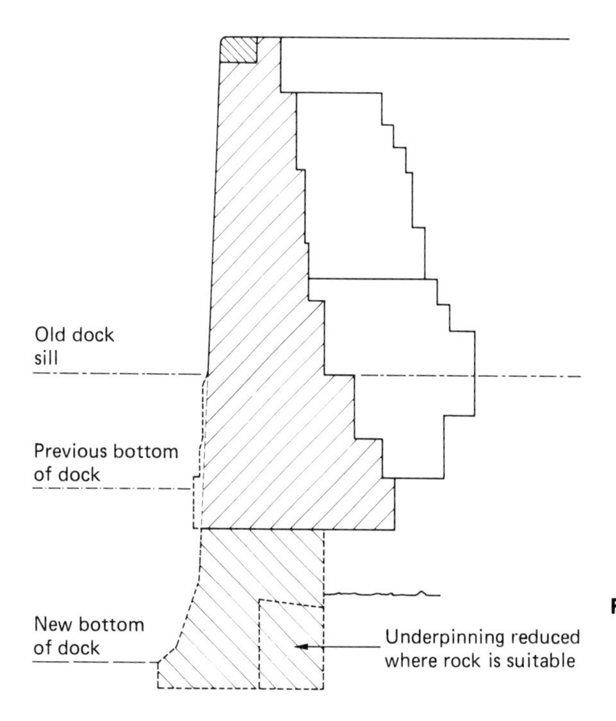

Figure 1.117 Section of the east wall of the Sandon Dock, Liverpool, showing underpinning carried out in 1903
Source: Mersey Docks & Harbour Company.

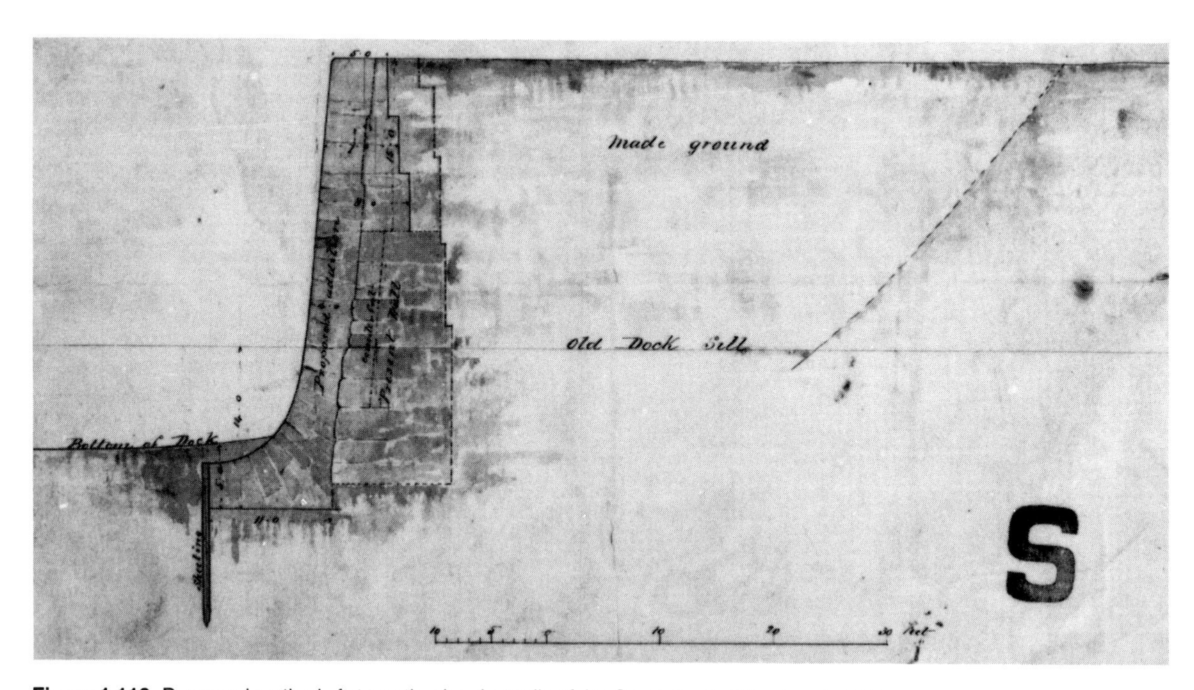

Figure 1.118 Proposed method of strengthening the walls of the Great Float, Birkenhead, 1858
Source: Mersey Docks & Harbour Company.

water and backfill, so it is prudent to review the kind of alterations made.

Various methods were used to strengthen walls so the berth in front could be deepened. In Birkenhead, a project was undertaken in 1858 to strengthen the wall of the Great Float by adding a new skin, new toe, and sheet piling. On the other side of the River Mersey, the East Wall of the Sandon Dock was underpinned in 1903 (Figures 1.117 and 1.118).

In Port Glasgow various types of waterfront wall were strengthened to provide for deeper water. The curved wall shown in Figure 1.119 was strengthened by the addition of a new sheet-piled facing, tied back to the original wall and running through to a mass concrete counterweight. Similar strengthening was carried out on the wall shown in Figure 1.120 but in this case the tie rod was taken back to an anchor block. Another wall in the same location was completely buried by a more recently constructed wall (see Figure 1.121).

In some instances walls began to fail during construction and designs were adapted to prevent failure. Figure 1.122 shows a dock wall under construction at Sharpness where the design has evidently been modified to stabilize an unstable portion of the structure.

Where the wall had already been constructed but it was decided to further strengthen the structure a number of novel solutions were used. Figures 1.123 and 1.124 show two completely different methods in Gloucester. In the first a large weight has been added to the back of the wall to bear on an oversail, presumably to act as a counterweight, and in the second counterforts have been added which extend well below the base of the original wall.

In some cases the complete failure of a section of wall has forced the engineer not only to rebuild the fallen section, but also to strengthen that part which did not collapse. For instance, in Belfast[23] a collapsed section (Figure 1.125) was built to a different design (Figure 1.126) and the portion of wall still standing was tied back and encased in a new structure (Figure 1.127). In Limerick[24] the dock wall shown in Figure 1.128 collapsed due to a toe failure (the timber grillage gave way between pile heads). This failure was caused by a combination of heavy rain and very low water. The engineer's original proposed reconstruction (Figure 1.129) was considered to be too expensive and he was forced to use the modified design shown in Figure 1.130.

Collapsed walls were not always rebuilt. In Barrow[25] an open-piled quay was built on top of the collapsed portion of an old waterfront wall (see Figure 1.131).

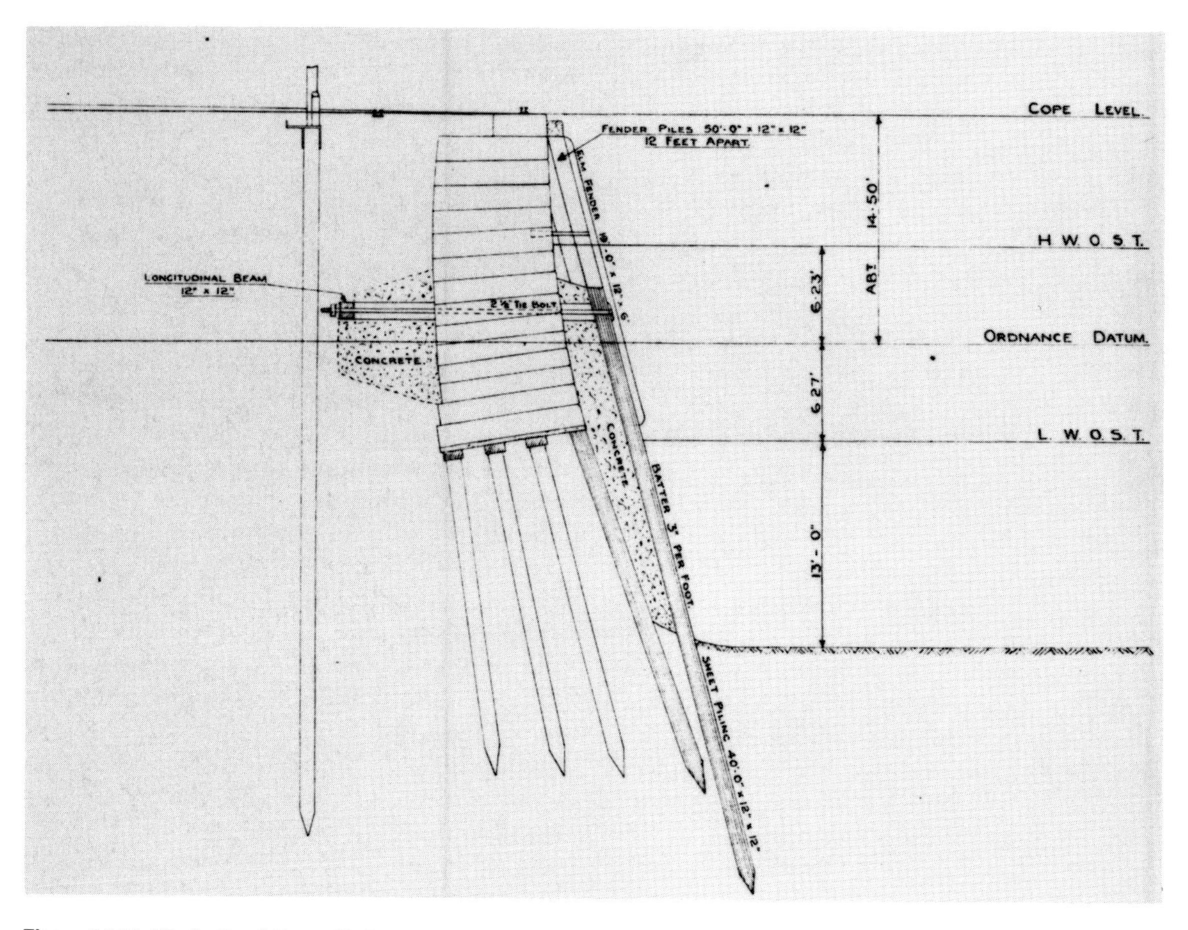

Figure 1.119 Windmillcroft Quay, Clyde, constructed in 1838, strengthened in 1884
Source: Clyde Port Authority and Glasgow District Council.

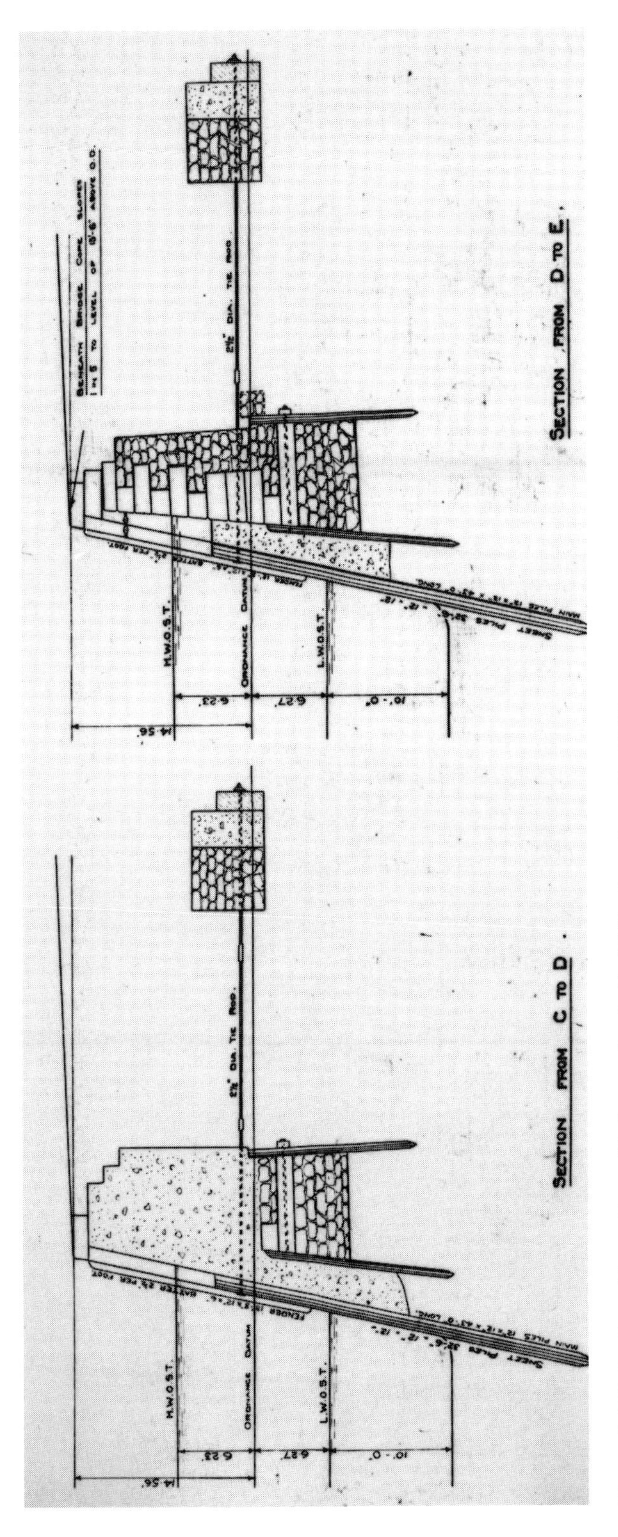

Figure 1.120 Customs House Quay, Clyde, constructed in 1852, strengthened in 1887
Source: Clyde Port Authority and Glasgow District Council.

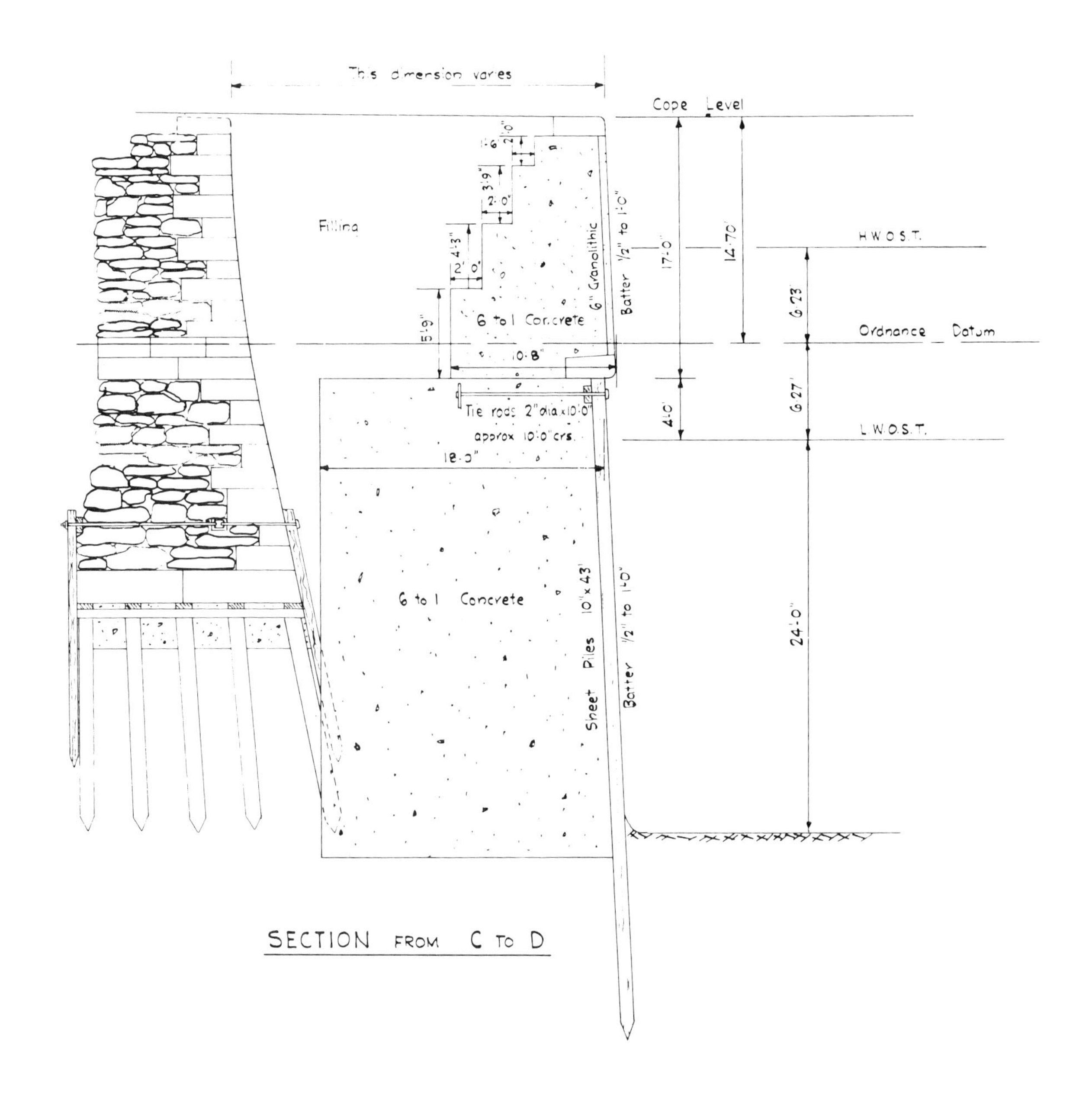

Figure 1.121 General Terminus Quay, Clyde, original wall constructed in 1849–50, reconstructed in concrete in 1932–4
Source: Clyde Port Authority and Glasgow District Council.

Figure 1.122 Photograph showing construction work on a dock wall in Sharpness, 1874. Propped walls indicate that some construction problems had been encountered
Source: British Waterways,

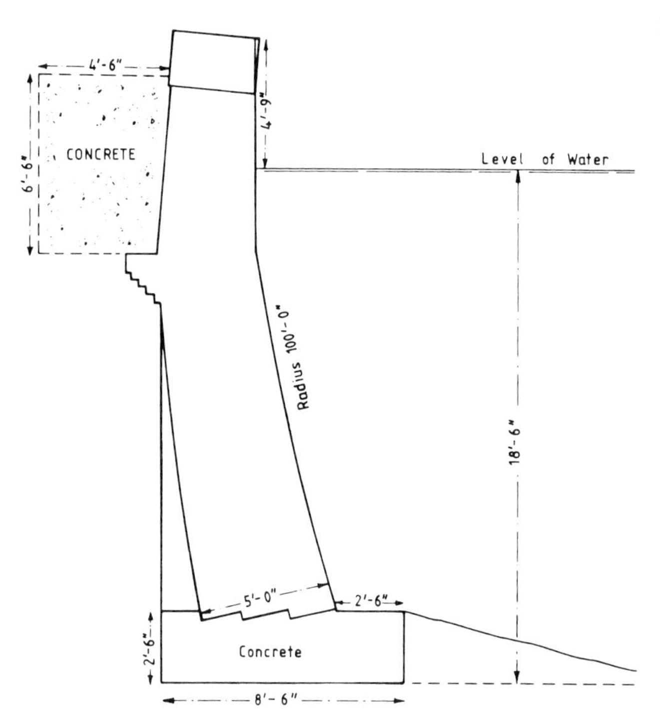

Figure 1.123 Section of a dock wall in Gloucester, 1852, showing the addition of a large concrete block above the oversail
Source: British Waterways Archives, Gloucester.

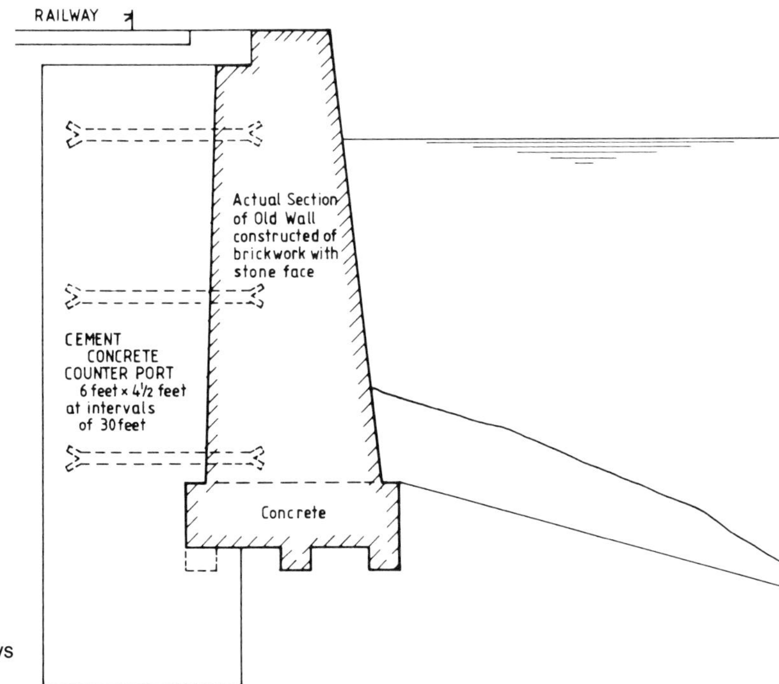

Figure 1.124 Section of a dock wall in Gloucester showing a counterfort which has been added, and extended below the original foundation level, 1908
Source: British Waterways Archives, Gloucester.

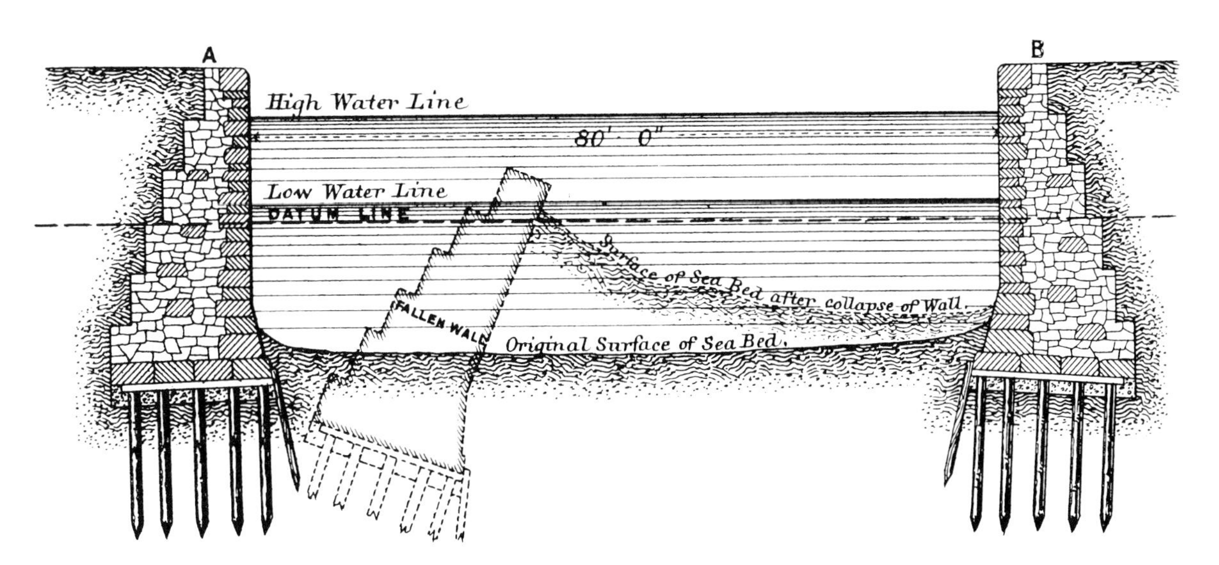

Figure 1.125 Illustration showing the collapse of a section of dock wall in Belfast, 1878
Source: Salmond, T.R., 'The river Lagan and the Harbour of Belfast.' ICE Proc., Vol. LV, 1878/79. Part 1, Plate 3, Figure 8.

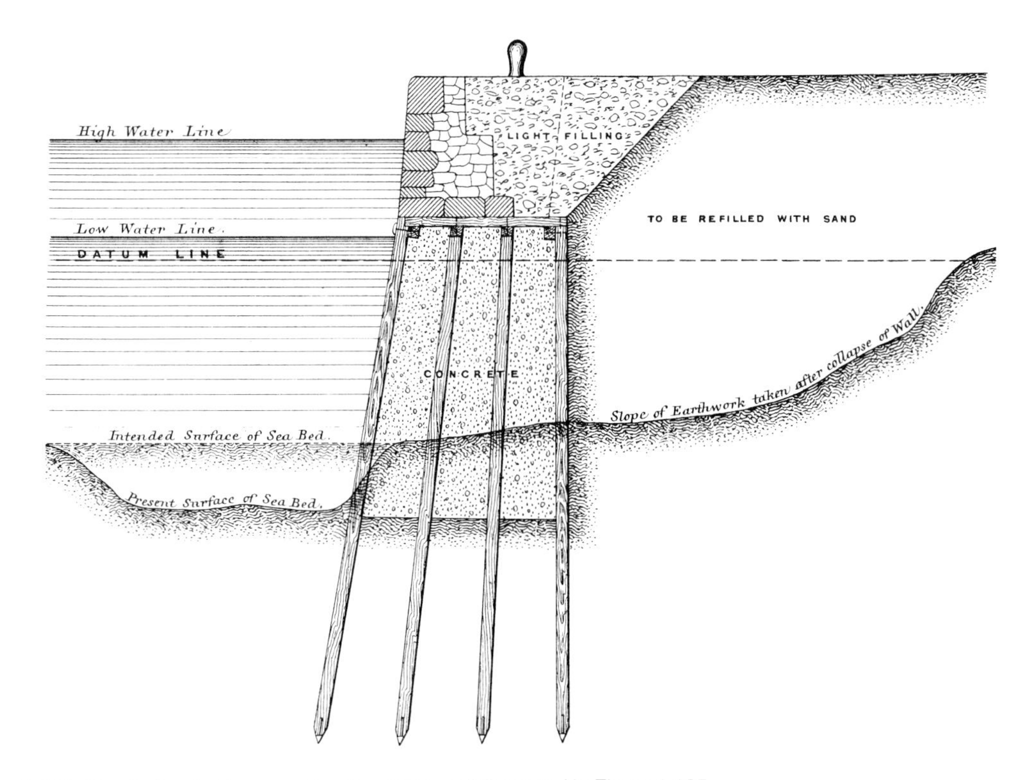

Figure 1.126 The design for the reconstruction of the wall illustrated in Figure 1.125
Source: Salmond, T.R., 'The river Lagan and the Harbour of Belfast.' ICE Proc., Vol. LV, 1878/79. Part 1, Plate 3, Figure 9.

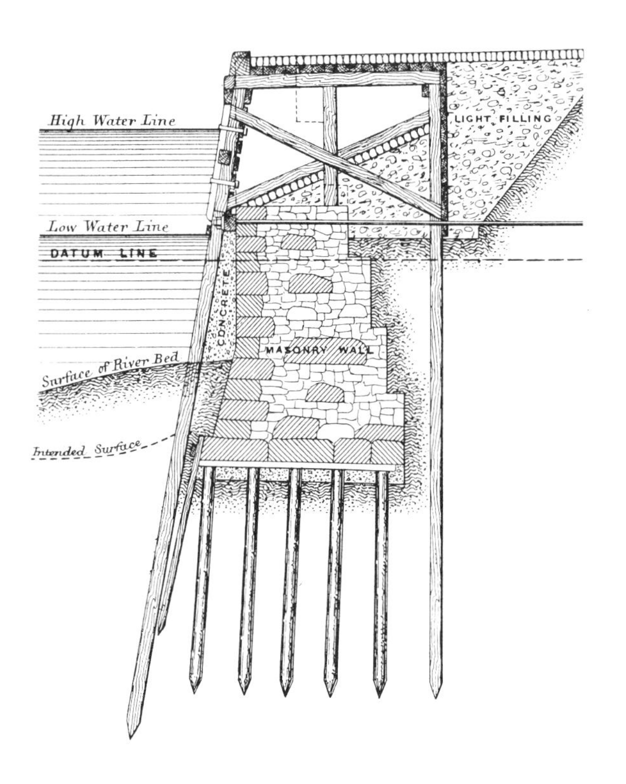

Figure 1.127 The design for the strengthening of a section of the dock in Belfast (see Figure 1.125)
Source: Salmond, T.R., 'The river Lagan and the Harbour of Belfast.' ICE Proc., Vol. LV, 1878/79. Part 1 Plate 3, Figure 10.

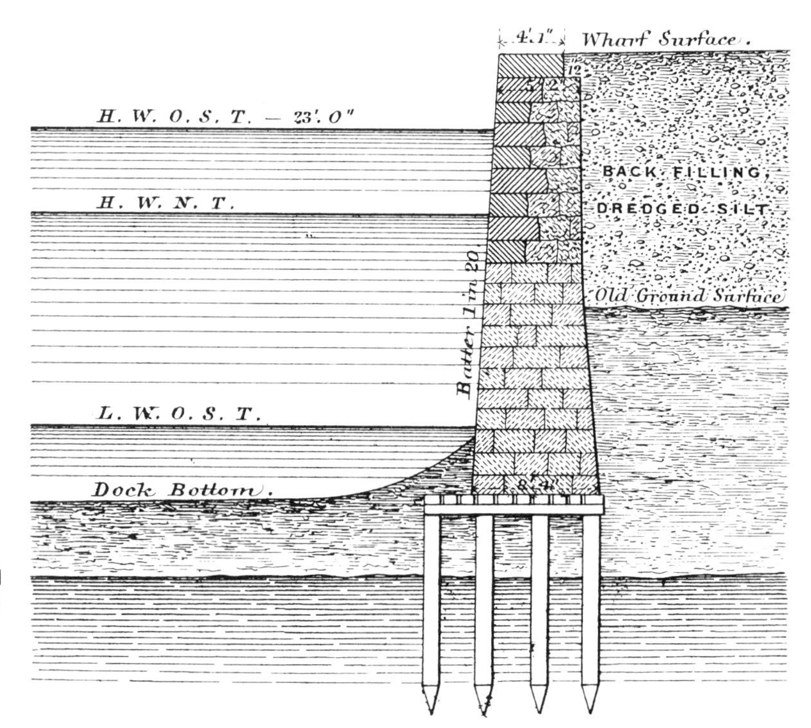

Figure 1.128 Section of the dock wall in Limerick which collapsed in 1887
Source: Hall, W.J., 'On the failure of the Limerick Dock Walls and the methods adopted for reconstruction and repairs.' ICE Proc., Vol. CIII, 1890/91. Part 1, Plate 9, Figure 2.

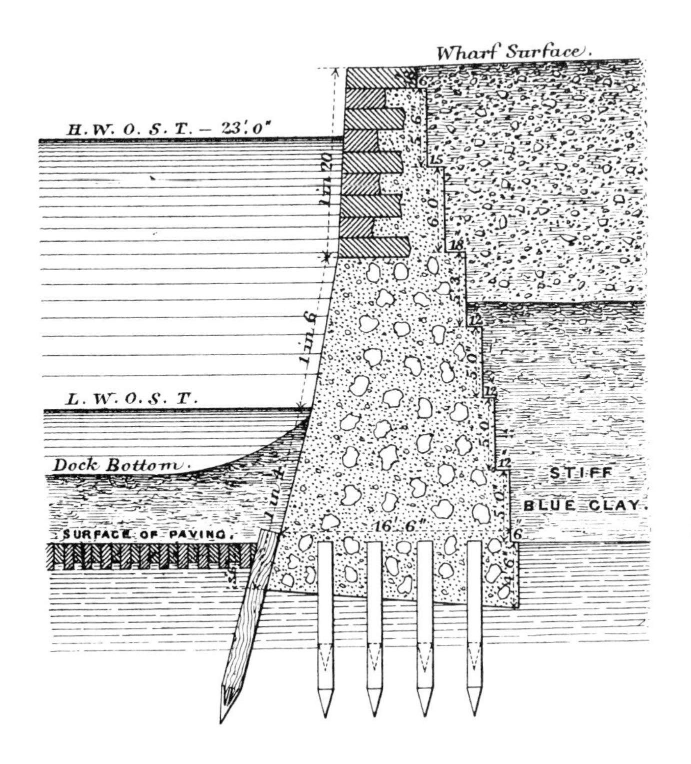

Figure 1.129 The proposed design for reconstruction of the dock wall which collapsed at Limerick (see Figure 1.128)
Source: Hall. W.J., 'On the failure of the Limerick Dock Walls and the methods adopted for reconstruction and repairs.' ICE Proc., Vol. CIII, 1890/91. Part 1, Plate 9, Figure 5.

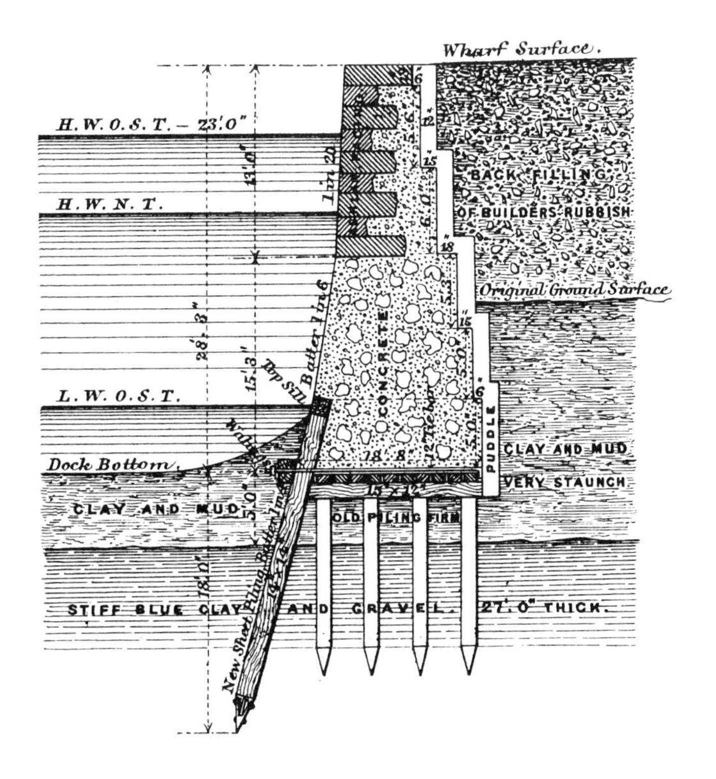

Figure 1.130 The accepted design for the reconstruction of the Limerick Dock wall (see Figures 1.128 and 1.129)
Source: Hall, W.J., 'On the failure of the Limerick Dock Walls and the methods adopted for reconstruction and repairs.' ICE Proc., Vol. CIII, 1890/91. Part 1, Plate 9, Figure 4.

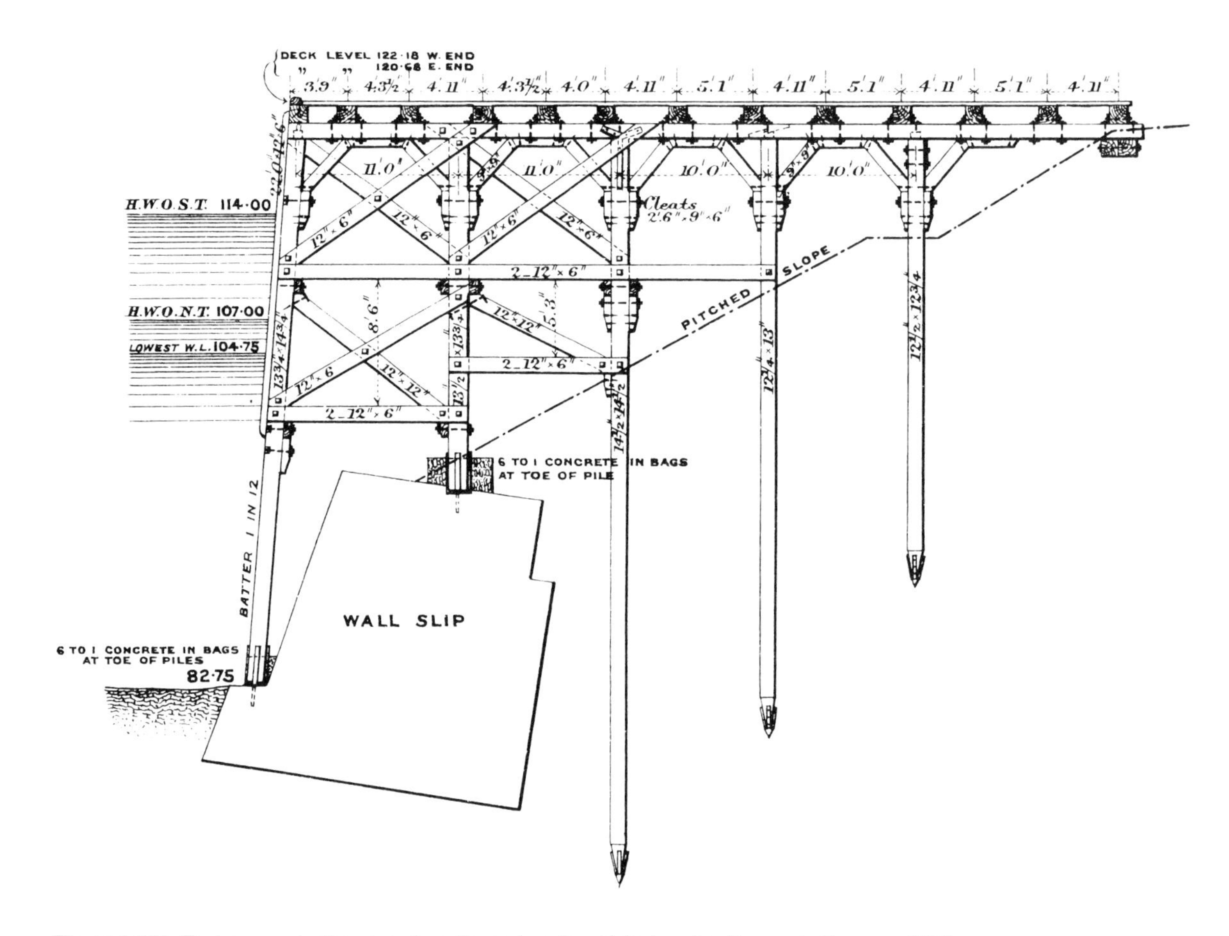

Figure 1.131 Timber quay built on top of a collapsed portion of dock wall in Barrow-in-Furness, 1901
Source: Savile, L.H., 'Lowering the sill of the Ramsden Dock, Barrow-in-Furness', ICE Proc., Vol. CLVIII, 1903/4. Part 4, Plate 4, Figure 4.

Chapter two

Review of circumstances affecting wall performance

2.1 Introduction

Wall performance may be affected by a variety of circumstances relating to loading, environment and durability. To make an assessment of the condition of an old waterfront wall, with a view to determining an appropriate level of maintenance or rehabilitation, it is important that these circumstances are understood. This chapter reviews the circumstances which frequently lead to distress in a wall, and sometimes even failure.

The circumstances have been categorized as follows:

1. exceedence of original design loads;
2. man-made changes to wall and environment;
3. natural changes in the environment;
4. deterioration of the structure of the wall;
5. geotechnical.

2.2 Exceedence of original design loads

2.2.1 Excess horizontal or vertical live loading

The horizontal or vertical live loadings assumed in the original design of a wall are often unknown. It has never been normal practice to mark on walls the maximum permissible loadings, and drawings of walls (when available) only occasionally show them. The survival of original calculations is even less common. It is therefore usually necessary to analyse the structural stability of old walls to determine permissible loading.

Overloading of walls as a cause of problems is believed to be under-reported, particularly where a wall has moved and later stabilized. Examples of excessive loading are as follows:

1. the stacking of dense bulk materials on the apron too close to the cope of quay walls (see Figure 2.1). Quay walls are unlikely to be designed for a vertical surcharge on the backfilling of more than 35 kN/m^2 (3 $tons/yd^2$) which is equivalent to about 2 m of sand or a lesser height of a denser material. Many walls will have been designed for a lower loading. It was the stacking of 916 tonnes of coal on a wharf in Limerick in 1849 which caused it to move outwards by 250 mm.[24] Additional details of this failure are given in Section 1.8. It is also reported that overloading has been caused by the stacking of containers too close to a wall;
2. the use of heavy mobile cranes and other cargo-handling equipment, particularly large forklift trucks;
3. the loading and vibration from modern traffic on top of and behind retaining walls and bridge abutments;
4. the application of high bollard pulls to quay

Figure 2.1 Material stacked close to a quay wall
Source: P.F.B. Tatham.

walls by ships or by bucket and cutter suction dredgers. Low walls and points which do not have the means to distribute bollard pulls, such as unrestrained ends of walls, are particularly vulnerable.

2.2.2 Ship and other impact loads

Ship impacts usually cause damage to the cope of a wall, but vessels with bulbous bows can cause serious damage below the water, particularly where ships are turned in confined areas. The damage is usually in the form of loosening of the individual components of the wall, and cracking. Ships with projections, such as ferries with belting, can also cause damage. Walls without fendering are clearly more vulnerable than those with fendering, which have been designed for the type of ship currently in use. A particularly serious result of damage caused in this manner is illustrated by the large hole punched in a dock wall at Liverpool (see Figure 2.2). In this case the problem has been exacerbated by the fact that a service duct runs along close to the back of the upper face of the wall.

Another ship impact problem is that caused by deep-draughted vessels in canals. These craft sometimes ground on the inverted masonry arches between bridge abutments or lock walls, and the damage caused by this eventually leads to a breakdown in the integrity of the arch and consequential movement of the propped walls.

Other impact loads may be applied to a wall by the misuse of cargo-handling equip-

Figure 2.2 A quay wall showing the effects of ship impact, Liverpool
Source: Mersey Docks and Harbour Company.

ment, the impact of flood debris, etc. In the past much damage was done below water level by the impact of bilge keels on vessels rolling at berth. This had the effect of prizing pieces out of the wall.

Another ship-induced load is the abrasion caused by the ship and its mooring lines on a wall. In areas where the stone or concrete is relatively soft, such as is the case with concrete containing chalk aggregate, abrasion can produce severe grooving in the wall. Wires can also penetrate joints in the coping, thereby causing further damage as a vessel rolls.

2.2.3 Increase in height without strengthening of wall

Occasionally the tops of walls have been raised in the past without any strengthening of the remainder of the structure. This may have occurred as part of a redevelopment or when railways were introduced. Such wall raising reduces, very rapidly, the factors of safety against overturning and sliding, for the following reasons:

1. The horizontal force against the back of the wall increases as the square of the height;
2. the disturbing moment increases as the cube of the height.

Wall raising for flood defence purposes is often carried out without raising the level of retained material behind the wall. In such cases the stability of the wall is less severely affected, but loads may still increase due to:

1. the additional water pressure due to the higher water levels in front of the wall;
2. dynamic and impact loading from the water flow and debris in this flow;
3. possible overtopping, with associated back pressures on the top part of the wall and surcharge on the ground behind the wall.

2.3 Man-made changes to wall or environment

2.3.1 Excavation at toe

A common cause of wall distress and some failures is excessive excavation of the ground at the toe of a wall. Often this is caused by over-dredging. Successive maintenance dredging operations may slowly reduce the level of soil at the wall toe. The effect of this dredging will not necessarily be obvious for a number of reasons:

1. silt may obscure the level of firm material;
2. echo sounding can give a false reading close to a vertical wall;
3. some dock areas were designed with a dock bed sloping away from the wall toe. This makes the control of the level of maintenance dredging more difficult.

There have even been cases where the rock under a wall has been fragmented by adjacent underwater blasting and the degradation of the foundation has resulted in the collapse of the masonry wall above.

In rivers and flood channels it is often necessary to deepen a waterway to increase the flow capacity to alleviate flooding in other areas. When this is carried out there is a danger of removing material from in front of the toe of old walls, piers and abutments of bridges, leading to instability, undermining or outflanking of the structures. Excavation below the level of the toe reduces the bearing capacity under the toe as well as the resistance to sliding and overturning.

2.3.2 Cutting of tie rods and damage to anchoring

Gravity walls, by definition, do not rely on tie rods. However, there are walls (e.g. Blyth Harbour, Southwold Entrance and Princes Dock, Glasgow, see Figures 1.46, 1.54 and 1.48) which have all the appearance of normal gravity walls but actually rely on tie rods. There are other instances of gravity walls which have been refaced with anchored sheet pile walls which depend on tie rods. Such tie rods are vulnerable during redevelopment – particularly where, as in one instance (see Figure 2.3), the tie rod was anchored to the foundation of a building which was subsequently demolished.

Walls with long tie rods (25 to 35 m) are particularly vulnerable. In one recorded case the rods were made 36 m long in order to reach good holding ground for the anchors. Subsequently there was some movement of the wall and ensuing investigations revealed that the rods had been cut at a point 22 m from the wall, during some unrelated trenching works, because it was not realized that the rods were connected to the wall.

In other examples bollards have been founded on separate blocks which are anchored to deadmen and are therefore vulnerable. Tie rods may also suffer from corrosion.

2.3.3 Increased erosion due to propeller wash

Erosion of the seabed and walls by propellers has become more common in recent years, particularly due to the effect of bow thrusters which can project a jet of water at up to 8 metres per second at right angles to a quay wall. Walls may be undermined and the body of the wall eroded. An example of this occurred at St Mary's Quay, on the Isles of Scilly (see Figure 2.4).

Erosion by propellers is particularly severe when the under-keel clearance is small or where

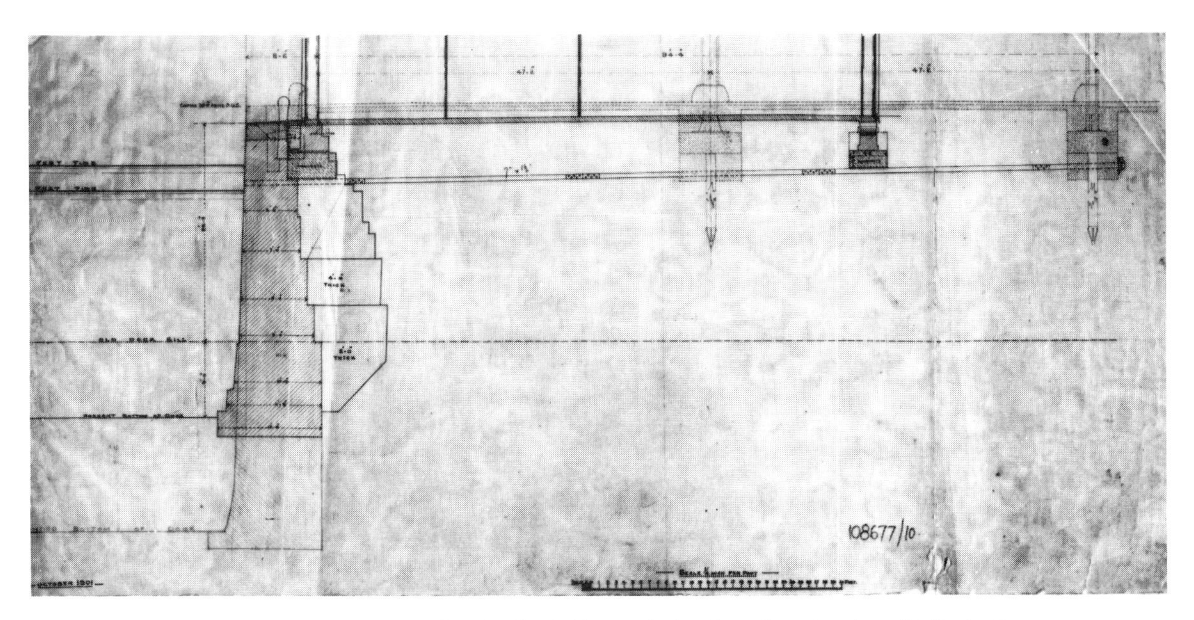

Figure 2.3 Cross-section of the south wall at Sandon Dock, Liverpool, showing a tie rod anchored to the footing of a building
Source: Mersey Docks and Harbour Company.

Figure 2.4 Propeller wash erosion at St Mary's Quay, Isles of Scilly
Source: Beckett Rankine.

the vessel has twin screws. A useful guide to this subject has been produced for the British Ports Federation by BHRA.[26] The same guide also points out that scour may be caused by a number of other mechanisms, as follows:

1. run-up or breaking of ship-induced waves;
2. the movement of water around a vessel in a narrow channel. In canals this may lead to undermining of bridge abutments and canal walls;
3. the movement of the propeller jet towards the channel bed due to the effects of squat in shallow water;
4. under-keel currents caused by vessels being moved laterally towards a berth by a tug;
5. water movement under a vessel caused by surging or rolling, due to long-period waves in the harbour.

Modern canal boats, which often have a high blockage factor (and therefore create more disturbance of the water) tend to cause more erosion problems than conventional canal barges.

2.3.4 Effect of new adjacent structures and redevelopment

New structures are sometimes built very close to waterfront walls. The effects of this can be as follows:

1. the new foundations might affect the loading on the old wall;
2. the backfilling might be removed to allow a basement to be constructed;
3. drainage paths and services might be interrupted;
4. active timber and iron ties might be severed (see Section 2.3.2 above).

Such new works can affect any type of waterfront wall but the effects are particularly important in the case of flood defence walls where safety and continuity are vital.

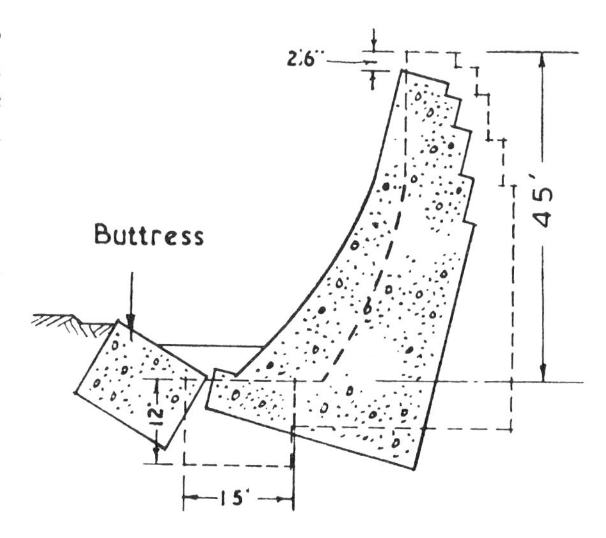

Figure 2.5 The failure of the Empress Dock, Southampton
Source: CP2, I. Struct. E.

There are also a number of indirect effects of new developments, such as:

1. the increased exposure of a wall and its site to currents, in cases where the flow of a river has been modified by new structures;
2. the increased loads and vibration due to increased road traffic levels caused by a new development in the vicinity of the wall;
3. the alteration of beach levels in front of a wall due to the construction of nearby harbour or sea defence works (see Section 2.4.1).

2.3.5 Interference with water levels

Changes in the water levels both behind and in front of a wall can have a considerable effect on the behaviour of the wall. If the wall has not been designed to accommodate a hydrostatic head difference, sudden changes may induce immediate failure. The following points should be noted:

1. a sudden drawdown of the water level in front of the wall may be necessary for maintenance or may happen by accident. (An example is the Forth and Clyde canal, which had to be partially emptied to remove pollution. Even though the water level was lowered over a period of days some of the walls collapsed.);
2. walls of canals and impounded docks are vulnerable where they rely on only one or two sets of lock gates to maintain the level of water;
3. many impounded docks have been built with a puddle clay lining immediately behind the back face of the dock wall. This makes them particularly susceptible to failure when the water level is drawn down (see box).
4. walls founded on clay are particularly at risk due to the slow rate at which water levels equalize on each side of the wall, and the slow dissipation of excess pore-water pressure under the wall, which reduces its effective weight and stability;
5. the conversion of an impounded dock to a tidal basin will have the same effect as a sudden drawdown;
6. the water level behind a wall may be accidentally increased by draining buildings or paved areas to new soakaways behind the wall;

The failure of the Empress Dock, Southampton, and the East India Dock, London

The Empress Dock in Southampton was constructed in 1888 and had shown signs of weakness at that time. A number of buttresses had been placed in front of the toe of the wall, to resist horizontal sliding. In spite of this the wall failed when the water was lowered in the dock (see Figure 2.5).

The East India Dock in London was emptied in 1943 to facilitate the construction of caissons for the Mulberry Harbours for the D-Day landings in Normandy. The walls, which were originally constructed with a curved brick cross-section (see Figure 2.6), failed by horizontal sliding. After failure the maximum extent of movement of the coping was about 0.96 m downwards and the toe moved about 2.13 m forwards (see Figure 2.7).

7. surface water drains, foul sewers or water supply pipes can be broken and leak behind the wall, thus raising the water level and the load on the wall;
8. the groundwater behind a wall can be

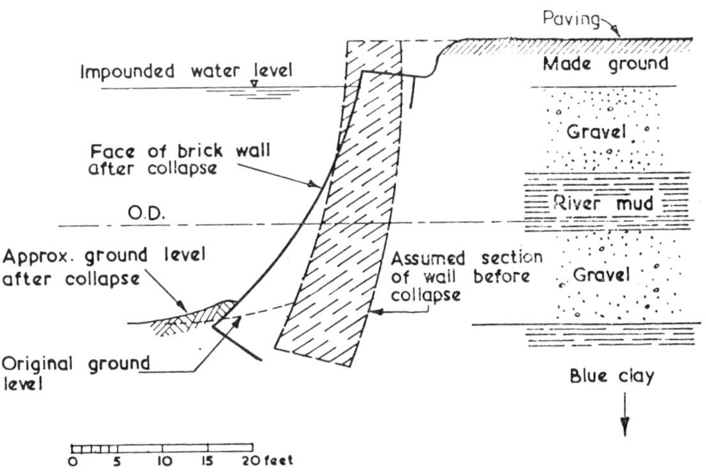

Figure 2.6 The failure of the East India Dock wall, London (see Figure 2.7)
Source: CP2, I. Struct. E

Figure 2.7 The East India Dock wall after failure (see Figure 2.6)
Source: Courtesy of 'Museum in Docklands', PLA Collection.

affected by installation of a barrier to the flow of water or the piercing of an existing barrier;

9. contractors' operations, such as the discharge of construction pumps or the construction of slurry walls, may raise the water level behind a wall;
10. some walls depend on drainage systems behind and through a wall (see Figure 1.47) to limit the maximum difference in level between the water at the back of the wall and at the front. If the drainage system is blocked (for instance, by uncontrolled grouting), a large hydrostatic load will be applied to the back of the wall.

It should be noted that interference with the drainage regime is not a new phenomenon. Indiscriminate culverting of streams in the nineteenth century is still, in some places, affecting the performance of old waterfront walls.

2.3.6 Bomb damage

Some walls were damaged by bomb blast during the 1939–45 war and have not been fully repaired. Most of these are in typical target areas, such as docks which were in use at that time. The effects of bomb blasts can be seen in the cracking of concrete and masonry. An effect which is possibly more important is the subsequent degradation of mortar joints which have been damaged by the shock. A possible clue to detecting bomb blast is the presence of characteristic shrapnel holes in or near the wall.

2.3.7 Mining subsidence

Mining subsidence can lower a wall relative to the water levels so that the top of the wall has to be raised (and the wall strengthened) to maintain the original freeboard. Sea walls may also suffer subsidence and this may have serious consequences with respect to maximum surge levels. Subsidence can also crack and tilt the wall.

2.3.8 Earthquakes

Earthquakes, which are rare in the UK but quite common in some areas of the world, can affect the stability of old waterfront walls. Not only do the earthquake forces themselves impose additional loads on the structure, but there is often the risk that retained material, or other soil in the vicinity, may suffer from liquefaction due to the vibration of the ground. In such circumstances large ground movements can occur which impose severe loads or even change the position of the whole structure.

2.4 Natural changes in the environment

Most of the waterfront walls under consideration have been standing for over 50 years, many are over 100 years old and an appreciable number are very much older. During the course of their lives there may well have been fundamental natural changes to the environment, both locally and on a global scale, which may now be affecting their performance.

2.4.1 Wave scour, current erosion and littoral drift

A common cause of deterioration of all types of waterfront walls is the erosion of the ground level in front of the wall by wave and current scour, or littoral drift. Wave scour arises from the interaction between the incident wave on a wall and the reflected wave from it. The severity of the scour is dependent on the shape of the wall, the type and profile of the beach or seabed, and the height, direction and period of the incident wave. Many coast protection walls have anti-scour toe protection of concrete apron slabs, sheet pile cut-offs or rock revetment to control erosion.

The use of 'Galveston type' walls (see Figure 2.8) has been shown to be more effective than other profiles in reducing the wave reflection and resulting scour (see para. 4.4.3 of CIRIA Technical Note 125[20]). Maximum wave scour will occur during storms which produce the highest waves at the site in question.

Current scour occurs when the flow velocity is high enough to erode the bed material. High velocities can be caused by river currents, tidal currents, or eddies round obstructions such as bridge piers. Extreme events such as a 1-in-500-year river flood can cause unprecedented scour in front of old walls (see box).[27]

Three recent examples of bridge scour

During the period 1987–9 there were three widely reported instances of scour occurring to British Rail bridges. In 1987 an accident, involving fatalities, occurred when a bridge near Glanrhyd, South Wales, failed due to the effects of scour. In early 1989 the Inverness Viaduct had to be closed after it had partially collapsed and in September 1989, a large scour hole was noticed in the Girvan Railway Viaduct on the Glasgow to Stranraer Line.

All these incidents, which appear to have been the result of the rapid onset of scour, occurred in spite of the British Rail system of regular inspections. The Girvan Viaduct, in which one fifth of the base of the pier had disappeared, had been inspected underwater two years previously.

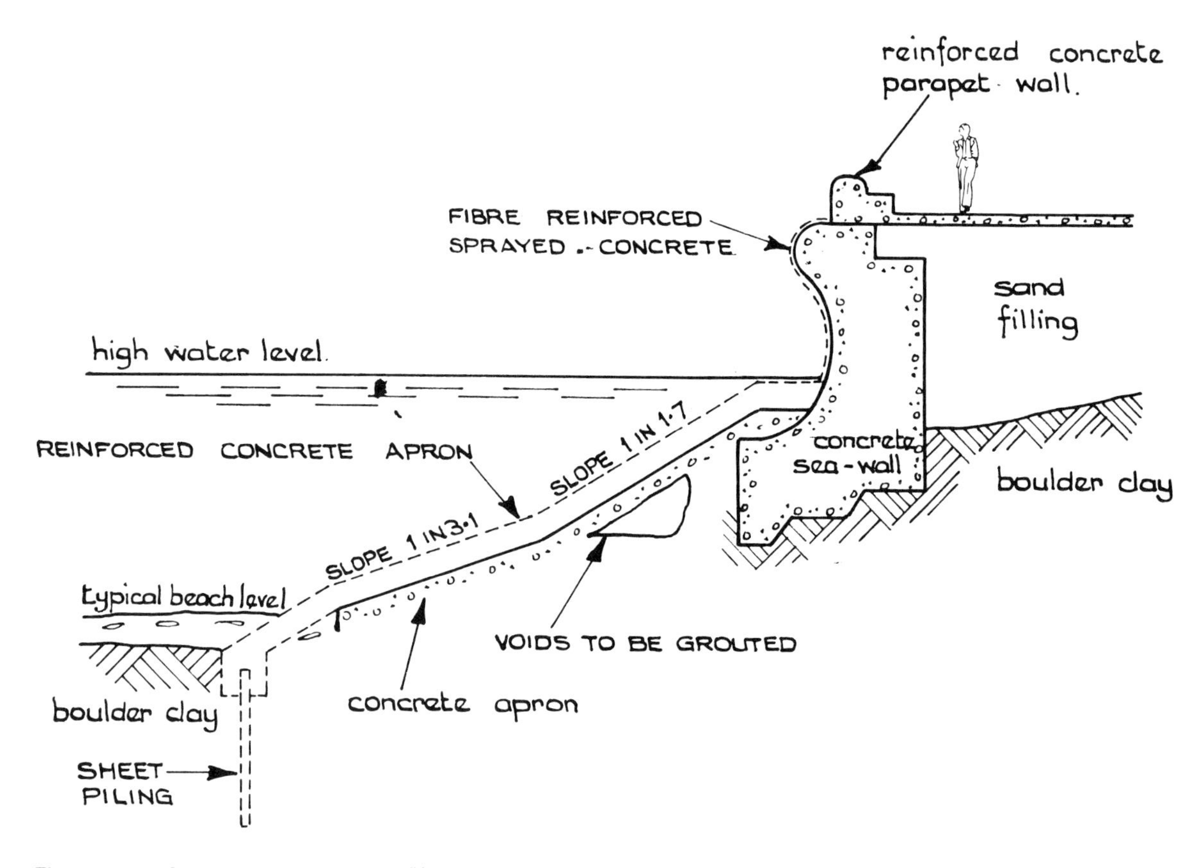

Figure 2.8 A 'Galveston' type sea wall at Blackpool
Source: Blackpool Borough Council.

Littoral transport is the movement of beach material either along or perpendicular to the shore and is caused by waves and currents. It can result in substantial accretion or erosion. Accretion of the beach is usually beneficial to the sea wall, but erosion will eventually undermine it. The results of erosion can be seen in Figure 1.102 where the toe of the wall has had to be reconstructed and extended on a number of occasions.

Changes in the rate of littoral drift, leading to drawdown of the beach, may occur for a variety of reasons. The change in beach or bed level may be quite sudden. The following points should be noted:

1. long-term changes in the direction of winds and waves may alter the coastal regime. (The coastline between Morecambe and the Wirral, on the north-west coast of England, is currently showing signs of beach depletion for this reason.);
2. removal of material from the beach or seabed, up-drift of the wall, may cause drawdown in front of it. (A current example of this is Skinnington on the east coast of England where beach levels have fallen

because British Coal and British Steel have stopped tipping unwanted material up-drift of the site.);
3. the construction of breakwaters and groynes can cut off the flow of material along the coast, causing erosion. Development of the seafront is a prime cause of this;
4. the removal or degradation of offshore bars may increase wave energy at the beach or alter the wave climate, thereby producing increased erosion;
5. offshore dredging can have long-term effects on beach levels, particularly if carried out close to the shore. Dredging farther offshore may affect beaches if the wave regime is altered by the removal of a bar.

The movement of littoral material is a complex subject. Many parts of the UK coastline can be divided into 'cells' in which all sediment sources and sinks are inter-related with the sediment travelling along the beach. (A sink is a point at which sediment is lost from the beach.) MAFF have funded a study to determine the extent of these cells, in preparation for detailed investigations at specific locations. (See also Section 5.8.2.)

2.4.2 Natural changes in water level

Change in the water level behind a wall may occur due to:

1. natural change in the groundwater levels;
2. overtopping by waves in storm conditions;
3. heavy rain or floods.

The water level in front of a wall can change in the short term because of surges caused by wind or atmospheric pressure, or because of changes in river flows.

Long-term changes are caused by phenomena such as the greenhouse effect, which is presently understood to be causing a global rise in sea level. Estimates of the magnitude of such effects are constantly changing as research proceeds.

2.4.3 Long-term changes in ground level

Due to past geological events, and in particular the last Ice Age, there is slow long-term movement of the land in certain parts of the country. In general in the UK there is a tendency for the land in the south and east to sink and for that in the north and west to rise. The effect of this movement is to alter the level of water in front of some waterfront walls. Figure 2.9 shows the changes of water level which have occurred in various parts of the UK in recent years.[28]

2.4.4 Increase in exposure to waves

Walls can become more exposed to waves on account of the following circumstances:

1. an increase in the water depths in front of the wall described in Sections 2.3.7, 2.4.2 and 2.4.3 so that waves of greater height can reach the wall without breaking;
2. the erosion of sandbanks or other coastal mechanisms which increase the exposure of the wall to waves;
3. the damaging or removal of a breakwater or similar structure;
4. an increase in wave energy due to greater water depth after dredging an approach channel.

An increase in the height of waves reaching the wall may lead to wave erosion at the toe, damage to the face and structure of the wall, or overtopping.

2.4.5 Exposure of back of wall to waves

Waves may overtop a wall and penetrate the surface of the backfilling, thereby causing the backfilling to be washed out. Sometimes the

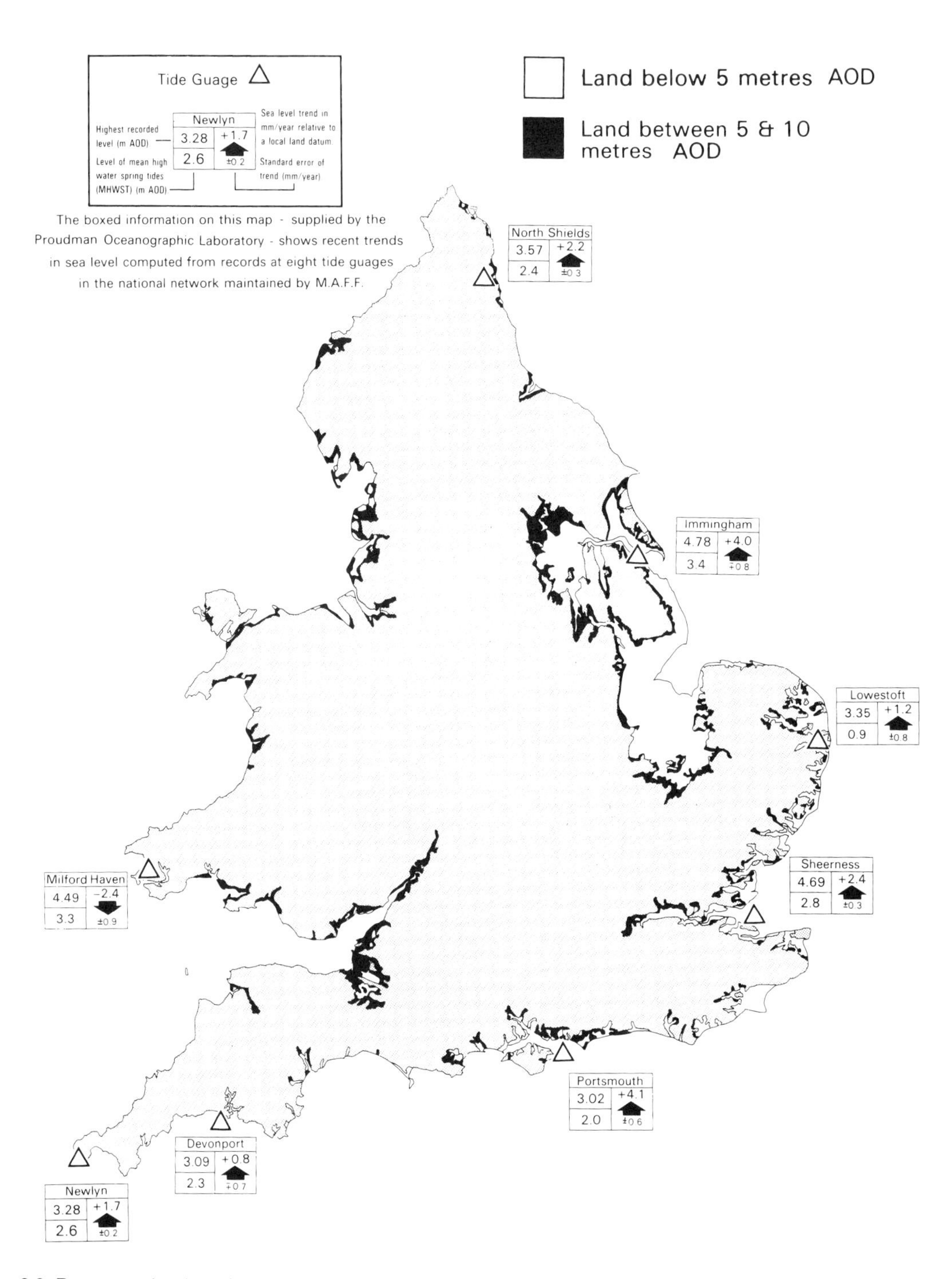

Figure 2.9 Recent sea level trends in England
Source: Whittle, I.R., 'The Greenhouse Effect. Lands at Risk. An Assessment.' MAFF Conf. of River and Coastal Engineers, 1989. Crown Copyright, reproduced with the permission of the Controller of HM Stationery Office.

front face of the wall is then pushed forward by the water pressure acting on the back of the wall (see Figure 7.28).

A similar situation can arise in the types of breakwater where a front masonry wall is backed by permeable rubble (see Figure 1.91). If the breakwater is overtopped, and if the rubble backing does not have a durable impermeable surface, the masonry wall can be subjected to high water pressure, leading to eventual deterioration of the wall.

2.4.6 Outflanking of ends of wall

Ideally, walls subjected to wave attack should be designed so that they end where the shore is shielded from wave attack or the material of the shore is not erodable by waves. Alternatively, the ends of the wall should have 'return walls', i.e. the wall should extend at right angles into the land.

Where the ends of a wall do not conform to one of the above methods of end protection the shore beyond the end of the wall can erode away so that the wall is outflanked and destroyed from each end. Outflanking of walls has occurred on the Holderness coast at Hornsea and on some waterway walls where return walls have not been built.

2.4.7 Effect of vegetation

The growth of vegetation in and around old walls can have a serious effect on the integrity of the structure. The growth of bushes and saplings can rapidly disrupt a wall and render it vulnerable to water penetration and frost action.

Vegetation may also affect the loading on a wall, by the application of 'wind rock', and the roots of plants may accelerate the deterioration of mortar (see Section 2.5.2). Another characteristic of vegetation is its effect on the water content of the adjacent soils (see Section 2.6.2).

2.5 Deterioration of the structure of the wall

2.5.1 Introduction

The general decay of the structure of the wall is one of the most commonly cited reasons for walls requiring repair or rehabilitation. There are a number of underlying causes for this decay and these are reviewed below.

2.5.2 Deterioration of mortar

Deterioration of mortar is one of the most important causes of wall degradation and leads to many of the other more obvious signs of wall distress. The causes of mortar deterioration in waterfront walls include the following:

1. leaching by water of the soluble components of inappropriate permeable mortars;
2. attack by chemicals in the water in front of the wall, in water percolating through the wall, or leached out of brickwork;
3. attack by chemicals released from growing plants;
4. erosion by abrasive materials in the water;
5. displacement by the roots of vegetation;
6. freeze/thaw cycles acting on porous mortar;
7. overstressing of mortar arising from movements or overloading of the wall.

Deterioration of mortar leads to loosening of brick and stonework in the structure. This in turn leads to loss of brick and masonry by wave forces, floods or even the action of vandals. One point which should be appreciated is that mortar joints under load will initially deform, but at a certain stage the joints will fail by crushing, thereby accelerating the degradation of the structure.

2.5.3 Deterioration of brickwork

Most brickwork deteriorates over a period of time. The principal causes of brickwork deterio-

Figure 2.10 Wyndford Lock, on the Forth and Clyde Canal, showing degradation of stonework due to wave attack and continual wetting and drying
Source: British Waterways.

ration in waterfront walls are as follows:

1. freeze/thaw cycles acting on poor-quality brick;
2. sulphates in certain types of brick reacting with the mortar;
3. restraint caused by insufficient movement joints;
4. in canals and docks, the combination of wave attack and continual wetting and drying.

In particular it has been noted that engineering bricks to BS 3921[29] are prone to granulation and exfoliation as a result of exposure to freeze/thaw conditions.

2.5.4 Deterioration of stonework

The principal causes of stonework deterioration in waterfront walls are as follows:

1. attack by acid in rainfall or urban atmospheres;
2. freeze/thaw cycles;
3. attack by chemicals released by plants;
4. in canals and docks, the combination of wave attack and continual wetting and drying (see Figure 2.10);
5. ageing of the stone, which changes its crystalline structure and affects its strength.

2.5.5 Deterioration of concrete

Concrete is affected by the same mechanisms that cause deterioration of mortar (Section 2.5.2). It can also deteriorate because of:

1. alkali aggregate reaction between cement, water, and certain types of aggregate;
2. erosion of weak, poorly made concrete, particularly that placed under water;
3. reaction to poor aggregates taken from local sources;
4. continual wetting and drying;
5. corrosion of iron or steel embedded in the concrete;
6. degradation of high alumina cement in the concrete.

2.5.6 Abrasion

Abrasion is the erosive effect of beach gravel being thrown at the wall by waves. Considerable local damage may be caused by abrasion to timber, concrete and masonry structures (see Figure 2.11). The effects of abrasion are described in the CIRIA Report *Seawall Design Guidelines*,[30] and should be taken into account when repair or reconstruction work is being designed (see Chapter 7).

Figure 2.11 Abrasion by sand and shingle of the sea wall at Sheringham, Norfolk
Source: H.G. Richings.

2.5.7 Loss of facing stones

Facing stones, particularly of granite, have frequently been used to improve the durability of breakwaters and sea walls against abrasion, and to protect the front face of quays against damage by vessels. However, the bonding of the facing stones to the wall sometimes fails under severe loading, such as wave attack or ship impact. (Both types of failure has been reported from Liverpool.) The reason for the failure may often be attributed to poor keying of the facing stones into the main structure (see Figure 2.12).

Under wave attack a concealed separation of the facing stones from the core of the wall may occur, which in time leads to the formation of a fissure or 'chimney'. In such circumstances the loss of a few stones from the face of the wall, which would not normally arouse suspicion, can lead to the sudden collapse of a considerable area of the facing stonework.

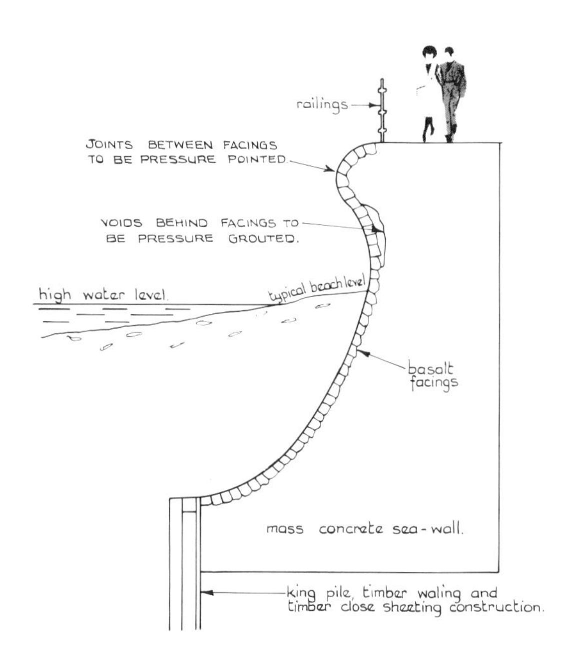

Figure 2.12 Illustration of sea defence works between Dean Street and Chapel Street at Blackpool showing effects of poor keying of facing stones
Source: Blackpool Borough Council.

2.5.8 Decay of timber

Timber used as bearing piles, sheets piles, or foundation grillages, which remains fully submerged at all times, does not usually deteriorate, although some softwoods are prone to marine borer. Local settlement is experienced where block masonry walls constructed in the late 1700s and early 1800s are founded on timber grillages. Where it can dry out, timber may well decay and this can be the cause of unexplained settlements and cracking, particularly where it is not known that timber forms part of the foundation.

Timber inserts in the structure of the wall may also decay if exposed. Timber rubbing strips were often used to give some protection to the face of quay walls and invariably are either broken away by ships or eventually rot, thereby exposing the quay wall to direct impact damage by ships.

Since timber is prone to decay when it is exposed to air, lowering water levels (by such actions as river works, natural drought or the drawing down of a dock) may have a serious effect on the timbers making up the foundations.

2.5.9 Loss of wedges

A few walls were constructed as wedged dry stone walls. The wedges were driven into the joints to fix each individual stone into the body of the wall. The walls of the Cob in Lyme Regis were constructed in this way. Waves may loosen and remove the wedges so that stones are no longer fixed and can fall out. Progressive deterioration of the wall then follows.

2.5.10 Cracking of gravity walls

Cracking of concrete, stone masonry and brickwork walls is a symptom of other defects such as:

1. overstressing of the soil under the foundation;
2. differential settlement;
3. partial failure by overturning or sliding;
4. mining subsidence;
5. thermal restraint;
6. ship damage;
7. settlement due to decay of timber foundations;
8. shrinkage;
9. breakdown of the bonds of surface renders and facings;
10. local pull-out of bollards;
11. expansion of corroded steel inserts.

Once cracking has happened, progressive failure may occur such as loss of filling behind the wall, loss of hearting within a breakwater, and breakup of the wall as it becomes more vulnerable to wave attack.

However, the cracking of walls is not necessarily a cause for concern. Many cracks are formed at an early stage in the life of the structure and, having provided a necessary method of stress relief, do not thereafter give rise to accelerated deterioration. It is common to find this type of crack in walls adjacent to places where local restraint is provided, such as at the corners of a dock.

2.6 Geotechnical

2.6.1 General

As old waterfront walls have, by definition, been standing for many years, it might at first sight seem unlikely that geotechnical problems would be a major concern. However, soil behaviour can be affected by a number of factors which change over the years. Many of these are discussed in BS 8002[31] and are summarized below.

2.6.2 Clays

The following characteristics of clay are important in relation to the performance of walls:

1. under sustained shear stress, the strength of overconsolidated clay diminishes with time or progressive strain. A slip failure may occur many years after construction;
2. clay will swell and shrink with changes in moisture content, caused, for example, by absorption of water by tree roots. Pressure of retained material is also affected. Changes in moisture content of clay in wall foundations can cause movement and cracking of a wall;
3. clay under load may continue to consolidate over a long period, leading to wall settlement. This applies particularly where the stress under the toe of the wall is high;
4. as clays are relatively impermeable, rapid reduction in water level will lead to excess pore pressure under the wall so that the resistance to forward movement is reduced.

2.6.3 Sands and silts

Where sand is used as backfilling, the active pressure applied to a wall will be increased if the sand is subject to vibration, for instance by traffic. Sands and silts, especially when loosely packed and saturated, may also be susceptible to liquefaction when subjected to an appropriate shock or other input of energy, such as seismic effects.

Silts and fine sands with a relatively low permeability can be effectively weakened by groundwater movement and changes in porewater pressure. High upward flows of water through sand can cause 'piping' so that the bearing capacity is reduced. Silts can suffer from frost heave where the water level is just below the silt.

2.6.4 Softening of rock

Rock under a wall may weather and soften, particularly if it is exposed to water, alternate wetting and drying, or freeze/thaw cycles when wet. The softening could lead to a reduction of bearing capacity under the toe of the wall. Locks on the River Trent, for example, which are founded on marl and mudstone, have suffered from softening and subsequent erosion. The same weathering process will cause problems with a rock face behind a skin wall if the skin wall does not adequately protect the rock from wetting and drying.

2.6.5 Inadequate factors of safety in original design

Although Coulomb published his theory of earth pressure on retaining walls in 1773, for over a hundred years after this, many wall designs were based on rules of thumb. Thus in 1881 Sir Benjamin Baker in his ICE paper 'The actual lateral pressure of earthwork'[32] derided theoretical calculation and stated that in his experience 'the width of the base should be one third of the height for a wall built in ground of average character'.

Even where Coulomb's theory was used, there were many problems in its application such as the following:

1. the properties of clay were not understood;
2. inadequate factors of safety against overturning were used. For instance, Romilly Allen,[33] writing in 1876, recommended that the resultant should be within the middle half as opposed to the middle third of the base. (It was French practice to keep the resultant within the middle three-quarters of the base.);
3. properties of soil were not adequately investigated and measured;
4. methods taking account of 'designed' backfilling (see Section 1.2.2) and of the effects of counterforts were not available. Even now difficulties are posed in analysing these.

Methods of assessing the factor of safety against a deep slip were not developed, so there were several wall failures during or shortly after construction (see Section 1.8). Many others have been left with a low factor of safety.

The result of such problems is that when old waterfront walls are investigated and analysed by conventional modern techniques, it is not unusual to find they have factors of safety below those recommended in modern codes of practice, and sometimes below unity.

Chapter three

Management of maintenance and rehabilitation

3.1 Introduction

Due to their nature and situation, old waterfront walls are subjected to a greater degree of wear and tear than their land-based equivalents. Not only do they have to withstand the destructive forces of the weather but they must also contend with the vagaries of the water in which they stand, and the effect of this water on the ground on which they are founded. In addition to this they must suffer the uses and abuses to which man puts them.

The fact that so many of these walls are still standing to this day probably reflects more on the quality of the original materials used in their construction, and the fact that unsound walls had to be replaced shortly after building, than on the maintenance they have received in their lifetime.

The maintenance of any structure is carried out to protect the investment which it represents, and its ability to provide the service for which it was intended.[34] All too often the objectives of maintenance are expressed in terms of the physical condition of the structure rather than in the context of the function which it is required to perform.

The important facts to be identified are thus not only the ownership of a wall, the responsibility for its upkeep and its present state, but also the use to which it is being, or is to be, put.

3.2 Maintenance and rehabilitation policy

It is important to recognize that maintenance, in this context, is an all-embracing term which includes all the operations necessary to keep a wall in a defined state. Maintenance includes:

1. complete reconstruction;
2. rehabilitation;
3. repair;
4. observation (inspection and monitoring);
5. demolition (or allowing the wall to degrade).

It should be noted that inspecting a wall on an occasional, but regular, basis is a valid form of maintenance. What is important is that there is a system in place which actively considers the state of a wall and defines the owner's response to its upkeep. The alternative to this is to ignore maintenance until a crisis forces management to take action. This may well be too late (see Section 3.3) to avoid extensive rehabilitation works.

The owner of an old waterfront wall, or the authority responsible for its upkeep, should take steps at the outset to evaluate the role of the structure in its present situation, and decisions should be made which define the level of maintenance to be provided and the method of carrying it out. In other words, a maintenance

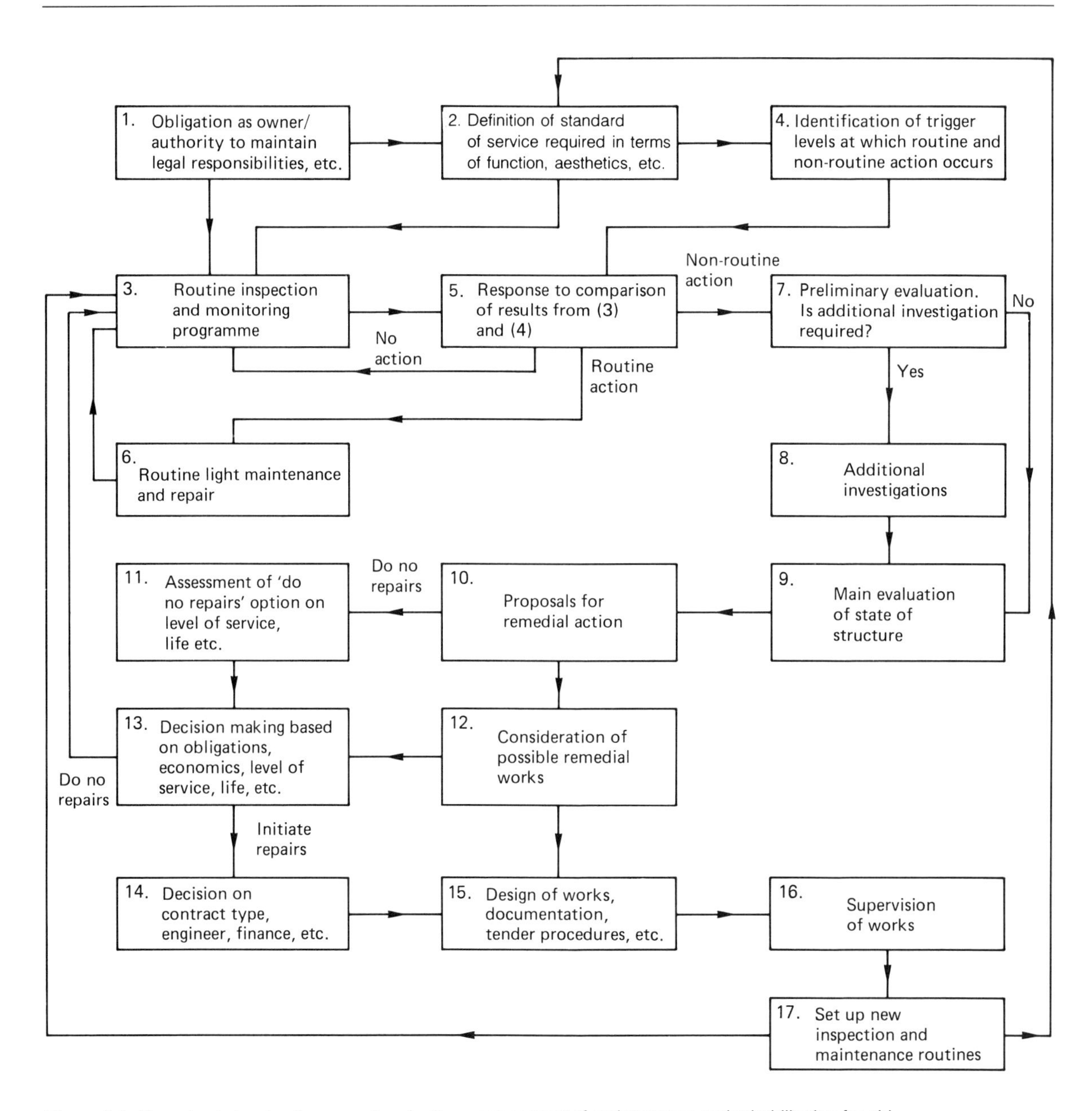

Figure 3.1 Flow chart showing the procedure for the management of maintenance and rehabilitation for old waterfront walls
Source: Livesey Henderson.

policy should be defined. Guidance on the development of a maintenance policy is given in BS 8210,[35] which, although it refers to buildings, is also applicable to waterfront walls.

The flow chart in Figure 3.1 illustrates in a simple way the steps to be taken to carry out evaluations and to implement procedures leading to appropriate maintenance and rehabilitation. The main items in this flow chart are described in more detail in Sections 3.4 to 3.10 below.

3.3 The effects of lack of maintenance

A crucial factor in determining appropriate levels of service, and indeed justifying the application of maintenance or rehabilitation, is an assessment of the consequences of lack of maintenance. In some cases the effect will be a deterioration of the standard of service provided, a reduction in safety standards and, finally, the collapse of the structure. In others there will be no appreciable change in the level of service prior to sudden failure. It is this latter case which is inclined to lull the owner into ignoring maintenance until it is too late.

Figure 3.2 is a diagrammatic representation of this degradation process. It is clear from this figure that maintenance and rehabilitation are closely linked with the life of the structure and its standard of service. It is also apparent that maintenance of a wall to keep up its appearance can prolong its life and postpone the need for major rehabilitation.

One of the most important factors in evaluating a wall's performance and establishing a maintenance programme is determining the 'critical point of disrepair' (see Figure 3.2). This is the point at which the onset of progressive failure occurs. If the wall is repaired before this point is reached, the life of the structure may be extended indefinitely. If repairs are not carried out, major loss of use and reconstruction will be inevitable if the structure is to be returned to its functional level. The 'critical point of disrepair' is dependent on the wall type, its environment and its function (see Chapter 6).

3.4 The obligation to maintain

There are a number of reasons for maintaining a structure and very few for not maintaining it at all. The reasons may be broken down into four main categories.

1. Legal — When the owner/authority has a statutory obligation to preserve, maintain or repair the structure for reasons of aesthetics, function, safety, etc.
2. Commercial — When it is in the owner/authority's best interests to maintain the structure to ensure that it functions efficiently.
3. Amenity — When it is in the best interests of the owner/authority to maintain the structure for amenity purposes.
4. Environmental — The moral obligation of all owners/authorities, and the engineers working for them, to attempt wherever possible to enhance the environmental quality of their structures.

3.4.1 Legal considerations

The following legal considerations may determine an obligation to maintain an old waterfront wall:

1. statutory requirements that structures should be safe for the personnel who use them and those that pass or visit them, including any relevant local regulations;

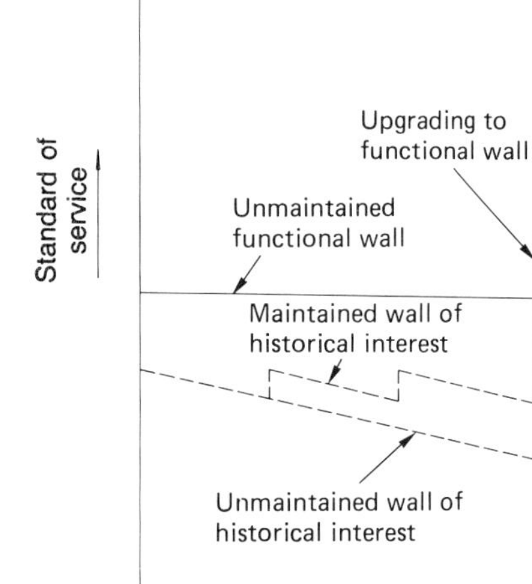

Figure 3.2 Graphs illustrating the relationship between standard of service and service life, for various types of wall and maintenance strategy
Source: Livesey Henderson.

2. the statutory obligation of landowners to preserve rights of way, such as footpaths, towpaths, promenades, etc;
3. the requirement for organizations which have been set up by Act of Parliament to maintain their structures and to fulfil certain functions (such as flood defences, coast protection, harbours to provide storm protection and specified water depths, canals and waterways to provide minimum widths and depths, etc.);
4. obligations written into leases and other legal agreements;
5. the requirement to maintain structures of historical interest in an adequate state of repair, including 'listing' of buildings and structures, ancient monuments, and other special local regulations laid down in conservation areas and in local or national planning legislation.

Table 3.1 UK organizations providing guidance on legal obligations

	Geographic region			
Area of interest	*England*	*Wales*	*Scotland*	*N. Ireland*
Docks and harbours	DTp	WO:TPD	DTp/SDD/DAFS	DENI
Sea defences	MAFF	WO:WD	RHC/DAFS	DENI/DANI
River walls	NRA	NRA	R&IC	DENI
Bridges	DTp	WO:HD	R&IC/SDD	DENI
Environmental	DoE	WO:EPD	SDD	DENI
Safety	HSE	HSE	HSE	HSE

Key to abbreviations:
DTp: Department of Transport
MAFF: Ministry of Agriculture, Fisheries and Food
NRA: National Rivers Authority
DoE: Department of the Environment
HSE: Health and Safety Executive
WO: Welsh Office
TPD: Transport Division
WD: Water Division
HD: Highways Division
EPD: Environmental Protection Division
SDD: Scottish Development Department
DAFS: Department of Agriculture and Fisheries Scotland
R&IC: Regional and Islands Councils
DENI: Department of the Environment Northern Ireland
DANI: Department of Agriculture Northern Ireland

A list of the organizations which can provide further guidance on the more important legislative obligations relating to old waterfront walls is given in Table 3.1.

Ownership and responsibility, as defined by local by-laws and Acts, and the permissive powers of regional authorities, may often be split. It is not uncommon to find, for instance, that a privately owned wall is functioning as an essential component of a major flood defence scheme. Rehabilitation of such a structure may entail redefining responsibilities and apportioning the costs of the work between interested parties.

3.4.2 Commercial considerations

Irrespective of the legal considerations mentioned above, most organizations will wish to keep their waterfront walls in a state of good repair for commercial reasons. They will wish to ensure that the walls remain suitable for their intended function and that they do so at the least possible cost. The most cost-effective way of preserving the required standard of services is to have a programme of maintenance and rehabilitation. (See Section 3.5.)

3.4.3 Amenity

In many instances, and particularly when structures are being incorporated into a new development, it is in the best interests of the owner/developer to maintain the old waterfront walls in a state of good repair. This is most important when the success of the development depends on the aesthetic qualities of the site. Even when commercial influences are less important there is often pressure from local groups to keep certain areas in pristine condition.

3.4.4 Environmental

It is now accepted that everyone has a responsibility to try to ensure that whatever they do is environmentally acceptable. In terms of dealing with old waterfront walls this goes farther than complying with EC Directive 85/337, 'The assessment of the effects of certain public and private projects on the environment',[36] concerning environmental assessment. It implies that all engineers and managers who are directly responsible for the maintenance and rehabilitation of these walls have a moral obligation to be aware of the environmental effects of any works which they are planning, or the effect of allowing a structure to fall into disrepair.

3.5 Defining standards of service

Standards of service are the standards of performance to which the wall should conform. They may be defined by reference to:

1. the function of the wall;
2. the loads to which it may be subjected;
3. the physical environment in which it must survive;
4. acceptable tolerances in its performance;
5. its appearance;
6. a combination of any of these.

Thus for a quay wall one might expect to specify:

1. the size, type and draught of vessel to be accommodated;
2. maximum vertical and horizontal loads to be applied to the wall, including the stacking of material on the apron;
3. the range of water levels to be accommodated both in front of and behind the wall;
4. permissible settlement of the wall and the apron;
5. permissible horizontal movement of the wall.

Breakwater and sea defence walls might require specifications relating to sea states, areas of protection, and design flood levels and risk of exceedence.

Definition of standard of service is essential in that it forms the basis on which:

1. inspection and monitoring programmes are developed;
2. the 'trigger levels' at which specific actions are initiated are identified;
3. decisions are made concerning the economics of various options for maintenance and rehabilitation.

In some instances it may be necessary to determine the loads a particular wall is able to take before setting the standard of service. It is important that the levels adopted are both suitable and attainable. In other cases it may be necessary to take account of the changing use of a wall or to define a standard of service which relates to a future use of the wall.

If a wall is exposed to wave attack or current scour, it is important to compare these with the conditions for which it was designed. Significant departures from the design conditions can then be counteracted accordingly.

3.6 Inspection and monitoring

An essential part of the maintenance procedure is the programme for inspection and monitoring. Inspection is required for the following reasons:

1. to establish a baseline set of conditions against which future inspections will be evaluated;
2. to identify changes in the appearance, position and surroundings of the walls which may require attention in order to maintain a specific level of service;
3. to establish the state of the structure and its surroundings after significant events which might have had an effect on its integrity and appearance.

Monitoring is usually implemented so that the behaviour of a specific area of the wall can be examined in detail. Types and methods of inspection and monitoring are covered in Chapters 4 and 5.

It cannot be stressed too heavily that all organizations should have a regular and systematic inspection procedure. Anything else will lead inevitably to unnecessary expenditure or loss of wall function. *Ad hoc* or 'crisis' inspection is no substitute for a well-managed inspection procedure. The minimum requirements are:

1. planned, fixed-interval inspections;
2. adequate and consistent methods of recording results;
3. a management which acknowledges the value of regular inspections.

It should be noted that the initiation of an inspection and monitoring programme does not necessarily lead to increased capital spending on the wall. On the contrary, there are occasions when the regular inspection of a structure may show that some apparent defects are not increasing in magnitude and are not detrimental to the integrity of the wall. Relatively inexpensive superficial inspections carried out on a regular basis are very valuable and may, in the long term, lead to substantial cost savings since, in certain circumstances, they may eliminate the need for other costly investigations.

Programmes for inspection and monitoring are developed in the context of the function and type of wall as well as the level of service required, which in turn will depend on the obligation to maintain the structure. Naturally, the owner's organization is taken into account in the planning of an inspection programme, and the times and methods adopted are tailored to suit the organization. The owner and/or authority responsible for its upkeep is, after all, the person with the most experience of the wall's behaviour.

3.7 Investigations

Investigations into the state of an old waterfront wall and its surroundings are required for a variety of reasons. The investigations may be part of the inspection process or may be needed on an *ad hoc* basis, as follows:

1. to establish a baseline condition for standards of service and routine inspections;
2. to determine the state of a specific area of the wall which is showing signs of distress;
3. to establish the ground conditions behind or under a particular length of wall for stability analyses;
4. to determine the internal state of the wall when tell-tale signs indicate that material is being lost;
5. to provide information for the design of repairs or rehabilitation;
6. to provide information on whether strengthening of the wall is required for a proposed change of use.

Whatever the reason for carrying out the investigations it is important that great care is taken in selecting the correct method, and planning it in the most cost-effective manner.

3.7.1 Planning investigations

It is likely that a number of different types of investigation will be needed and there may be several ways of achieving the desired results. The following points should be noted:

1. it is important to establish precisely what the unknowns are and what is to be achieved by carrying out the work;
2. consider the benefits of carrying out investigations in stages so that each can benefit from the one before;
3. when dealing with old structures there are considerable benefits to be gained from detailed historical research, often in locations remote from the site such as the Institution of Civil Engineers, the Public Records Office, etc.;
4. consider the use of one investigation to provide information on a number of different aspects. For instance, one borehole may be used to sample the internal material in a wall, detect where the bottom of the wall is founded and provide geotechnical information on the substrata;
5. consider the reliability of the investigation method being proposed and whether the results will be representative of the whole wall;
6. compare the benefits of an investigation which provides information on causes of distress and methods of repair with those of one which only gives information on the former.

3.7.2 Selection of method

The various investigation techniques available are described in detail in Chapter 5. A number are well established but others are relatively new, or are old methods applied to new conditions. In the latter case it is important to establish whether the methods proposed are really suitable for the site and whether they will give the desired results.

There have been instances in the past when inappropriate methods have been selected and the results have been disappointing, or useless. Where possible, it is advisable to obtain details of previous instances where the technique in question has been used to enable a judgement of its usefulness to be made.

Before embarking on an expensive and complex method of investigation it is as well to recall the wise words of Ralph B. Peck (1972):

> An instrument too often overlooked in our technical world is a human eye connected to the brain of an intelligent human being. It can detect most of what we need to know about subsurface construction. Only when the eye

cannot directly obtain the necessary data is there a need to supplement it by more specialized instruments. Few are the instances in which measurements by themselves furnish a sufficiently complete picture to warrant useful conclusions.

3.7.3 Implementing investigations

The method of implementing the investigations will depend to some extent on whether the authority responsible for the wall has its own engineering division. Even if this exists, it is advisable to consider employing engineers who are familiar with the problems of old waterfront walls, since a number of the aspects to be considered are highly specialized. Supervision of the investigation requires experienced staff who can modify the scope of the work as it proceeds. One of the most important duties of the specialist engineer is to identify and interpret all the relevant information which the owner already has in his records.

Figure 3.3 illustrates a typical process of implementation in stages where the engineer is reviewing progress and modifying the scope of work to suit. The analysis of the first stage of investigations is carried out before the scope of the second stage is determined. In some cases further stages may be appropriate but usually cost-effectiveness reduces markedly after Stage 2.

If specialist engineers are engaged to carry out the first-stage study, this work may be done on a lump sum basis. This first stage will define the scope of the detailed investigations, which can then be quantified and costed with reasonable accuracy. If the first stage shows that the detailed investigations should be carried out by engineers with a different field of experience, the opportunity exists for them to be engaged at this time.

It will be appreciated that where the scope of work is to be varied during the investigation an appropriate form of contract must be used. To some extent, all investigation contracts need to be flexible to allow for the circumstances which they reveal. In the case of waterfront walls this is especially so, since not only is the back of the wall obscured by the fill material and the foundations hidden, but the front face is also wholly or partially covered by the water. Investigations may reveal large areas of facing which are in need of repair, voids in the wall and damage to the wall toe. In these cases it would be cost-effective to extend the investigations to identify all the areas in need of repair.

The most suitable contract forms are those which fix a unit rate for carrying out a specific item of work but enable the number of items to be varied to suit the circumstances. Most specialist contractors recognize the variable nature of this work and are prepared to accept flexible contract forms. It is suggested that the budgets for both engineer's and contractor's work should be increased by around 15% to take account of these contingencies.

3.8 Evaluation and decision-making processes

The evaluation of the state of an old waterfront wall and the decision-making process needed to determine the most appropriate course of action are fairly complex. In principle, they are similar to those required to assess the foundations of bridges sited in, or alongside, water. The French Ministry of Transport has published a guide to assist in this task[37] and Figure 3.4 is taken from it, with some alterations to suit waterfront walls. It should be noted that the procedure in this guide allows for situations where:

1. the initial studies indicate that there is insufficient information to analyse the state of the wall and that further investigations are necessary;
2. the cost of remedial works to return the structure to its defined standard of service

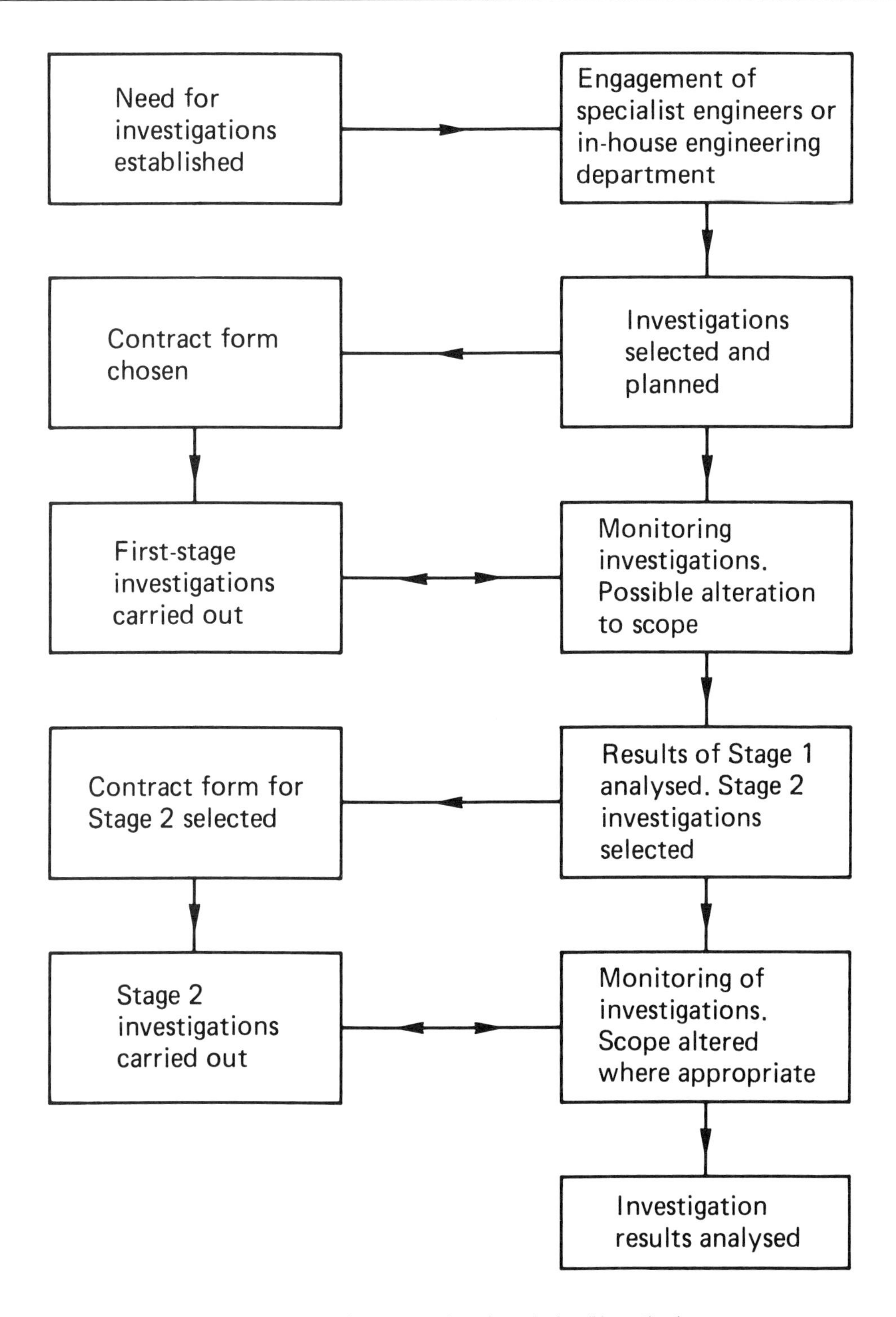

Figure 3.3 Flow chart illustrating the stages in the implementation of a typical wall investigation
Source: Livesey Henderson.

are too high and alternative standards of service are considered.

Figure 3.4 is a more detailed explanation of the operations to be carried out in boxes 9 and 13 of Figure 3.1. The following points should be noted:

1. the analysis of a wall's condition (box 4) may be very complicated due to the number of causes and effects possible. Chapter 6 gives guidance on this aspect of the work;
2. the standard of service defined in box 7 would normally be the same as that established for the inspection procedures (see Section 3.5). However, when rehabilitation of the structure is being considered the standard of service may be increased;
3. details of methods used for maintenance and rehabilitation (box 9) are described in Chapter 7;
4. economic studies (box 10) should take account of the fact that there may be special funds and grants available to assist in the construction of certain types of works. A representative selection of the UK agencies which administer these funds is given in the box. District and County Councils will normally be able to provide guidance on this subject;
5. economic studies may be augmented in some cases by use of specialist techniques such as those developed at Middlesex Polytechnic[38] for flood damage cost assessment. There are also guides to investment appraisal published by the British Treasury.[39,40]

A representative list of UK agencies which may fund work on old waterfront walls

Ministry of Agriculture, Fisheries and Food
Welsh Office
Department of Agriculture and Fisheries, Scotland
Scottish Development Agency
Department of Agriculture Northern Ireland
English Heritage (and equivalents)
Historic Buildings Trusts
National Rivers Authority
Department of Transport
District and County Councils
Scottish Regional and Island Councils
English Tourist Board

3.9 Maintenance finance

BS 8210[35] gives sound advice relating to maintenance finance. The following paragraphs are a précis of the guidance given.

1. the planning and control of finance is an important aspect of maintenance management not only for the control of maintenance, but also to demonstrate that the owner is getting value for money and that the maintenance proposals justify the funds requested;
2. budget proposals should include repair/replace decisions, the optimization of programmed planned maintenance and lifecycle costing. It is important that budget proposals are presented to management in a way which will identify the benefits to accrue from the funds required;
3. the detailed maintenance programme will involve decisions regarding optimum repair reaction times and the most appropriate method of execution (direct labour or contract: the best type of contract). This will lead to the need for budgetary control during the course of the year;
4. a technical audit should be carried out to assess the benefits from expenditure in the previous year. It could also make recommendations for improving benefits;

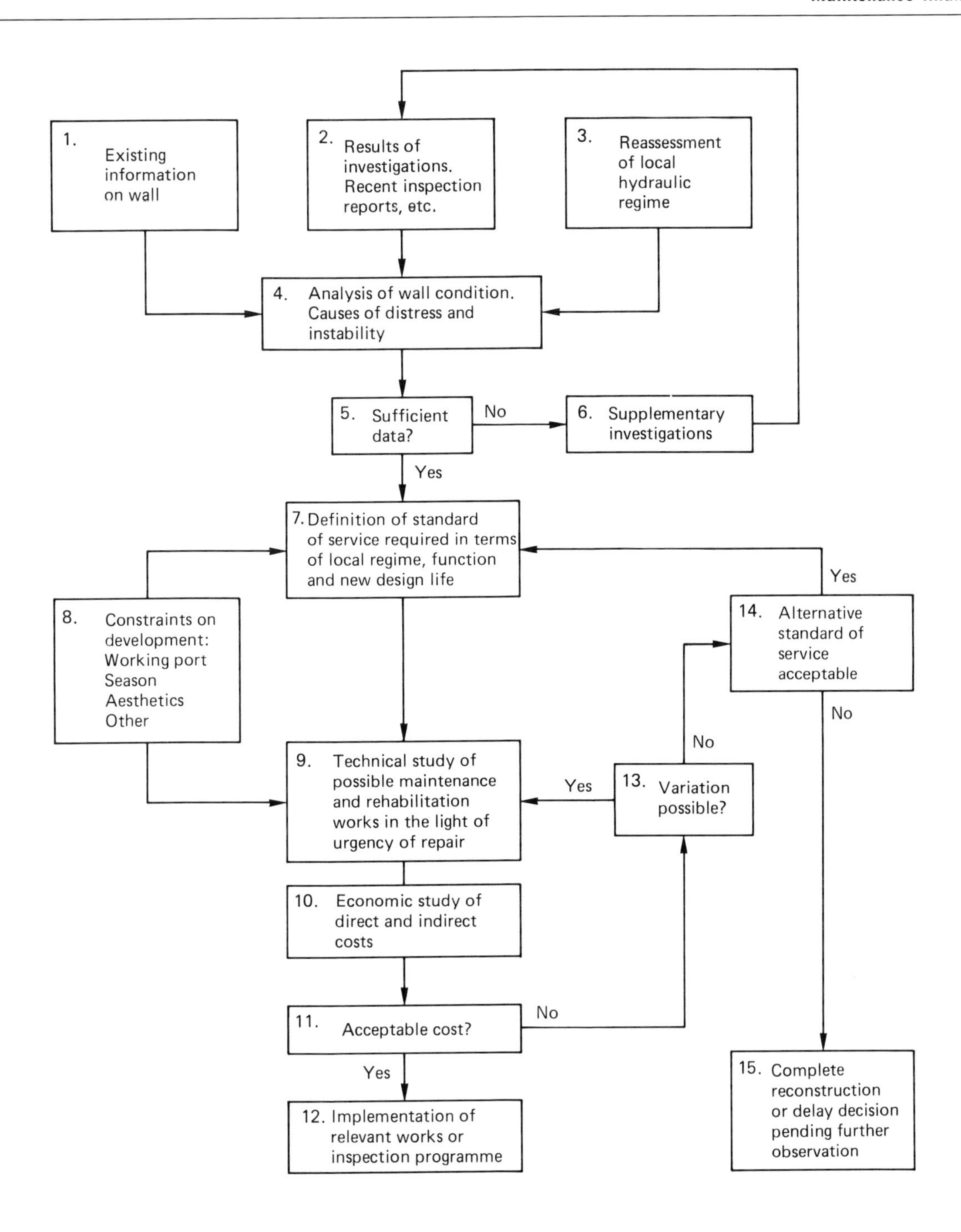

Figure 3.4 Evaluation and decision making in the maintenance and rehabilitation processes. (After the French Ministry of Transport)
Source: Livesey Henderson.

Table 3.2 Grading of repair and rehabilitation work

Grade	*Type of repair*	*Examples of repair*
1	Superficial repair	Refacing, pointing, etc.
2	Superficial/structural repair	Joint filling, void filling, resurfacing, etc.
3	Mainly structural repair	Replacing bricks and stonework, refilling, replacing ties, etc.
4	Structural strengthening (original structure preserved)	Underpinning, adding ties, adding toe protection piles, replacement of backfill, etc.
5	Structural strengthening (parts of original structure replaced)	Replacing internal structure, adding concrete surrounds, etc.
6	Complete rebuilding	

Note: The order in which the repairs are carried out will have to be determined by the owner/authority responsible for the wall. In some cases repairs in grades 3–6 will be carried out as emergency works.

5. it will be obvious that owners, or their executive management, need to be directly involved with many of the aspects of maintenance finance described, particularly those which will have a direct impact on the owner's primary activity. It is reported that some owners divide their budgets into three portions; an investigation fund, a project preparation fund, and a fund for current projects.

3.10 Implementation of repairs and rehabilitation

The approach to implementation of any repair and rehabilitation work will depend on the magnitude and complexity of the work to be carried out; as determined by inspection and investigation. Work may be graded according to its significance as shown in Table 3.2. Low grades indicate non-urgent, non-structural work while high grades indicate work which is likely to have a considerable effect on the stability and life of the structure.

It would be usual to find that the lower-grade work was carried out by the owner's maintenance personnel (where these exist) unless the work involved specialist techniques. The higher grades would most likely be the subject of a conventional works contract of some type (see Section 3.10.2).

3.10.1 Contract period

Many waterfront walls are not accessible during parts of the day or season for operational or environmental reasons. It is important that any constraints on working are identified, and included in the contract documents, before repair or maintenance work is put out to tender. The following questions will need to be answered:

1. does the maintenance work have to be fitted into periods when the wall is not being used (i.e. will the contractor suffer frequent interruptions)?
2. will the work have to be carried out in winter when there are fewer tourists/visitors about?
3. will the work have to be split into more than one part for technical or budgetary considerations?
4. will the work be affected by environmental constraints such as tidal variation, low river flow, summer sea conditions, etc.?

3.10.2 Contract type

The contract types selected for the work will depend on a number of factors, including the work's magnitude, urgency and definition. The following list may be helpful in the choice of the most appropriate contract form.

Nature of work	*Contract work*
Emergency work	Cost plus contract
Total quantity of work defined	Fixed-cost contract
Some quantity of work defined	Consider the use of two contracts, or variable elements within a fixed-cost contract
Type of work defined, but not quantity	Fixed unit price contract, variable quantities
All work undefined	Fixed time rate plus materials cost, or fixed time rate until work is defined.

It should be noted that these contracts are intended to leave the client with the risk of estimating the quantity of work, while the contractor takes the risk for his own performance. If the contractor is forced to estimate the quantity of work, the result will inevitably be a higher price to cover the risk. If a contractor underprices and is unable to carry out the work, it can lead to legal action or bankruptcy.

Many organizations have arrangements with contractors on the basis of 'term contracts', where the rates for certain types of work are agreed in advance. These are particularly useful for diving work or minor repairs. They also enable the organization to mobilize a contractor rapidly in the event of an emergency. However, it is still essential to provide the contractor with a clear specification for the work to be carried out.

3.10.3 Selection of contractors

Repair work on old waterfront walls is often complex, delicate and carried out in inhospitable conditions. To be successful it requires the engagement of experienced contractors who are familiar with the maritime and river environment. Considerable care should be exercised in the selection of contractors. In some cases it may be beneficial to nominate subcontractors for specific parts of the work if a particular proprietary method of repair is deemed to be the most effective. However, in this case it is prudent to carry out a trial of the method, where possible, to be certain of its efficacy. Where grant assistance is being sought for a scheme it is likely that a precondition will be that competitive tenders are to be obtained.

3.10.4 Control of maintenance works

There are two aspects of the control of maintenance works which should be addressed:

1. the actions of identifying, initiating and completing maintenance works should be adequately recorded, so that any recording system or database is kept up to date;
2. the maintenance work itself should be adequately supervised.

Figure 3.5 shows the procedure adopted by one owner who is responsible for coast protection work. Similar procedures are being set up by other coastal authorities. The important point to note is that there is a clearly defined system in place which ensures that the repair work is carried out, and documented.

Arrangements for the supervision of repair work will vary according to the nature of the work. It will be important to ensure that the quality of the work is, as specified, of the highest standard. This will mean providing continuous, high-quality supervision by staff who are sufficiently well qualified and experienced to be able to cope with whatever problems may arise. For work in enclosed docks or canals, where it may be necessary to lower water levels, monitoring of these levels and the behaviour of the walls will be essential, from both a safety and an

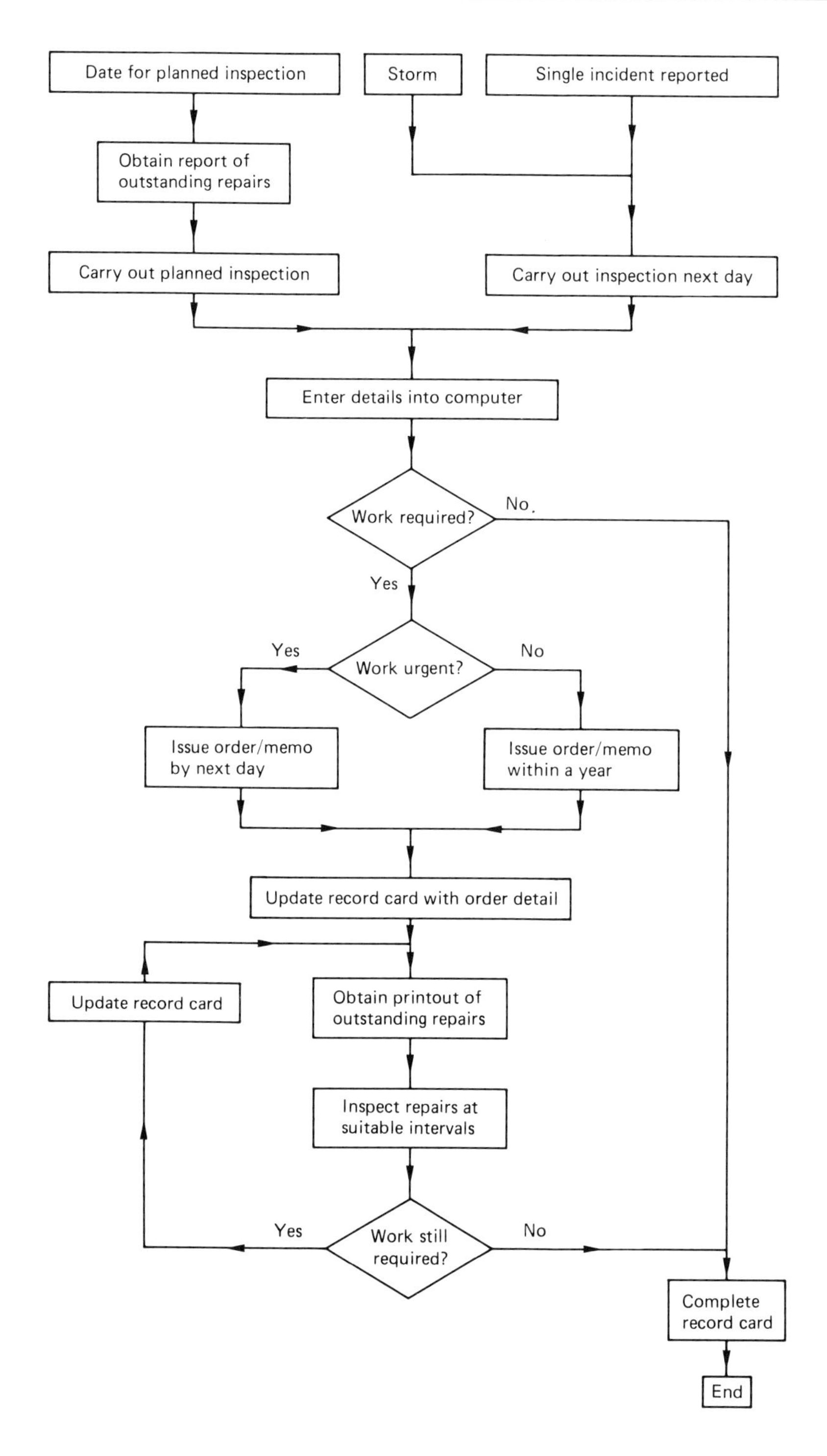

Figure 3.5 Maintenance procedure flow chart for coastal structures (as used by Wirral Borough Council)
Source: Wirral Borough Council.

operational viewpoint.

3.11 Post-contract actions

The action of repairing and maintaining a structure may have a number of significant effects:

1. the standard of service may have been raised or lowered as a result of the works;
2. the pre-contract investigations may have revealed facts about the wall and its environment which were previously unknown;
3. the maintenance works themselves may have revealed additional information about the configuration and state of the wall;
4. the engineering evaluations may have established that the structure is either more, or less, stable than originally thought.

Information which is gained during the course of the maintenance contract should be used as the basis for a post-contract review. The following points should be considered:

1. has the standard of service been altered by the works or information gained from the works and, if so, is the owner/authority aware of this?
2. has all relevant information about the wall and its environment been assembled in a form which will act as the baseline for future inspections and monitoring?
3. have inspection and monitoring procedures been reviewed, and, where necessary, revised, in the light of the information gained from the works?
4. have suitable means of storing and retrieving this information been agreed?

Where a Development Corporation, or its residuary body, has jurisdiction in a particular waterside area it may be necessary to ensure that owners or developers of properties with waterfront walls take the correct action in maintaining them in an acceptable state. In such circumstances it would be prudent to develop a standard procedure to be followed by all developers. One such document being developed by the London Docklands Development Corporation covers:

1. the setting of acceptable structural and quality standards;
2. the carrying out of baseline structural and aesthetic surveys;
3. the agreement of a repair programme to improve the state of the structures to the standards set;
4. the agreement of a future maintenance programme to maintain these standards.

Chapter four

Inspection and monitoring

4.1 Introduction

Inspection is the basic means by which the maintenance engineer is kept aware of the condition of the assets entrusted to his care.[8] The type and regularity of inspection required depend on the type of structure and the use to which it is put. Monitoring is the act of taking a series of measurements at regular intervals at a specific location on a structure, to assist in the engineering evaluation of a defect identified during the inspection process.

Old waterfront walls, being a family of structures with an immense variety of uses and forms, do not conveniently fall into a simple inspection category. It is therefore necessary to examine their function and importance more closely before making decisions which relate to their inspection and monitoring.

It is first necessary to review the objectives of an inspection programme. These are listed by Sowden[8] as:

1. affording assurance of structural safety;
2. providing for the economic management and control of operational serviceability by
   (a) identifying the need for preventive action
   (b) detecting incipient defects at an early stage
   (c) monitoring the development of those defects, in order to determine the urgency for, and the nature of, corrective action
   (d) compiling quantitative records of deterioration on which to base maintenance planning;
3. providing feedback to improve standards of durability in design and construction and of cost effectiveness in maintenance;
4. providing data to assist in formulating future maintenance strategies;
5. checking on changes in service conditions;
6. ensuring continuing compliance with social and legal obligations and guarding against political and professional embarrassment.

The manner in which an inspection and monitoring programme fits into the overall management of maintenance and repair is set out in Chapter 3.

4.2 Procedure for implementation of programme

The following procedure is suggested for identifying the scope and nature of a proposed inspection and monitoring programme:

1. review the importance of the structure in terms of its function, level of service and the consequence of its deterioration or failure;
2. assess the risk of deterioration or failure in terms of normal and extreme events;
3. on the basis of the items above, determine an appropriate inspection programme, if

necessary varying the frequency of inspection for different parts of the structure;
4. identify inspection methods which are appropriate for the wall or walls in question;
5. determine whether the inspections are to be carried out in-house or by specialist organizations, and ensure that the inspectors are trained to do the tasks assigned to them;
6. establish a suitable method and format for recording the results of the inspection including long-term storage and future access.

These procedures are discussed in more detail below.

4.3 Function of the wall

Clearly, the function of a wall will have a considerable bearing on its importance, and this will be reflected in the frequency and type of inspection adopted. A breakwater, protecting a harbour of refuge (see box), or a river wall forming part of a flood defence are likely to be judged as having greater significance than an old dock wall in a disused port. Likewise, the deterioration of a pier used by a few fishermen might be considered to be of low priority unless the pier is protecting a marina, or a site of historical significance or amenity.

A useful way to evaluate the importance of a wall is to consider the effects of its deterioration and failure. This is already a method adopted in the assessment of coast protection and flood defence works, where the loss of property and damage occurring due to failure of the defences is evaluated on a cost basis (see Section 3.8 and reference 38).

Where waterfront walls form part of a working port or bridgeworks the effects of failure are fairly easy to establish. The effects of deterioration on a structure of high amenity value may be more difficult to ascertain, but these are more likely to be covered by some statutory obligation to preserve them in a state of good repair. Statutory obligations are also covered in Section 3.4.1.

Inspection procedure at Holyhead

A good example of the attention paid to a breakwater protecting a harbour of refuge is to be found at Holyhead. At this port the owners, Sealink (UK) Ltd, carry out a comprehensive inspection every four weeks. This inspection is undertaken by an experienced supervisor.

The long breakwater is marked at 30 m intervals to assist in the location of defects and the inspection, which includes the measuring of rock mound profiles to check on the security of the wall foundations, is reported on 'by exception' (i.e. only changes to the structure or its foundations are noted).

For smaller authorities, who may find it difficult to justify expenditure on inspection, there may be good reason to take specialist advice as to the need, or otherwise, for this addition to their annual budget, particularly if the risk of failure is low (see below) or the consequences of failure may not justify expenditure.

4.4 Risk of deterioration and failure

No matter how important a wall is judged to be, the level of inspection selected must be a reflection of the rate of natural deterioration of the structure, its susceptibility to extreme loading and climatic conditions, and its propensity for sudden total failure or part irreversible loss of function. Although natural deterioration may be easy to assess, the susceptibility of a wall to extreme loads and the probability of it sustaining permanent degradation or failing are complex issues. If the structure is of importance these characteristics should be assessed by specialists, who will also be able to advise on the best type of inspection.

4.5 Inspection programme

An inspection programme should identify the types of inspection to be implemented and their frequency.

4.5.1 Types of inspection

Most organizations with an established inspection programme use some form of three- or four-tiered system. These can be described in the following terminology:

1. superficial inspections;
2. general inspections;
3. principal inspections;
4. special inspections

(Note: 1, 3, and 4 are terms used in the Organization for Economic Cooperation and Development (OECD) Bridge Inspection Report[41], 2 was introduced in the Department of Transport Bridge Inspection Guide[42].)

Superficial inspections

These take place regularly as part of the routine work for one or more members of the wall owner's staff. Personnel who work on or near a particular length of wall or structure note and report any defects, changes or unusual features. In some cases a chargehand, foreman or ganger is allocated the task of carrying out a regular walk-over inspection. This type of inspection is carried out daily, weekly or at a longer frequency, depending on the type and extent of structure involved, its use and the importance of its location.

General inspections

General inspections are more formal and detailed than superficial visual inspections and take place at greater intervals, say every two years or so. They are a combination of a general visual inspection and detailed inspections of representative areas and areas of particular note or concern. Monitoring of specific locations may also be carried out. General inspections are normally made by trained technical staff from the employer's organization or an outside organization.

Principal inspections

Principal inspections include a detailed examination of all aspects of a structure including any areas under water or with difficult access. The inspections are carried out at intervals of between two and ten years depending on the organization, the type and age of the structure, and the regularity and detail of the other inspections. Such inspections are carried out by a team of qualified engineers (and divers, where appropriate) who are fully conversant with the type of structure and the defects which may occur. Large organizations may manage to staff a suitable team with in-house personnel but smaller companies may need to engage a firm of consulting engineers and divers to carry out a principal inspection.

Special inspections

Special inspections are carried out on an *ad hoc* basis following severe adverse conditions such as floods, storms, exceptionally high or low tides, ship impact or when any of the other regular inspections indicates a major cause for concern. These inspections usually concentrate on one particular area or region rather than providing overall coverage like the regular inspections. The special inspection should also be carried out by qualified personnel.

4.5.2 Frequency of inspection

The frequency of inspection will depend on the vulnerability of the structure and the severity and variability of the conditions to which the wall is subjected. The effect of deterioration will also have a bearing on the inspection frequency.

A large breakwater, sea or flood defence under continual attack by the sea, whose collapse would or could endanger human life or cause considerable damage to or loss of property, should be inspected on a frequent basis. An old dock or lock wall in relatively benign conditions, whose demise would be an aesthetic rather than a catastrophic loss, would not warrant such frequent inspections.

The frequency and timing of inspection may be event-dependent, such as a special inspection after a severe storm, or it may depend on the ease of access to the area to be inspected. Low waters of equinoctial spring tides and drought conditions are particularly good times to inspect the lower courses of masonry and concrete in waterfront walls.

Many walls deteriorate at extremely slow rates and it would be impractical and unjustifiable to inspect these at frequent intervals. In such cases it may be better to devise an 'inspection by exception' system where the occasional superficial visual inspection is only supplemented by more detailed inspections when a change in the wall is detected. This method is adopted for many canal walls, where settlement of the wall or collapse of the

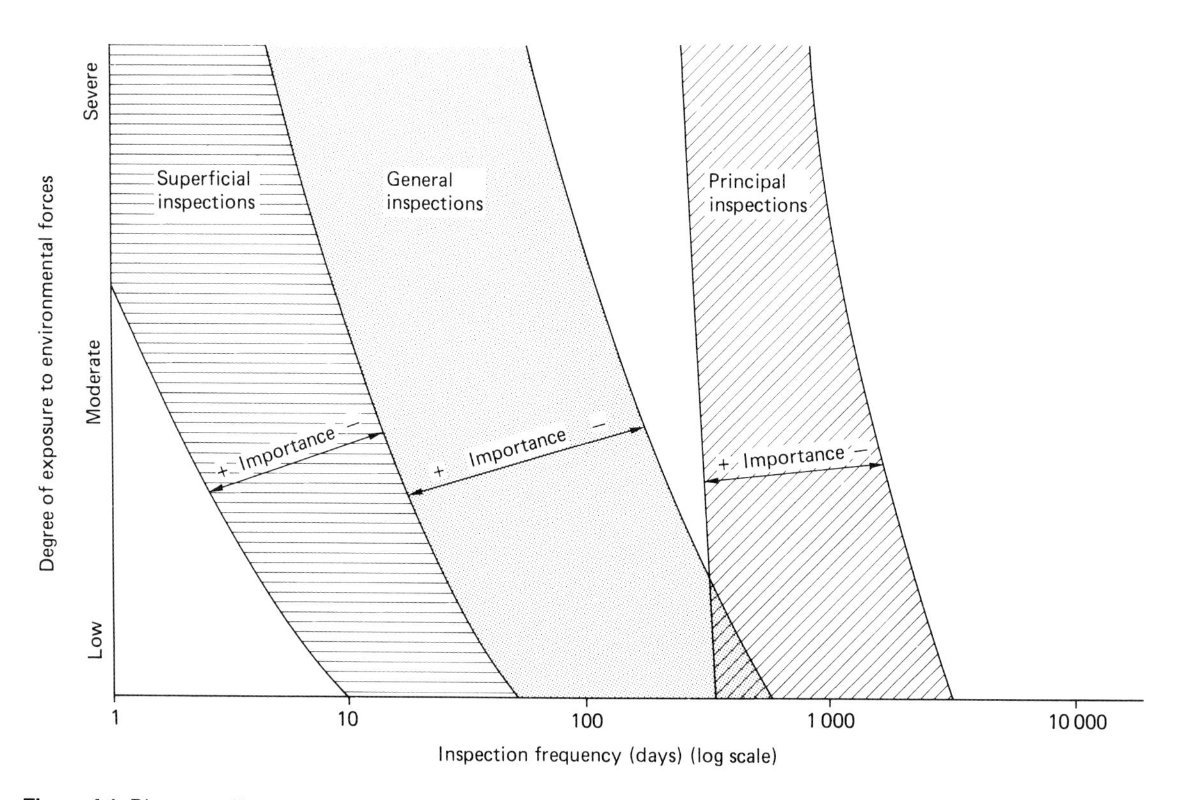

Figure 4.1 Diagram to illustrate the relationship between type of inspection, frequency of inspection, degree of exposure to environmental forces and importance of structure
Source: Livesey Henderson.

filling behind alerts the inspector to wall problems.

Organizations such as county councils, with responsibility for a considerable number and variety of old waterfront walls, may well find it difficult to justify initiating regular inspections on structures which have never been regularly inspected in the past and whose condition is currently of no great concern. One such organization, which commenced with no records of old waterfront walls, has adopted the following procedure when an old waterfront wall gives cause for concern:

1. the County Engineer is advised that a wall has been identified which is deteriorating or on the point of collapse;
2. repair or rehabilitation work is carried out;
3. the structure is recorded and taken into the inspection system;
4. a general inspection is carried out every two years thereafter.

It is not possible to lay down hard and fast rules regarding the optimum frequency of inspection for any particular category of wall. Walls with similar physical characteristics occur in widely differing locations and their assessed importance to their owners may also differ markedly. The tabulation of recorded examples in generic terms has proved to be difficult. Figure 4.1 indicates in general terms the spread of recorded inspection frequencies for all types of waterfront wall.

Figure 4.1 should not necessarily be used as a guide for determining appropriate inspection frequencies. There is evidence to suggest that many organizations carry out informal superficial inspections without recording them. What is important is that inspections are made in a methodical fashion and take adequate account of the importance of the asset being inspected.

4.6 Inspection report format

Many organizations, and especially the smaller ones, have suffered considerable loss of information over periods of years due to the non-formalization of inspection reporting. This is an unnecessary and costly waste of funds, particularly for those organizations which could ill afford the expense in the first place. It is essential that inspection reports carried out at different times are comparable, since this will save both time and money when it is decided to take more advanced measures to study or repair the wall.

The format of an inspection report will vary according to the nature of the inspection, the grade of personnel carrying out the inspection and the wall being inspected. However, all routine reports will include a number of common features:

1. the report will be recorded in the same format on each occasion;
2. preprinted forms will be used to assist in the recording process and to act as an *aide- mémoire* to the reporter;
3. walls and portions of walls will be clearly identified by name and reference;
4. means will be provided for any defects to be located with reasonable accuracy;
5. a consistent method of describing defects will be laid down;
6. the date, time of inspection and name of the inspector will be recorded.

Figure 4.2 shows a defects report form used by Wirral Borough Council in their coast-protection maintenance system. Although this form is designed for a computerized database, and is supported by other forms, comprehensive lists of structure, materials and damage codes, etc., it illustrates how a relatively simple form can be devised for defect inspection.

WIRRAL BOROUGH COUNCIL
COAST PROTECTION OFFICE

BOROUGH ENGINEER'S DEPARTMENT
DEFECTS REPORT FORM DRF 2

Date..............Taken by.......................Card No. [| | | | |]

Incident reported by ....
WBC Staff = D [ ] Councillor or MP = C [ ] Member of Public = P [ ]

Coastal Structure.................................. Wall Code [| |]
Approximate Location............................ From Ch [| | |]
To Ch [| | |]

(Identify Slipways, Accesses or Road Junctions)

Nature of Defect

Structure.. ST [|]

Material... MA [|]

Damage... DA [|]

Condition.. CO [|]

Date surveyed by CP Engineer [|] Y [|] M [|] D Entered on computer [ ]

Remarks...

....Order No. [| | | | |]

Figure 4.2 Defects Report Form, as used by Wirral Borough Council
Source: Wirral Borough Council.

In addition to the basics mentioned above there will be other facts to be reported in certain types of inspection:

1. details of any severe floods, storms, wave conditions, etc. which have occurred since the last inspection;
2. details of the surrounding hydraulic regime and any apparent changes in the environmental conditions;
3. details of any noticeable change in service conditions;
4. an assessment of the condition of the wall according to a condition coding system (see below);
5. an assessment of the circumstances causing any new or deteriorating defect;
6. an assessment of the priority which should be given to repairs (see below).

It may also be of value to take photographs of any defects noted, since this assists in the reporting process and makes comparison from one inspection to the next easier.

4.6.1 Repeatability

One of the most important aspects of the inspection of old waterfront walls is that it must be possible to repeat an inspection and be confident that any small changes in the condition of the wall will be detected by the comparison of two consecutive inspection reports. To achieve this, the location of existing defects and their severity must be reported reliably and preferably by the same inspector on each occasion.

Accuracy of location can be enhanced by marking the wall with unobtrusive distance marks in convenient locations or the use of non-corrodible pins driven into the structure. It is important to ensure that the method selected is not liable to be vandalized.

4.6.2 Description of defects

A full description of the defects which occur in old waterfront walls is given in Chapter 2. For a superficial visual inspection there are a limited number of defects which can be iden-

Table 4.1 Alphanumeric condition code suggested by the UK Department of Transport

Extent:	A.	No significant defect
	B.	Slight, not more than 5% affected (of area, length, etc.)
	C.	Moderate. 5% to 20% affected
	D.	Extensive. Over 20% affected
Severity:	1.	No significant defects
	2.	Minor defects of non-urgent nature
	3.	Defects of an unacceptable nature which should be included for attention within the next two annual maintenance programmes
	4.	Severe defects where action is needed (these should be reported immediately to the Engineer) within the next financial year

tified (although clues to their causes may also be apparent, see Section 4.7). and these may be listed as follows:

1. cracking;
2. distortion, misalignment and tilting;
3. bulging;
4. loss of mortar or wedges;
5. loss of facing;
6. spalled or eroded surface;
7. erosion, undermining or scour.

Methods of measuring the severity of these defects are given in Sections 4.7, 4.8 and 4.12.

4.6.3 Condition codes

Condition codes are sometimes used if a wall owner has a lot of walls, or walls of considerable length. They are a convenient way of describing the state of a wall for inclusion in a computer database or other system of recording where an overall assessment of the state of all the assets is needed. Many of the coastal authorities and British Waterways either have, or are currently setting up, such systems.

Condition codes are normally devised to suit the type of structure being inspected and the end use to which the overall assessment is to be put. In their Bridge Inspection Guide[42] the UK Department of Transport suggests an alphanumeric scale to denote the extent and severity of defect (see Table 4.1).

However, the extent of the defects of waterfront walls may be difficult to determine and, even if known, may be of little value to the engineer. It may therefore be more appropriate to use a simpler format to denote the need for repair, as used by some county councils.

Code	*Action*
1	Requires immediate action.
2	Repair as soon as possible.
3	Repair before the next inspection.
4	Keep under observation.

In France a classification system is used which attempts to relate the code to the nature of the defect[43] (see Table 4.2). Both this coding system, and those above require the inspector to be capable of making an engineering judgement on the severity of the defect and the need for repair. The codes would not therefore be used as a method of recording a superficial inspection carried out by untrained personnel, unless the inspection was rapidly followed up by an experienced engineer to record the nature of the defects found.

Condition codes may also be devised for a specific type of material. For instance, the following brickwork code was used in a recent diving survey of the Surrey Docks, London.[44]

Code	*Description of brickwork.*
1	Good, minor pointing only required.

Table 4.2 Classification system used by the French Ministry of Transport

A.	No defects
B.	Defects without important consequences other than aesthetic
C.	Defects which indicate the risk of abnormal development
D.	Defects which indicate developing deterioration
E.	Defects which clearly show a change in the behaviour of the structure and which bring into question its durability
F.	Defects which indicate the approach to a limit state, necessitating a restriction of use or taking out of service

2 Bricks eroded but firm, mortar broken down to a maximum depth of 25 mm.
3 Bricks eroded and loose in places, mortar broken down to a depth greater than 25 mm.
4 Areas of missing bricks to depths indicated on the sketches.

4.7 Inspection methods above water

4.7.1 General visual inspection

A close visual inspection of a wall will often reveal early signs of major distress as well as defects to be remedied during routine maintenance. The following points should be noted as being indicative of some problem:

1. ponding on the apron or top surface of the wall (differential settlement, rotation, loss of backfill);
2. steps between adjacent wall units (differential settlement, rotation);
3. cracks in the apron (differential settlement, forward movement);
4. difficulty of operating travelling cranes due to misalignment (movement of backfill);
5. step between wall and apron (loss of backfill, forward movement);
6. repeated repair of same crack (continuous movement);
7. seepage of water through cracks or joints above water level (loss of hearting);
8. unusually green area of foliage beside canal wall (seepage through wall);
9. canal boats unable to come alongside wall (loss of depth due to movement of lower part of wall or wall material fallen on to bed).

Much information on the state of a wall may be gained from viewing the front, either from a boat or by walking along the beach or river bed at low water. In some cases a small hovercraft may be the most effective form of transport.

4.7.2 Detailed inspections

For general and principal inspections a detailed survey of the features of the wall is required. For these detailed surveys it is necessary to locate any defect found and to note its characteristics.

For location purposes each length of wall has a unique chainage referenced to a recoverable origin. For ease of reference, identifiable, fixed objects such as bollards, corners of quay walls, ladders, etc. are referenced prior to the inspection. These can then be used as local reference chainages. It is important to use two reference objects since it is possible that one object may have been moved or replaced in the period between inspections.

Features to be observed during an inspection include known and new defects and early signs of distress. The types of feature to be detected will depend on the wall, its possible mode of failure and its function. Features can be classified into the following categories:

1. physical damage, including abrasion;

2. loose, projecting or missing bricks or blocks;
3. cracks in mortar, blocks, bricks or concrete;
4. disintegration of wall materials;
5. total or differential settlement;
6. deviation from the original lines of the structure by translation, rotation or bulging.

The more complex of these features are described in greater detail below.

Cracking

Cracks indicate that the structural strength of a particular element has been exceeded. It is essential to determine whether the crack is static or growing, whether it continues through mortar and blocks or through mortar only, and whether the crack is a potential source of further deterioration.

Cracks through mortar only are usually indicative of local stress relieving whereas those which extend through a number of blocks and mortar joints are of greater structural significance. In the latter case the length, width and orientation of the crack needs to be measured and monitored. The monitoring of cracks is covered in Section 4.12.4.

Settlement

Both total and differential settlement can be measured by standard levelling techniques. Reference is made to a fixed, identifiable benchmark which is well outside the zone of influence of the forces affecting the area under consideration. The benchmark must be permanent and non-moving. The use of brass studs and washers is a common method of forming a benchmark if an Ordnance Survey benchmark is not available.

All levelling points are normally fixed, permanent and easily identifiable, with all-round vision and good accessibility where possible. They are placed at about 30 m centres on long walls and there is usually a minimum of two points on short walls. The accuracy of levelling over this distance will be of the order of 1 mm.

Translation

Translation of a wall can be detected by measuring the offset distance of a number of points on the wall from a fixed baseline outside the zone of influence. The points on the wall can be the same as those established for levelling. The baseline may contain intermediate points within the zone of influence whose movement will indicate a general translation of the ground behind the wall (see Figure 4.3). The accuracy of measurement in this case will be of the order of 5 mm.

In docks, where there may be walls facing each other, measurement of the separation will indicate whether translation is taking place. This measurement will be made easier by the use of electronic distance measuring equipment (EDM). However, it will still be necessary to determine which wall is moving. The accuracy of measurement will be limited to about 10 mm. It is essential that a record is kept of the water level in front of the wall when measurement of translation is made (see Section 4.9).

Inclination

A change in the inclination of a wall indicates rotation. Inclination may be measured with a plumb bob or an inclinometer (also known as a tiltmeter), which measures the change in inclination of points in or on a structure. The plumb bob is usually accurate enough for most purposes, but cannot be used on a windy day unless accompanied by adequate shielding and patience. If still water exists in front of the wall it may be used to damp the motion of the bob.

Inclination should be measured over as long a length as possible and the plumb bob should be suspended a short distance in front of the wall to facilitate the measurement of both positive

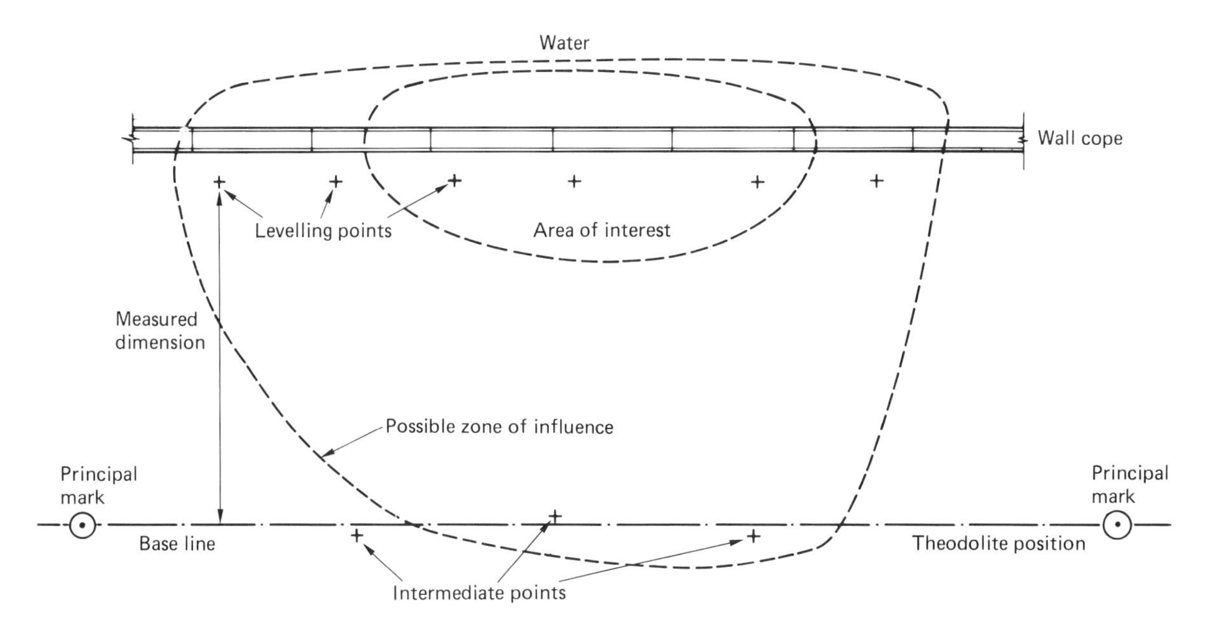

Figure 4.3 Illustrative example of baseline set up for monitoring wall movement
Source: Livesey Henderson.

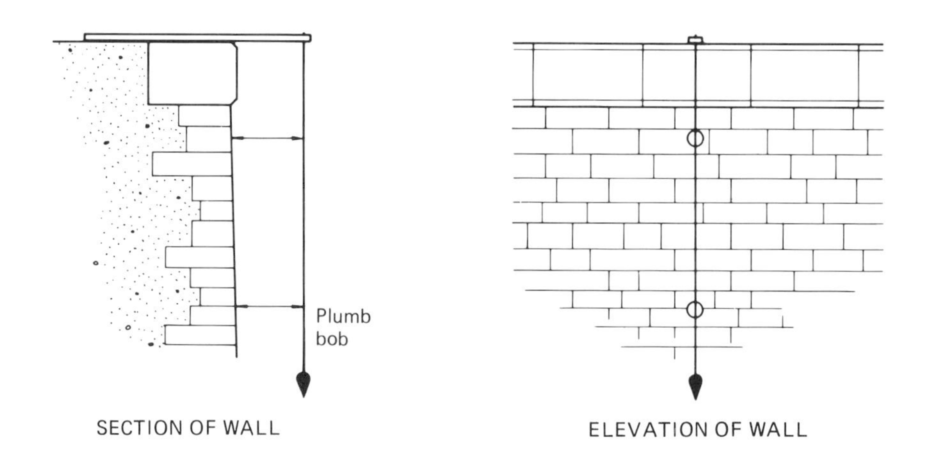

Figure 4.4 Method of measuring inclination of a structure with an irregular surface
Source: Livesey Henderson.

and negative inclinations. This method is not suitable for structures with irregular faces unless two patches of mortar are applied to the wall as shown in Figure 4.4.

4.8 Inspection methods below water

Detailed inspection of the whole of the submerged face of a wall is not usually cost-effective. It is more common for a general inspection of the whole wall to be carried out followed by a detailed examination of representative areas, or areas of special interest. The overall inspection can be carried out by the following methods:

1. team of divers
2. remote-operated vehicle (ROV)
3. sector scanning sonar
4. echo sounder.

Closed-circuit television (CCTV) can be used in conjunction with divers and ROVs, or manipulated from the surface. An advantage of the addition of CCTV is that it gives a permanent record and allows personnel on the surface to inspect the wall and co-ordinate the survey. A CCTV with its own light source often gives a better picture to the personnel on the surface than the diver has under water.

An example of a CCTV survey used to good effect in the inspection of the underwater portions of quays is given by Berry.[45] This survey, which was carried out in Massawa, Ethiopia, illustrates that, when visibility is good, long lengths of quay can be inspected with expediency by this method. Details of the use of divers, sector scanning sonar and echo sounders are given in Section 5.4.

For detailed inspection the chosen areas have to be cleared of marine growth and their location and level determined. Still photographs of defects and areas of interest should also be taken as a permanent record. Each photograph should be identified by its position, have a scale in the picture and an indication of its orientation. Clarity of the image may be improved by the use of fresh-water jackets between the camera lens and the objective.

For routine underwater inspection of walls, divers experienced in this type of work should be employed. For more complex problems it may be desirable to employ engineers with suitable diving experience.

4.9 Water levels and drainage

Differential water head in a wall can cause excessive loading, and the flow of water through a wall can wash out backfill and hearting (see Sections 2.3.5 and 2.4.5). When a wall is inspected it is essential to record water levels and drainage path observations. The main features to be measured are:

1. the water level in front of the wall;
2. the water level within and behind the wall;
3. the differential head between the water in front of and within the wall, and the rate of response of either level to changes in the other;
4. the porosity of the structure.

The water level in front of a wall may be measured either manually or by automatic means. Manual measurements can be made with a graduated pole or a dip meter suspended over the front of the wall from the cope. Automatic measurements can be made by floating ball or bottom-mounted pressure transducer tide recorders.

Water levels inside a wall are more difficult to measure than those outside. Without the use of predrilled boreholes it is only possible to gauge the water level inside by observing weepholes in the face of the wall.

With predrilled boreholes (see Sections 5.2.4 and 5.3.2) it is possible to use dip meters or piezometers to measure the level of water within the structure or behind it. Another method of determining the water level in this area is by the excavation of trial pits.

Monitoring of water levels is necessary to determine the maximum differential head between levels and the rate of response to changes in levels. In tidal areas the head and the response will vary according to the tidal cycle and hence measurements should be taken over a full tidal cycle at both spring and neap tides.

Differential head can be surcharged by waves overtopping the wall and by heavy rainfall. In these cases the porosity of the structure may be of interest. Porosity may be measured by means of falling head or constant head permeability tests as detailed in BS 1377.[46] All water level measurements should be recorded with the time and date of the measurement and a note of any significant event which has occurred in the immediate past, such as heavy rainfall, etc.

4.10 Site considerations

An inspection of a waterfront wall should include an examination of its surroundings and a record of any changes. The key points to be noted will depend on the type of wall being inspected but will include some of the following:

1. the level of water in front of the wall, particularly the highest and lowest levels;
2. the flow velocities and directions in front of the wall;
3. wave heights in front of the wall;
4. ground/beach levels at the toe of the wall;
5. alterations to the environment upstream of, or along the beach from, a wall, such as construction of groynes, bank erosion, etc.

Two of the most important changes which can occur, and frequently lead to problems, are the scour of the material in front of the toe of the wall and the change in beach level in front of a wall. Both these can lead to instability of the wall and eventual collapse if not detected at an early stage of development.

Scour can be measured by normal surveying methods where the bed is exposed at low water levels. However, this is rarely possible. For submerged beds it will be necessary to adopt one of the following methods:

1. ship-borne hydrographic surveying (there may be problems with side scatter of signals off the wall at close proximity);
2. sector or side-scan sonar from a vessel or from equipment cantilevered from a vehicle running beside the wall cope;
3. lead line from a boat, or from the cope of the wall.

Further details of the sonar techniques are given in Section 5.4.2.

The monitoring of beaches subject to wave action is complex. Beaches are frequently affected by both longshore transport of material and some offshore/onshore transport. The presence of groynes further complicates the picture. Beaches may be well stocked with material at one moment but depleted overnight by a severe storm. Large seasonal variations may therefore be imposed on much smaller long-term changes.

Long-term changes in beach level can only be determined by a regular long-term monitoring programme or by selective measurements of maximum and minimum levels. The regularity of measurement will depend on the length of beach, its type of protection and the mobility of the beach material. It has been suggested that the stereoscopic technique described in Section 4.12.5 might be useful for monitoring beach levels.

4.11 Change in service conditions

During an inspection it is important that the inspector notes any changes to the service conditions which the wall is having to withstand. Such changes might be:

1. the imposition of surcharge loads on the apron of a wall;
2. alteration of use, such as the mooring of large vessels against a quay which has been out of use;
3. increase of traffic loadings behind a wall due to temporary diversions or introduction of one-way traffic flows.

4.12 Movement monitoring

To establish a meaningful monitoring programme it is necessary to include the following steps:

1. establish the baseline condition of the wall;
2. identify methods of measurement of sufficient accuracy;
3. ensure that the methods are repeatable;
4. devise a method of recording and presenting data which will allow detection of significant movement.

4.12.1 Establishment of baseline conditions

Lines and levels shown on construction and 'as-built' drawings cannot be used as baseline conditions. It is thus necessary to carry out a detailed survey of the wall to determine its present shape, and this will include:

1. the line and level of the cope;
2. the inclination of the front face;
3. bulges or distortions to the face.

A survey of this nature not only sets out the baseline conditions but can also indicate which areas of the wall are more likely to change in the future and hence need to be monitored more closely. Reference positions should be chosen which are stable, since these can also be monitored for comparison.

The monitoring positions on the wall should be selected with care, since positions which are unlikely to show defects nullify the monitoring. Enclosed corners are inherently stable while open corners are not. Walls at midspan between counterforts are more likely to show signs of distress than adjacent to a counterfort.

Reliance upon a single datum is not recommended. Two or even three well-separated points should be used and the interrelationship between them should also be monitored. It is unlikely that they will all settle but, as an additional precaution, one of them should be regularly checked against a known datum well outside the zone of any possible movement.

4.12.2 Accuracy and repeatability of measurement

The method of measurement used must have an accuracy appropriate to the range of possible movements to be expected. For example, a series of consecutive readings of 26 mm, 29 mm and 35 mm, taken with an instrument which is accurate to 5 mm, may or may not be significant. However, if the accuracy of measurement is 2 mm there is definite evidence of a trend. In general, movements need to be monitored to an accuracy greater than 10% of the anticipated change. A useful guide to the instrumentation and methods available for monitoring deformation is given in Chapter 12 of Dunnicliff.[47]

With all monitoring, water levels and seasonal variations must be noted as they may affect the readings. To eliminate the former the same monitoring exercise should be carried out at high and low waters on each occasion until it is known whether the tidal state is affecting the readings. Temperature may also be an influencing factor.

Most measurements are subjective. This subjectivity can be reduced by repeating each measurement a number of times. This not only gives a better estimate of the reading but indicates the range of possible error. By having the same person to carry out the monitoring on

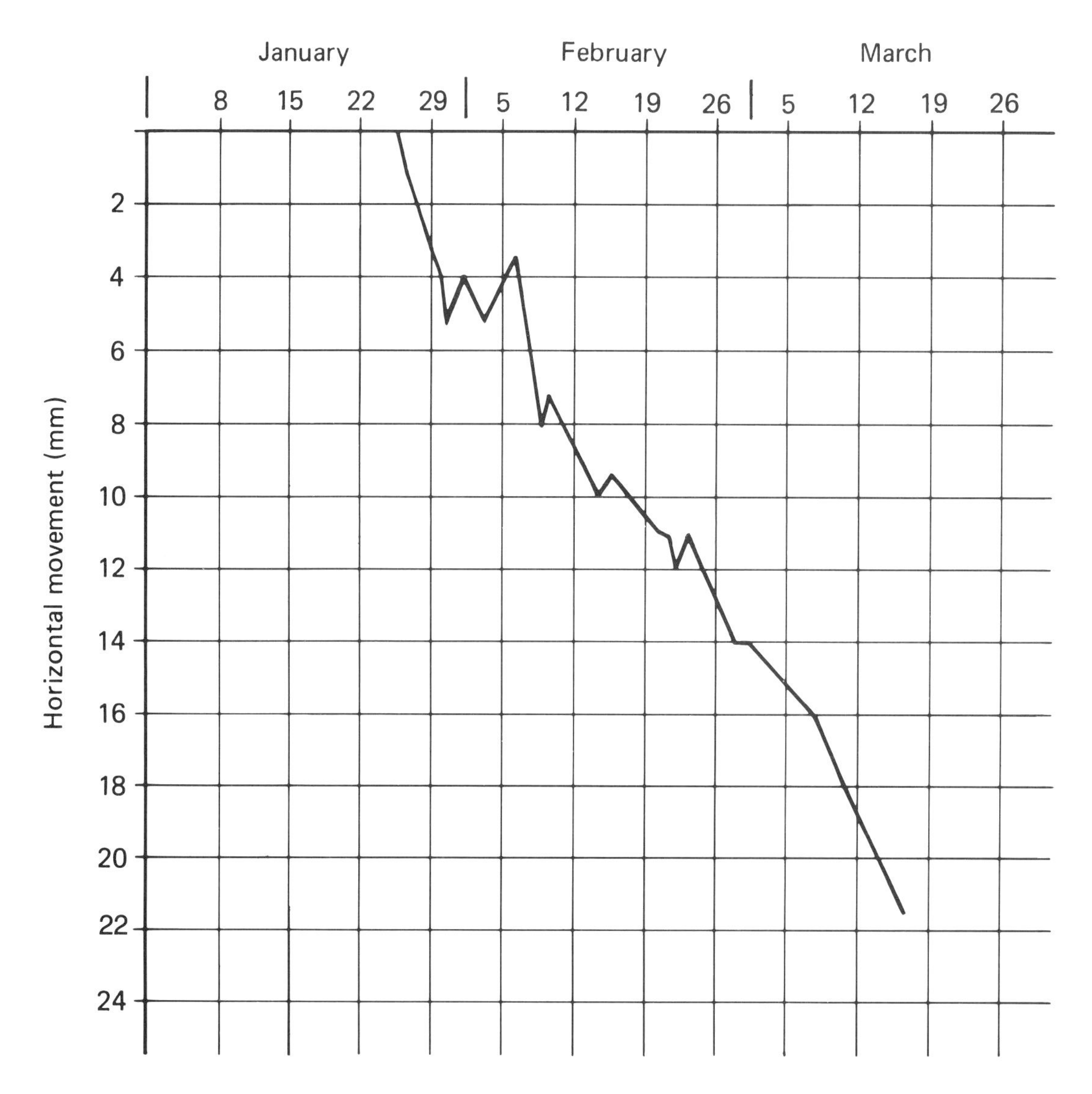

Figure 4.5 Example of the simple graphical presentation of wall movement
Source: Livesey Henderson.

each occasion it is possible to take into account any bias related to the measurements.

4.12.3 Recording information

Recording information on a standard sheet, or in a fixed format, allows the monitoring to continue unaltered by changes in staff, makes it easy to assimilate new data and facilitates computer processing of data. After each monitoring exercise the effect of the new set of data should be assessed. The following points may be helpful:

1. graphical presentation often helps to show up anomalies and trends. Figure 4.5 shows a

simple example, as recommended by English Heritage;[48]
2. regression analyses and moving mean analyses help to remove errors due to inaccuracies in the readings;
3. walls rarely move upwards or inwards;
4. doubtful sets of data should be collected again;
5. readings which are dictated to a second person should be repeated back.

It is very important to record ambient conditions, such as tide level, temperature, sunshine, etc. when noting a measurement. Walls may move appreciable amounts due to temperature change.

4.12.4 Cracks

It is necessary to monitor crack widths to determine whether cracks are stationary or moving. Crack width can be monitored by the following methods:

1. distance measurement by engineering calipers between brass studs fixed into the wall on either side of the crack;
2. use of an Avongard Tell Tale, which consists of two overlapping acrylic plastic plates attached to either side of the crack, marked so as to facilitate direct reading of any movement occurring;
3. use of vibrating wire strain gauges attached to either side of the crack, or specially adapted long gauges designed to span across courses of brick or stone. Such gauges can be linked to a signal-monitoring computer that can indicate changes of reading above a predetermined level.

Generally, the first method is less prone to vandalism and accuracies of up to 0.02 mm are possible. The use of glass telltales is not recommended as they can be accidentally broken and do not quantify the movement.

4.12.5 Stereoscopic combination

A monitoring method currently being developed by the University of Exeter[49] is the combining, stereoscopically, of an image in a photographic slide with the directly viewed scene of the same image. The method is capable of showing small changes in alignment and crack width by setting up the instrumentation at a remote location and viewing the scene against a slide taken on a previous visit. Accuracy depends on the geometry of the site and the type of movement being monitored, but can be as good as one millimetre in the right circumstances.

Design charts are given for this method to determine the accuracies which will be obtained in various circumstances. The accuracy will also depend on the stereo-acuity of the viewer.

Chapter five

Investigation techniques

5.1 Introduction

The most difficult parts of appraising an old waterfront wall are the determination of its shape and internal condition, and determination of the factors which influence its performance. The investigation methods selected should be suitable and cost-effective (see Section 3.7), and there should be a clear objective such as:

1. to investigate the causes of wall movement or other signs of distress;
2. to establish the load-carrying capacity of the wall;
3. to design rehabilitation works for upgrading the wall to take higher loads.

A comprehensive investigation is likely to involve techniques designed to provide information on the following:

1. the design of the wall and any history of repairs or failures, including those during construction;
2. the internal state of the wall;
3. the external state of the wall;
4. the state of the materials of which the wall is made;
5. the local drainage regime;
6. the geotechnical regime;
7. the local hydraulic regime;
8. the local sedimentary regime;
9. the present and future wall loadings;
10. any constraints to maintenance or rehabilitation.

Descriptions of the various techniques available are given below.

5.2 Determination of design and repair history

The design of the wall includes its shape, composition, strength, founding and backfilling. No reliable assessment may be made of the stability of the wall if this data is not all available. It is not sufficient to obtain the old construction drawings alone, as walls were frequently modified during and after construction.

Determination of the present shape of the wall will use several complementary sources. These are described below. Before any method of investigation is initiated it is prudent to review the numerous varieties of wall shapes which are described in Chapter 1, and to bear these in mind during the investigation.

5.2.1 Record drawings

The files, records or archives of the wall owner may contain original wall drawings and notes relating to its repair, modification or other historical events. Drawings tend to be of an outline nature and much of the detail is left to the person on the site. Although the drawings may give a general guide to the original design of the wall it is important to obtain corroborative evidence of this.

Some old construction drawings have the original notes of the designer written on them, showing where the 'as-built' structure differs from the drawings. However, few engineers or contractors were as conscientious as this and it is possible that the contractor may have altered the design, resulting in a change to the shape or materials. A good example of this is found at Sharpness (see box) where the contractor modified the design to suit his construction method and also had to alter the design because of failure during the construction.

Modifications to design at Sharpness

At Sharpness the original design proposed in 1874 is shown in Figure 1.44. The contractor modified this design in two ways to facilitate construction; by widening the upper part of the wall to allow the operation of a rail-mounted unit on top of the wall and by adding a considerable amount of brickwork to act as permanent shuttering. At one stage there was evidently movement of the part-constructed wall (see Figure 1.122) and the rear profile of the wall was altered to compensate for this. The props shown in the photograph are too heavy to be merely temporary works. They must have been installed to prevent further movement.

Walls were frequently changed in cross-section along a single length, to suit changing ground conditions, and some walls were so unsuitable that they were replaced or modified at a later date. The discovery of a set of old design drawings may therefore be more misleading than helpful, and it is essential that some cross-checking by other investigative methods is carried out.

Where a tidal river wall is being investigated it is likely that there will be previous walls behind the current wall and there are instances (see Section 1.5) where a number of walls may be superimposed.

5.2.2 Other historical sources

The history of civil engineering in Europe is well documented, and records of the achievements of European engineers in overseas territories are well covered. In the UK there are many books and reports documenting the building of ports, docks, canals and other structures with waterfront walls. In particular, the following publications are of great help at the commencement of a search:

1. A. W. Skempton. *British Civil Engineering Literature 1640–1840*. Mansell Publishing Ltd, 1987;
2. G. J. Jackson. *The History and Archaeology of Ports*. World's Work Ltd, 1983;
3. R. Morris and F. Morris. *Scottish Harbours*. Alethea Press, 1983;
4. D. Swann. 'The Engineers of English Port Improvements 1660–1830'. In the journal *Transport History*, 1968.
5. C. Hadfield. *British Canals. An Illustrated History*. David and Charles, 1974.

Each of these contains references to other books or publications which are generally more site-specific.

The main difficulty with historical research occurs prior to the publishing of the Institution of Civil Engineers proceedings in 1836. The publications mentioned above assist in covering some of the pre-1836 period. The ICE proceedings contain numerous useful papers, many of which are referenced in this book, and the ICE library has a computerized database which facilitates searches under both designer's name and site location.

Other national organizations which may hold records of interest to those investigating old waterfront walls are:

1. The Public Records Office, Kew, London;
2. The British Newspaper Library, Colindale, London;
3. The British Geological Survey, Keyworth, Nottingham;
4. The Telford Museum, Ironbridge, Staffs;
5. British Coal, London.

Local organizations may also be useful sources of information. The following may warrant investigation:

1. the county archives (in the County Records Office);
2. the local library;
3. the local museum;
4. the local authority;
5. local historical and geological societies;
6. the local newspaper;
7. old picture postcards.

5.2.3 Excavation

Excavation can be carried out in the apron area behind a dock, retaining or sea wall, or in between the back and front face of a pier or breakwater. This excavation would normally be in the form of a number of trial pits and trenches up to 5 m deep. Such excavations should not affect the integrity of the structure.

Excavations are limited by the groundwater level, since below that level the ground is liable to be unstable. Where the water in front of the wall is tidal the excavation should be started at high tide so that it can be completed at low tide. Trenching along the back of the wall is needed to explore for counterforts (maximum spacing usually 5 m) and tie rods. An excavation of this type will give the following information:

1. whether the wall has counterforts, and if so, how deep they are and what their spacing is;
2. whether there are any ties or old timbers linking the wall to the ground behind;
3. whether bollard blocks are built into the wall or are separate;
4. the form of construction and composition of the top of the wall below the coping stones;
5. the nature of the backfilling material;
6. the nature, in the case of a breakwater or pier, of the top surface and hearting.

The disadvantages of excavation are that it is depth-limited by expense, safety considerations and by the high groundwater levels found behind walls of this type, and it cuts off a large area of working space, an appreciable fact in a busy port or riverside location.

However, excavation gives a clear picture of the wall at the excavation site and allows large representative samples of wall and backfilling to be collected for testing. An example of the extent of detail exposed by excavation is given in Figure 5.1.

5.2.4 Use of boreholes and probes

Beyond the limits of economic excavation, a system of probing can be used to assist in the identification of the back profile of a wall. The use of probes is by no means foolproof. It is difficult to differentiate between large pieces of debris and wall components, and the probe may well glance off sloping or angled steps in the rear of the wall, giving a false impression of the shape of the structure. Backfilling is frequently of a coarse and granular material, with rubble and debris commonly used, and this makes penetration by any drilling method particularly difficult.

Various types of probe are available, from a manual percussive probe, to an electrically powered percussive drill with a 'flow-through' sampler. The latter not only defines the levels of hard strata encountered but also obtains a continuous 30 or 50 mm diameter core of material above.

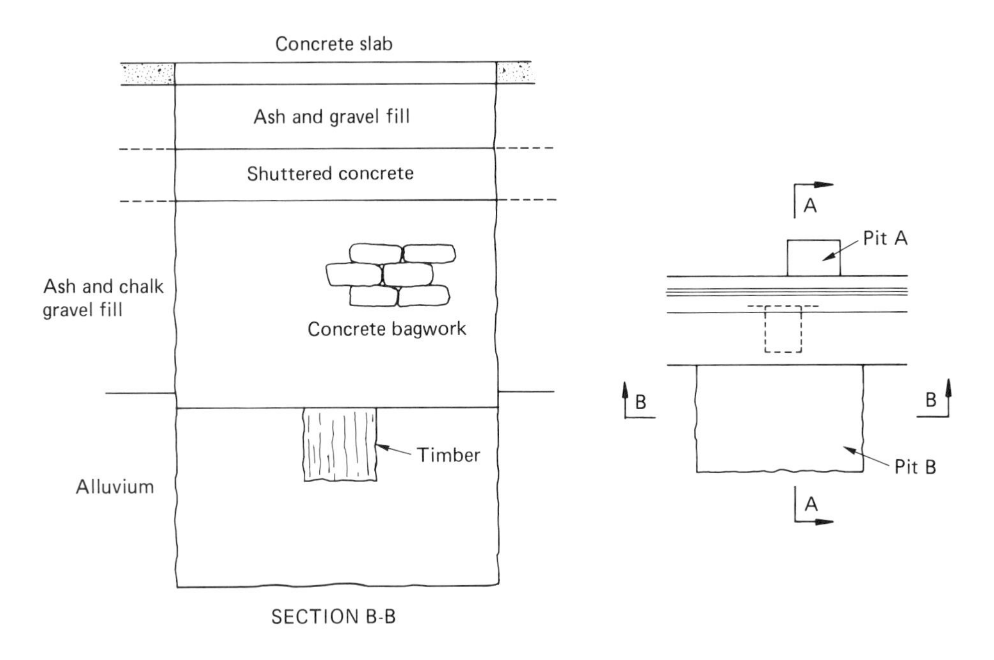

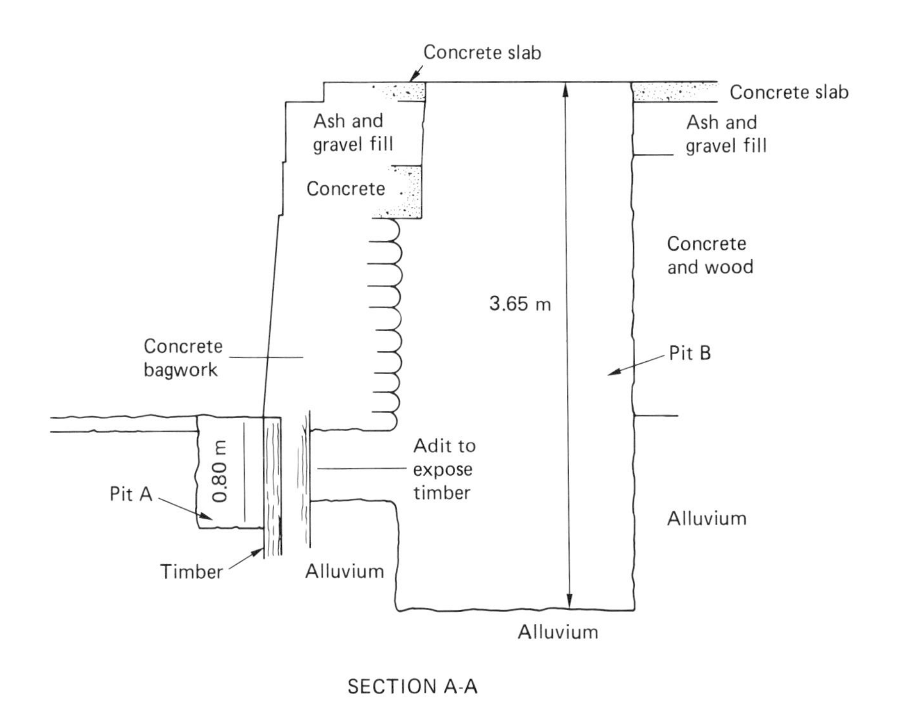

Figure 5.1 An example of the extent of detail which may be gained from excavation of a trial pit
Source: Sir Bruce White, Wolfe Barry and Partners.

Table 5.1 Relative costs of ground-investigation techniques (all figures in £s 1989)

	Probing	*Shell/auger*	*Rotary*
Mobilize rig – item	65–100	100–250	250–500
Establish rig – cost/hole	20–80	25–40	45–85
Investigation – cost/metre	8–10	10–13	31–44

Probes, however, are only suitable for investigation in homogenous and relatively soft materials such as clays or fine sands. Obstructions such as rocks or debris can cause refusal and thus give misleading information. Probes are also depth limited, dependent on soil conditions.

In materials which are predominantly soft but which contain occasional obstructions the use of a shell and auger (cable percussion) rig may be considered. This allows soft material to be continuously sampled as it is removed and the obstructions can be broken up by chiselling. Once the back of the wall is reached a short core can be taken by a rotary coring rig to prove the material.

The relative costs of probes, shell and auger, and rotary coring methods are illustrated in Table 5.1. To mobilize a rotary coring rig is roughly two or three times more expensive than a shell and auger, or probing, rig; and the cost per metre run is approximately three or four times as high.

5.2.5 Shape-detection problems

There are a number of special problems associated with the investigation of the base of a wall, and the toe and heel of the wall, if they exist. First, some walls have a curved back profile which is almost parallel to the front of the wall (see Section 1.2.2) and a borehole placed in the middle of the top of the structure would give a false indication of the nature of the base (see Figure 5.2(a)). To overcome this problem it is necessary to put down a number of boreholes and to be very careful in comparing the break-out levels with that of the toe level of the wall. An absence of designed backfilling might indicate this type of wall.

Second, some walls have a small heel at the rear of the base (see Figure 5.2(c)). This heel is difficult to detect but significantly increases the stability of the structure, due to the additional weight of the soil above the heel acting at a large lever arm about the centre of rotation at the toe of the wall.

Third, the toe of many walls is obscured by sediment and, in some cases, covered by debris and/or anti-scour armouring (see Figure 5.2(b)). To determine the extent of the toe it may be necessary to remove the material covering the toe or to probe with a steel rod. Removal of material can be achieved by the use of an air-lift, or by mechanical means from a barge or the shore. In the latter case it is important not to excavate to too low a level, which may endanger the stability of the wall by undermining.

Probing for the toe tends to be unreliable since it is not possible to be certain that refusal of the probe is caused by the toe or an obstruction. Some toes are extended forwards, at some time after the original construction, in a stepped fashion, and a probe may not detect this modification to the original design.

The front face of the wall should be carefully examined for signs of what the remainder of the wall might be like. The following should be looked for:

1. tie rod anchorages;
2. type and materials of the face of the wall including differences between the top and bottom of the wall;

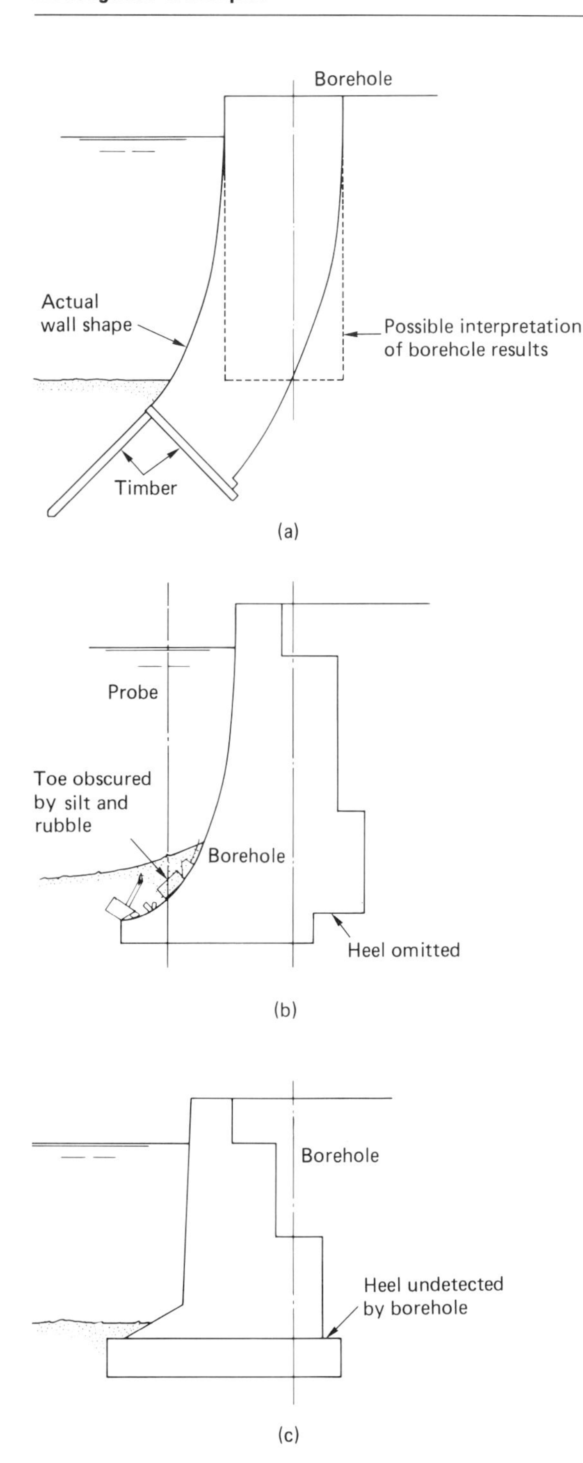

Figure 5.2 Wall shape detection problems for various types of structure
Source: Livesey Henderson.

3. drainage holes;
4. changes in founding levels;
5. structural features, such as arches;
6. timber piling.

5.2.6 Geophysical methods

Identification of the rear face of an old waterfront wall by geophysical methods (see Appendix B) may be possible but may be complicated by the lack of:

1. information regarding the shape of the interface, due to the loss of historical records;
2. contrast between the geophysical properties of the structure, the backfilling material and the original ground;
3. working space for the deployment of geophysical equipment on the ground due to the presence of buildings and dock facilities.

Close attention should be given to the type of information given in Chapter 1 on the internal construction features of a range of typical walls since the use of geophysical methods on many of these complex structures is likely to be impractical with the range of equipment and methods currently available. While to date no case histories have been found in the scientific literature relating to the application of geophysical methods to wall shape detection, all the methods described in Appendix B could be considered.

Choice of the best site-investigation method depends on close collaboration between the geophysical specialist and engineer/client. Use should be made of historical information, where it exists, to establish how the wall was designed and constructed so that the most appropriate geophysical method is selected and an assessment made of its likely degree of success.

The most likely approach at present is the use of ground-probing radar deployed on the front face of a wall. This would require the operation of the radar equipment in a vertical mode, and this has been shown to be a practical proposition. The main limitations of this method are:

1. attenuation of the electromagnetic energy by the internal fabric of the structure;
2. insufficient contrast in the electromagnetic properties of the materials on either side of the rear face;
3. lack of knowledge of the internal fabric of the wall leading to difficulty in calibration.

A further major problem associated with the use of ground-probing radar is that the electromagnetic signals are severely attenuated by sea water, either inside or outside the structure. This is a considerable disadvantage when the wall being investigated is partly submerged.

The seismic reflection method, with the seismic source and geophone array deployed on the front face of the wall, should also be considered. Like electromagnetic energy, seismic energy is attenuated during propagation. An additional problem is the lack of resolution when the normal low-frequency seismic sources are used. Further improvement of this method for wall investigation will require the development of a high-frequency seismic source, used in conjunction with suitable geophones, or accelerometers, and combined with modern data-processing techniques.

Ultrasonic reflection techniques have also been investigated but the results were not conclusive and no experimental work has been carried out.

Where boreholes have been used to investigate the internal fabric of a wall they may also be employed for cross-hole seismic or electromagnetic measurements. Of particular value would be boreholes on opposite sides of the anticipated position of the rear face of the structure. It may then be possible to predict the shape of the face between the boreholes by recording the variation of the compressional or shear wave velocity with depth.

Alternatively, it might be possible to position seismic sources in one of the boreholes with a geophone or hydrophone array and reflect seismic energy off the rear face. It has been suggested in Section 5.2.4 that horizontal probe holes could be used in this application. These would also be extremely useful for cross-hole seismic or electromagnetic surveys and the deployment of standard geophysical borehole logging tools.

In locations where the ground surface is clear, standard surface geophysical methods can be employed if there is a contrast in physical properties between the internal fabric of the wall and the geological material against which it abuts. The simplest method is a ground-conductivity survey which can be carried out rapidly. Data can be plotted and contoured so that changes in ground conductivity, related to changes on either side of the rear face of the wall, should be readily apparent.

Other surface geophysical techniques, such as seismic and electrical resistivity methods, may also be applicable. The use of a magnetic survey, to locate tie rods or vertical shuttering, should also be considered.

5.3 Determination of internal state

Many walls, retaining and free-standing, are composed of masonry or brick skins infilled with a hearting of a different material such as rubble, rubble concrete, etc. Loss or deterioration of this hearting will have a serious effect on the strength and stability of the wall.

Investigation of the internal portions of the wall is sometimes required to identify the nature of the hearting and its state, and to detect any cavities which may have been formed by loss of hearting through cracks in the outer skin of the wall. Investigation can be by means of:

1. excavation from the top of the wall;
2. drilling techniques;

3. remote inspection methods;
4. geophysical methods;
5. any combination of these.

5.3.1 Excavation

Excavation, as described in Section 5.2.3, gives a reliable picture of the composition of the upper layers of a wall and enables the investigator to determine whether there are any ties, cross walls, timbers, etc. within the wall. It may also detect whether there are any large voids in the core caused by leaching out of the hearting at lower levels. Excavation of this type has been carried out at Lyme Regis and West Bay in Dorset.

5.3.2 Drilling techniques

Drilling may be used to investigate the core of the wall if it is composed of sand, concrete, brickwork or stone. In these cases coring techniques may be used to sample the material and at the same time to determine changes in composition, voids and the level of the wall base. Sampling will allow the densities of the materials to be measured and their strength to be tested.

If the wall has been filled with coarse rubble, stone or debris, it is most likely that drilling methods will prove to be unsuitable since the hole will need to be cased and the drill string will become trapped in the loose material. Rotary percussive drilling with the 'Overburden drilling' system might overcome this problem but has the disadvantage that it is generally unsuitable for taking samples.

5.3.3 Remote inspection methods

Where the structure is delicate, listed, or the top is inaccessible; it may be difficult to obtain access to the core of the wall to inspect and measure voids. In such cases remote inspection methods can be used.

If a large enough hole can be made in the skin of the structure it may be possible to insert remotely controlled CCTV cameras. Where the entry hole must be small, such as through a borehole or a hole cored through the side of the structure, a borescope can be used. The borescope, which utilizes fibre optics, is inserted down a narrow hole and is able to give an all-round view of any cavity found. Photographs can be taken externally with a 35 mm camera attached to the end of the borescope and measurements made with the aid of this method. But lighting is a problem in large voids.

5.3.4 Geophysical methods

A description of the principal land geophysical surveying methods is given in Appendix B together with some practical case histories. The use of geophysical methods to investigate the internal structure of a wall has been very limited and confined mainly to the use of ground-probing radar, ground conductivity and seismic methods. Results of these methods to date indicate considerable potential but further research is required.

As seen in Chapter 1, the internal construction of many old waterfront walls is highly complex. Some information concerning this construction may be gleaned from historical records and any previous boreholes installed. It is essential that this information is made available to the geophysical specialist to enable the most suitable method to be selected for investigating the internal fabric of the structure.

It is likely that ground-probing radar will be used increasingly in the future to assess the internal fabric of old waterfront walls. One of the major problems that still has to be solved is the interpretation of the data. While the characteristic reflection pattern observed on the record from a cavity has been established already in other published work, this is not the case with many of the internal construction features illustrated in Chapter 1 of this book.

At present, although a large amount of data is generated during a radar survey, little work has been carried out on the calibration of the radar records against known targets within the wall. However, in time it is anticipated that the necessary experience to process and interpret radar data will be gained and this will enhance the use of ground-probing radar in this application.

Ground conductivity is directly related to the presence of water within the fabric of a wall. Its measurement is appropriate to the examination of a structure which has been exposed to an ingress of sea water, since water-filled zones will give rise to anomalously high conductivity values. The presence of iron rods and other metallic objects will lead to similar results, but in many cases this problem can be resolved by using a magnetometer at the same time. Again, calibration of the contour maps produced from a ground-conductivity survey against known targets in the wall is essential.

Seismic methods offer considerable scope in this application but very little practical work has been carried out to date. The transmission of a seismic pulse between opposite sides of the structure, or between boreholes within the structure, offers considerable potential for assessing the state of its internal fabric in terms of both its seismic properties and its derived dynamic elastic moduli. The computation of seismic tomography from the seismic data should also be considered in situations where good borehole control is available.

If boreholes are drilled into the wall then they may be used for a wide variety of geophysical logging tools to assess the internal fabric. Radioactive logs are of particular relevance and these include natural gamma (for lithology), gamma-gamma (for formation density) and neutron (for porosity). If the boreholes are fluid-filled it is possible to derive the dynamic elastic moduli of the internal fabric from the full wavetrain sonic log and the formation-density logs. Further details of geophysical borehole logging are given in Appendix B.

5.4 Determination of external state

The state of the external surfaces of old waterfront walls may be determined by visual inspection above water level, as described in Section 4.7, and under water by divers, Remote-Operated Vehicles (ROVs) and acoustic methods, as mentioned in Section 4.8. The use of divers and acoustic methods are described in more detail here.

5.4.1 Investigation by divers

Investigation by divers can be used for general inspections and for examining particular defects. Diving work tends to be expensive, because the number of diving personnel required to satisfy the safety regulations is high. The operations are frequently hampered by poor visibility, so it is more usual to employ divers for the detailed investigation of specific areas of a wall.

Where a large area of wall is to be examined it is important that the divers have a reliable location system. This should consist of a method of obtaining longitudinal chainage along the cope of the wall together with a method of measuring the vertical distance down from the cope level to the area of interest. There are two ways of achieving the latter; by the use of vertical, marked lines hung at one metre centres down the face of the wall, or by a light metal frame suspended over the side of the wall.

The following additional procedures are recommended:

1. there should be two-way communication between divers and personnel on the surface;
2. all divers should be equipped with means for removing fouling, such as algae, seaweed and crustacea, from the surface of the wall;
3. all reporting should be in a standard format to suit the type of wall being inspected. An example of a standard report sheet is given

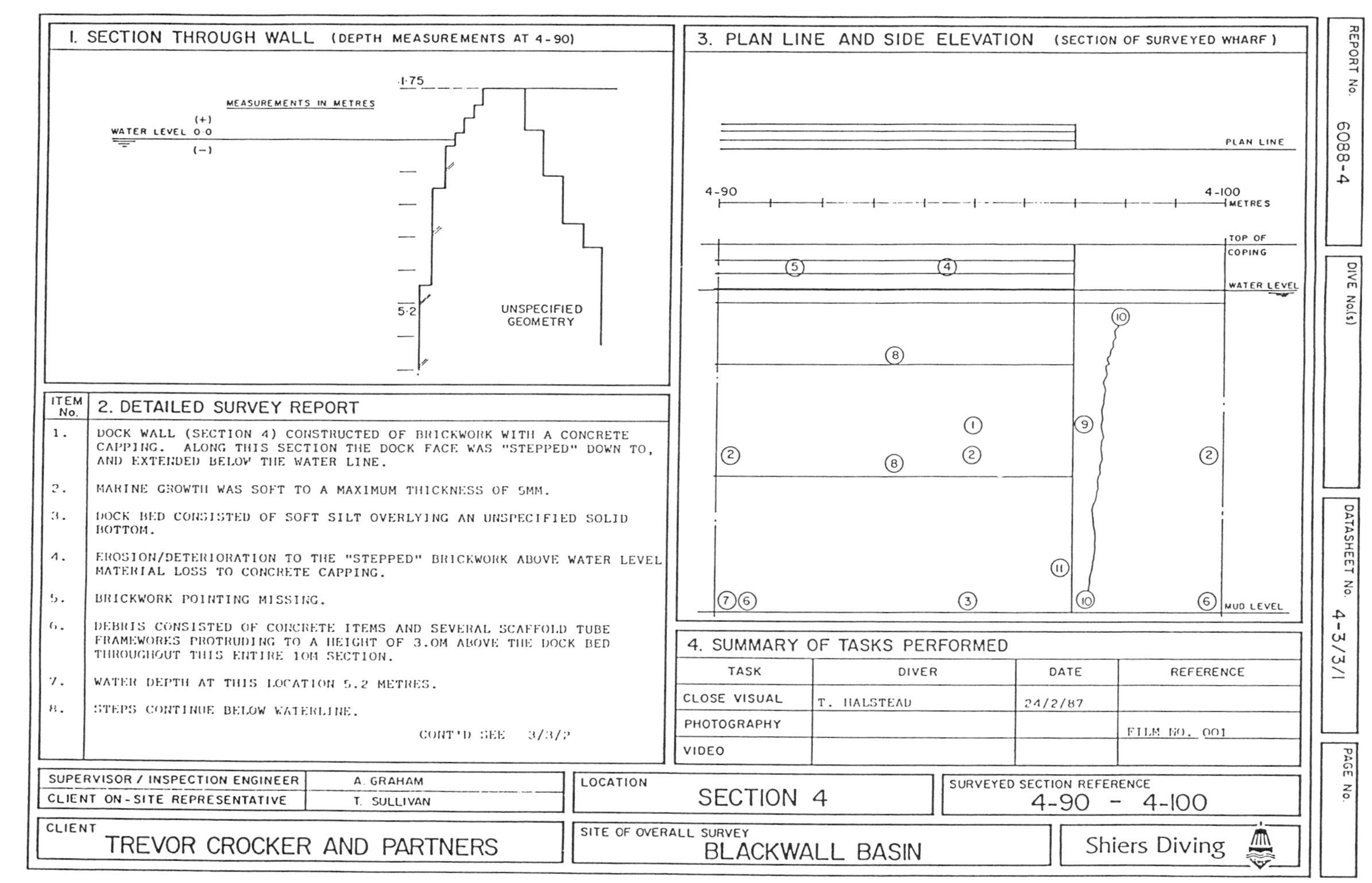

REPORT No. 6088-4	DIVE No.(s)	DATASHEET No. 4-3/3/1	PAGE No.

1. SECTION THROUGH WALL (DEPTH MEASUREMENTS AT 4-90)

3. PLAN LINE AND SIDE ELEVATION (SECTION OF SURVEYED WHARF)

ITEM No.	2. DETAILED SURVEY REPORT
1.	DOCK WALL (SECTION 4) CONSTRUCTED OF BRICKWORK WITH A CONCRETE CAPPING. ALONG THIS SECTION THE DOCK FACE WAS "STEPPED" DOWN TO, AND EXTENDED BELOW THE WATER LINE.
2.	MARINE GROWTH WAS SOFT TO A MAXIMUM THICKNESS OF 5MM.
3.	DOCK BED CONSISTED OF SOFT SILT OVERLYING AN UNSPECIFIED SOLID BOTTOM.
4.	EROSION/DETERIORATION TO THE "STEPPED" BRICKWORK ABOVE WATER LEVEL MATERIAL LOSS TO CONCRETE CAPPING.
5.	BRICKWORK POINTING MISSING.
6.	DEBRIS CONSISTED OF CONCRETE ITEMS AND SEVERAL SCAFFOLD TUBE FRAMEWORKS PROTRUDING TO A HEIGHT OF 3.0M ABOVE THE DOCK BED THROUGHOUT THIS ENTIRE 10M SECTION.
7.	WATER DEPTH AT THIS LOCATION 5.2 METRES.
8.	STEPS CONTINUE BELOW WATERLINE.
	CONT'D SEE 3/3/2

4. SUMMARY OF TASKS PERFORMED

TASK	DIVER	DATE	REFERENCE
CLOSE VISUAL	T. HALSTEAD	24/2/87	
PHOTOGRAPHY			FILM NO. 001
VIDEO			

SUPERVISOR / INSPECTION ENGINEER	A. GRAHAM
CLIENT ON-SITE REPRESENTATIVE	T. SULLIVAN

LOCATION: SECTION 4

SURVEYED SECTION REFERENCE: 4-90 - 4-100

CLIENT: TREVOR CROCKER AND PARTNERS

SITE OF OVERALL SURVEY: BLACKWALL BASIN

Shiers Diving

Figure 5.3 Example of a standard report form for use during a diver's survey
Source: DHV Burrow-Crocker Consulting.

Table 5.2 Typical diver's report/inspection result

Metres from 'Zero' datum	*Ref. No.*	*Defect type*	*Defect size and depth*
81–5	1	Crack, (vertical diagonal) within brickwork	5 cm wide at mud line deeper than 20 cm, misaligned by 2 cm. At waterline 20 cm wide. It starts immediately below limestone coping to a depth 4–45 m minus datum top
92–3	2	Outfall	At waterline 13 cm dia. approx
94–2	3	Hole (horizontal)	75 cm long × 15 cm wide × 10 cm deep × 3.5 m minus datum top
95–4	4	Hole (horizontal) with crack (vertical) and loose bricks	2.5 m long × 50 cm deep × 50 cm wide. Crack deeper than 15 cm × 50 cm long × 3 cm wide total distance from top to bottom of defect 2.3 m. 1.15 minus datum top to top of defect
103–3	5	Outfall	At waterline 13 cm dia. approx.
104–4	6	Hole (missing brick)	1.5 m minus datum top. One brick deep
108	7	Hole (horizontal missing bricks)	50 cm long × 30 cm wide × 25 cm deep × 1.75 m minus datum top
110–9	8	Hole (missing brick)	1 brick deep × 1.5 m minus datum top
113–1	9	Hole (horizontal, 2 missing bricks)	50 cm × 10 cm × 1 brick deep × 1 m minus datum top
114	10	Hole (missing brick)	1 × brick deep × 1.5 m minus datum top
135	11	Outfall	1 m wide × 6 m long stone faced with vertical dividers
140	12	Hole (missing brick)	1 × brick deep in first course below limestone coping at waterline × 1 m minus datum top

in Figure 5.3, and an alternative in tabular format in Table 5.2;

4. where possible, defects or unusual features should be photographed or videoed using low-light techniques as a permanent record;
5. light and simple apparatus should be provided for measuring the shape of the wall. Particular care is needed to define the exact shape of curved or sloping faces of walls. (An adjustable template may be useful for this task.);
6. an airlift pump will often be needed to remove debris from the wall toe as probing is unreliable due to the amount of hard debris which is usually found in canals, rivers and harbours.

5.4.2 Acoustic techniques

The front face of a waterfront wall, and in particular the area at the toe of the wall where a build-up of sediment is likely, may also be investigated with acoustic techniques. The same methods will identify areas where scour is occurring and where large cavities or recesses exist in the wall face. The techniques which may be used are:

1. echo sounding;
2. sector scanning sonar;
3. side scanning sonar.

Echo sounding

Reference should be made to the British Ports Federation's (BPF) publication *Evaluation of Echo Sounders for Hydrographic Surveying in Ports*[50] for selection of a suitable echo sounder and to BPF's Guidance notes in *Hydrographic Surveying in Small Ports*[51] for how to carry out the survey.

Echo sounders cannot normally penetrate silt and may give false echos near a wall. Other methods must therefore be used to measure the profile of the toe of the wall and the profile of firm ground immediately in front of the wall. Echo sounders may sometimes be used in the horizontal mode to give useful results.[52] The instrument is mounted on a vertical beam which is lowered to the dock or sea bottom at predetermined positions where vertical profiles of the wall are required. The transducer is moved up or down the beam to produce the vertical profile. This system, which is illustrated in Figure 5.4, requires modifications to be made to the pulse length, and screening and damping of the transducer, to achieve the desired results (also shown in the figure).

One limitation of the technique described above is that, for a general inspection, a judgement has to be made, on cost grounds, between the number of profiles taken and the likelihood of defects between the profiles remaining undetected. However, if a survey of this type indicates the presence of a high number of defects, then a more detailed type of investigation can be implemented.

Sector scanning sonar

Sector scanning sonar is usually deployed, in this application, from a winch mounted on a trailer which is towed along the top of the wall. The transducer is lowered into the water at the chosen chainage. A trace is produced showing the profile of the wall below water and the nearby sea or river bed. Voids under the toe can be measured provided the transducer is positioned at a low enough level. The transducer can also be turned through 90° to record horizontal profiles of the wall.

Figures 5.5 and 5.6 show a typical sector scanning sonar deployment for recording wall profiles and an example of the trace recorded. The detailed results shown in Figure 5.7 were obtained in Gloucester Docks. Another good example of the use of this type of equipment is shown in Figure 5.8, where the profile of a river has been obtained and the presence of undercutting on one bank is clearly demonstrated.

Side-scanning sonar

This type of sonar may be used in a variety of situations to examine the sea floor in front of walls. The combined use of a side-scan sonar and an echo sounder gives not only a visual indication of the seabed but also some indication of the dimensions of the sea floor features.

An interesting sonar record using a modern system is shown in Figure 5.9 for the sea floor immediately in front of the breakwater in Weymouth harbour. The excellent clarity of the rubble base of the breakwater should be noted, together with the fine detail of the sunken tender.

This example demonstrates the improvements made to side-scan sonar over the past decade. The use of high frequencies (675 kHz) and colour graphics gives the best final product and a good visual presentation to the civil engineer. However, simple systems may yield sufficient information since their sonar frequencies of around 65 kHz give a reasonably clear picture of the sea floor.

5.5 Testing of wall materials

Wall materials may need to be tested for a number of reasons, such as to determine:

1. the density of the material for stability calculations;
2. the strength of materials for calculations relating to crushing and cracking;
3. the type, extent and depth of degradation;
4. the physical and chemical composition of wall materials;
5. the porosity of material for assessments of the hydrostatic head due to differential water levels.

Figure 5.4 Deployment of dock wall profiling equipment and an example of the results obtained
Source: ABP Research and Consultancy.

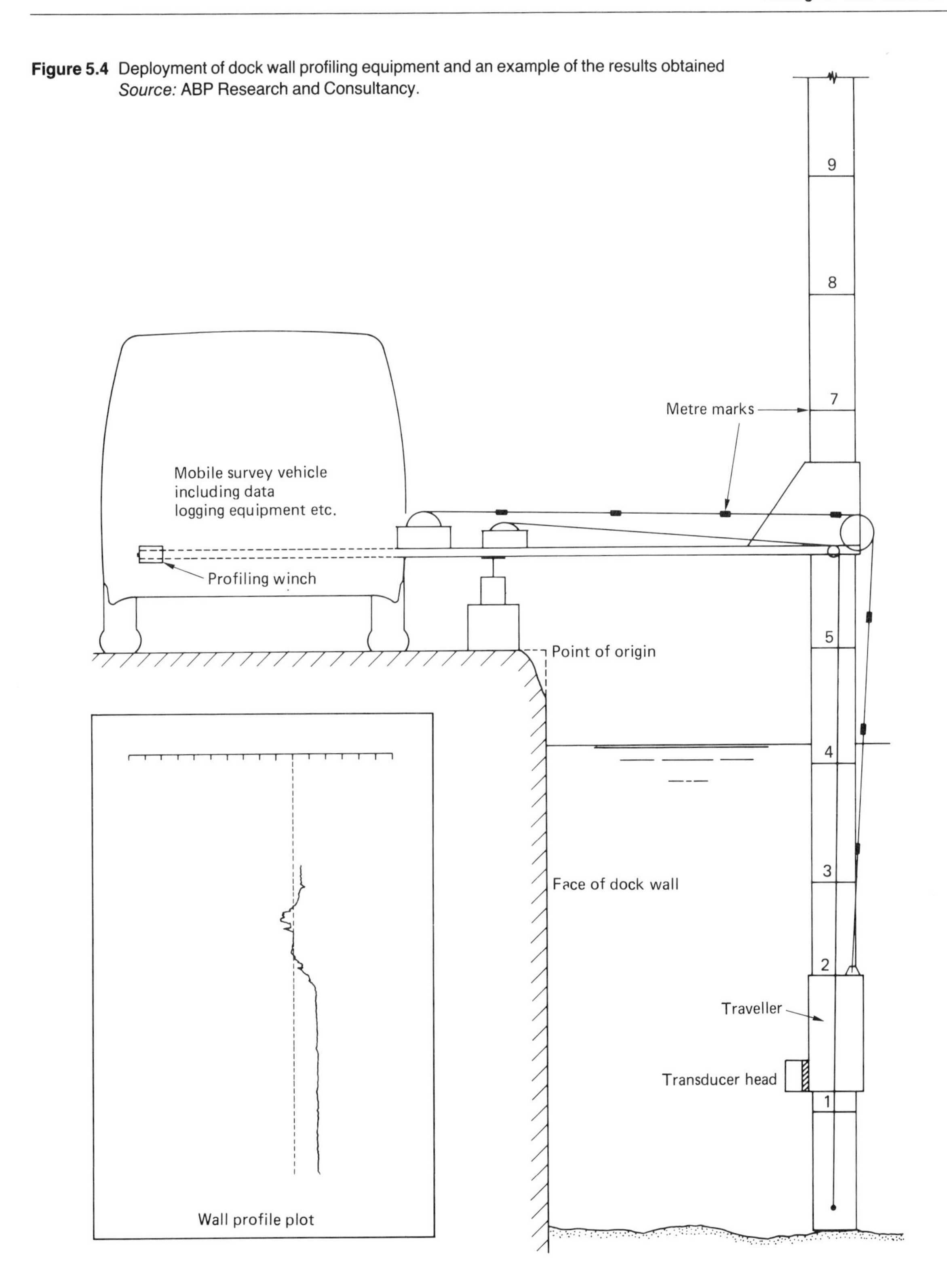

Figure 5.5 Sector scanning sonar equipment deployed for wall profile measurement
Source: SIMRAD Albatross Ltd.

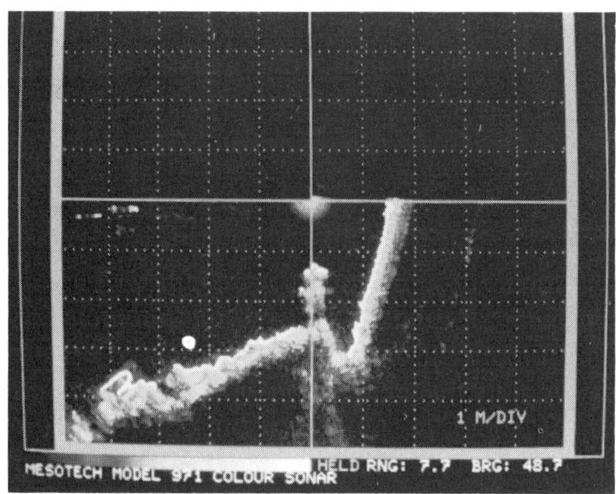

Figure 5.6 Typical harbour wall profile obtained from equipment shown in Figure 5.5
Source: SIMRAD Albatross Ltd.

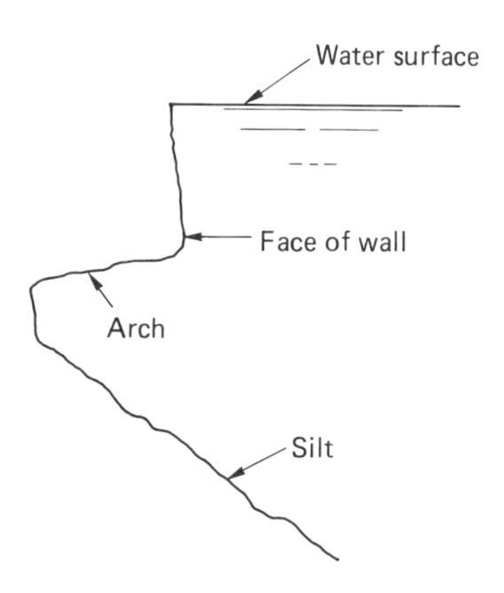

Figure 5.7 Profile of an arch wall in Gloucester Docks obtained with equipment shown in Figure 5.5
Source: Livesey Henderson.

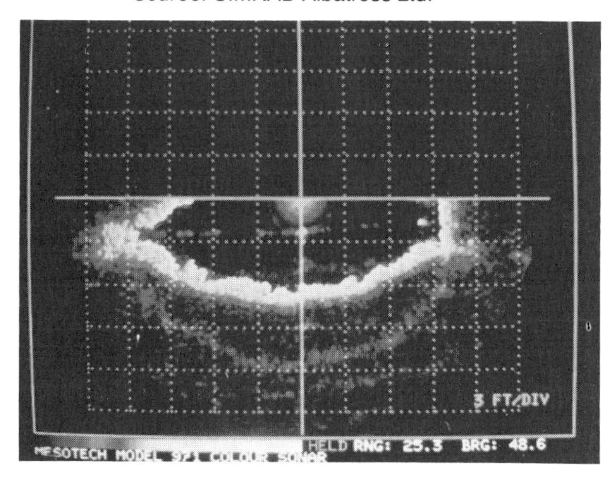

Figure 5.8 Profile of a river obtained with sector scanning sonar equipment, showing undercutting of one bank
Source: Livesey Henderson.

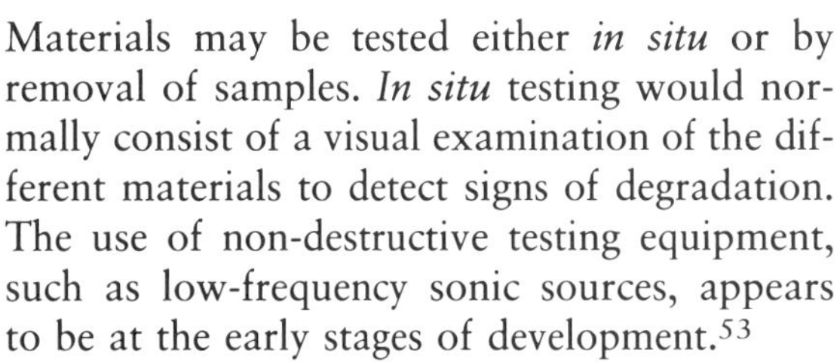

Materials may be tested either *in situ* or by removal of samples. *In situ* testing would normally consist of a visual examination of the different materials to detect signs of degradation. The use of non-destructive testing equipment, such as low-frequency sonic sources, appears to be at the early stages of development.[53]

Samples for testing at a laboratory would usually be obtained during excavation of a trial pit in the top or back of a wall, or from cores or bulk samples taken from a borehole.

5.6 Geotechnical aspects

It is frequently necessary to obtain information concerning the ground conditions beneath and behind a waterfront wall. The ground conditions beneath the wall will dictate the maximum

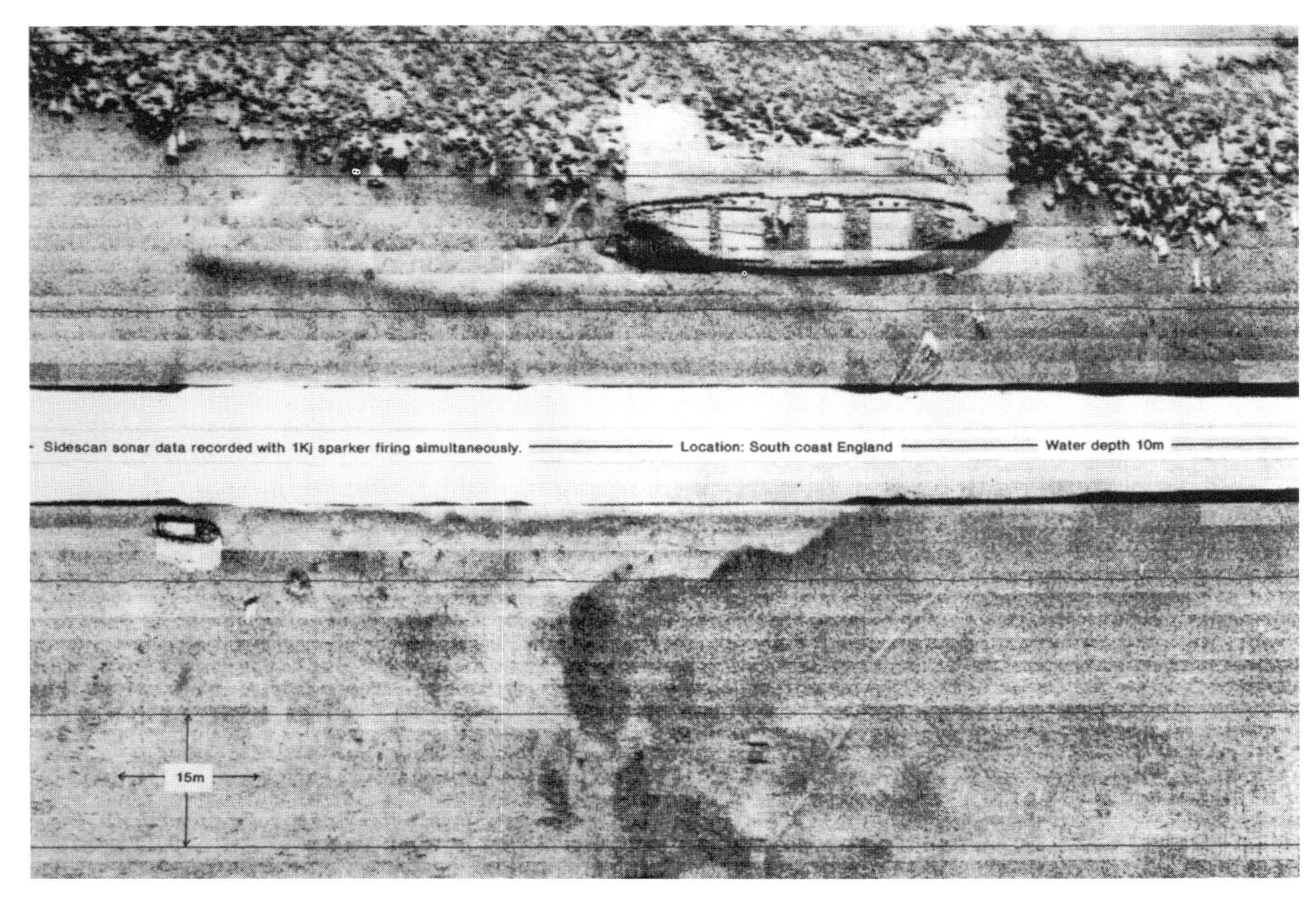
Sidescan sonar data recorded with 1Kj sparker firing simultaneously.
Location: South coast England
Water depth 10m
15m

Figure 5.9 Side scan sonar record showing wreck at toe of rubble mound in front of the breakwater at Portland
Source: British Geological Survey.

bearing pressure that can safely be withstood by the ground, as well as the wall's resistance to sliding and susceptibility to a deep slip failure. Information about the ground behind the wall will assist in determining the forces exerted on the wall by the backfilling.

Ground conditions under the wall may be obtained from a borehole installed from the top and carried past the base into the ground below. This is frequently combined with the process of obtaining information on wall shape and collecting samples of wall material. It should be noted that walls often have timber grillages and piles beneath them. Thus the material encountered by the borehole may or may not be representative of the foundation as a whole.

Backfilling material may also be sampled during installation of boreholes but in many cases, it may be simpler to excavate behind the wall. Once again the backfilling may vary with depth, so a sample from a shallow depth may not be truly representative of the lower levels. Such features as puddle clay behind the wall, the presence of buried temporary works and debris will all tend to complicate investigation of the backfilling. Further information on substrata may be gained from geophysical methods, and this may be an economic way of augmenting the information obtained from other investigation methods, particularly if geophysics is being used to define the shape of the wall or to identify voids or other features.

Whenever the installation of boreholes is contemplated it is well to remember that judicious positioning of the hole may enable it to be used for soil investigation, sampling, geophysics, measurement of water levels and, when the borehole is in the wall itself, for deploying an inclinometer. Where possible, therefore, boreholes should be capped for future use and not backfilled.

There is also an advantage in taking at least one borehole down to a depth well below the base of the wall and, if feasible, down to hard strata. This will greatly assist in correlating geophysical records with actual ground conditions and will provide data to check the factor of safety against a deep slip failure. Boreholes in front of the wall may be needed where a deep slip is possible or where materials in front of the wall may contribute to resisting horizontal sliding.

5.7 Review of local hydraulic regime

5.7.1 Cross-wall flow

Simple water level measurement, behind, inside and in front of a wall, is described in Section 4.9. There may be occasions when it is necessary to make a more comprehensive study of the flow of water through a wall and this may be achieved with a series of boreholes, forming a cross-section of the wall and the backfilling. Measurement of water levels in the borehole, and in front of the wall, throughout the tidal cycle will enable a picture to be obtained of the hydrostatic head acting on the back, or front, of the wall.

5.7.2 Cyclic water level changes

The water level in front of a wall will normally vary in a cyclical fashion according to the function of the wall and the hydraulic regime in which it is situated. Thus the water in an impounded dock and in canals will only vary by small amounts, but walls forming breakwaters, piers and river banks may be subjected to considerable changes in water level.

Generally, these water levels will have been measured by the local water or harbour authority, and it will not therefore be necessary to carry out long-term water-level measurements. However, it may be necessary to assess the effects of severe drought or floods, tidal surges and such phenomena on top of the normal cyclical effects.

5.7.3 Rapid water level changes

An important aspect of the assessment of old waterfront walls is their risk of failure, and this may in turn depend on the chances of a rapid drawdown of water in front of the wall and the behaviour of the wall in these circumstances. It is thus important to investigate the possible causes of a rapid change of level of this nature and its chances of occurring.

The possibilities of the following happening may need to be investigated:

1. lock gates failing, causing rapid drawdown of water;
2. any need to lower water at a slow rate for maintenance purposes;
3. any need to lower water levels at a fast rate (for instance, to flush out a pollutant);
4. rapid drawdown, or increase in level, due to seismic effects (tsunamis);
5. rapid drawdown, or increase in level, due to hurricane, typhoon, etc.

5.7.4 Water velocities

High water velocities in the vicinity of a wall may cause erosion of toe ground leading to reduction of wall stability. In many cases the wall has been designed for the berthing of sailing vessels and the water velocities arising from these manoeuvres have been insufficient to cause any problems.

Modern powered vessels, particularly those with bow thrust, create a flow of water at appreciable velocities which can cause scour at the wall toe. Roll on/Roll off (Ro/Ro) vessels are often introduced to port areas which contain old waterfront walls. It is thus desirable to estimate the velocity of water at the wall toe to check whether the toe ground will remain in place. The effect of propeller wash in shallow and confined areas has been studied by BPF[26] and their report *Propeller Induced Scour* gives guidance on methods of estimating water velocities caused by propeller wash.

Natural water velocities will also scour the toe ground in front of a wall. Most walls subjected to scour will have been designed to resist forces. However, when the hydraulic regime is altered in any way these velocities may increase, and it may be necessary to measure or predict the new velocities. In England and Wales the National Rivers Authority should be consulted on flood data and river flow records, and, in Scotland, similar information can be provided by the River Purification Boards.

Currents may be measured by direct-reading or self-recording current meters, and measurements should be taken over a sufficient period to assess the effects of spring tides, high flows associated with floods, etc.

5.7.5 Wave measurements

Wave attack is of particular importance when assessing the stability of breakwaters and sea walls. Wave measurement is difficult to carry out in the nearshore zone and in consequence it is normal for measurements to be taken in deep water, further off-shore. The effect of the shallowing water and any obstruction to the waves is estimated by the use of refraction and diffraction methods. The effect of bottom friction will also need to be estimated.

In an assessment of wave climate for the purposes of an old waterfront wall investigation it is normally necessary to identify changes in regime which have occurred. Thus the absolute values of wave height and period in the offshore zone may be less significant to the investigator than changes in seabed bathymetry in the nearshore zone, which will affect the waves reaching the wall. However, actual wave heights will be needed for checking the rock sizes and slopes used in the foundation mounds under breakwaters and for designing rock mounds to protect sea walls.

5.8 Review of local sedimentary regime

5.8.1 Investigation of ground-level changes at wall toe

One of the primary causes of wall failure is the removal of the toe ground and thus the measurement of the ground level at the toe of a wall is an important investigation. Where the wall is in a relatively benign environment, where natural fluctuations of the toe ground are unlikely, it is important to establish whether the ground is still at its design level or has been altered by dredging or other works. It will also be important to discover whether there have been any changes in the level during the wall's history.

Some walls are in an environment where the ground level is frequently altering, and in these cases it may be necessary to monitor the toe ground level over a period of time to establish maximum and minimum levels. It will also be of benefit to determine the causes of the fluctuations.

Measurement of toe ground levels may be carried out by various means. For impounded docks and other locations where the wall toe is permanently obscured it may be possible to determine the measurement from acoustic surveys of the wall face (see Section 5.4.2), or from measurements made directly with a hand line. (In docks, rivers and canals air lifting by diver may be needed to reveal the profile of firm ground as discussed in Section 5.2.5.) In tidal locations it may be possible to obtain access to the toe of the wall in the dry and measurements may then be made to a fixed point on the wall itself, such as the cope level.

5.8.2 Assessment of coastal processes

For walls in the coastal zone, the wave attack and the level of ground at the wall toe will be controlled to a great extent by the coastal processes in the area. These processes are usually:

1. wave-induced littoral drift;
2. onshore/offshore sediment movement;
3. sediment movement due to tidal currents.

Coastal processes are complex, and considerable amounts of data collection and analysis are required to determine the processes active in an area and quantify their effects. Generally, it would not be feasible to investigate these processes solely for the purpose of assessing their effects on an old waterfront wall.

However, if the coastal zone has been studied for other reasons, such as coastal defence, beach management, etc., then the results of these studies are of great value in the prediction of waterfront wall behaviour. In particular, it should be noted that the construction of new sea defence works, piers, groynes, etc. may have a considerable effect on the beach levels in front of a wall.

5.8.3 Assessment of river regime

River regimes will affect bridge piers and abutments, river walls and waterfront walls in estuaries. Although the analysis of a river regime is complex, the relationship between sediment types and scour/deposition velocities may be determined by examination of a site, and these in turn may be used to predict the effects of extreme flows arising at times of flood. The important aspects to be investigated are the effect of scouring currents on toe ground levels and the possible increase of currents due to changes in the river regime, construction in the river, etc.

5.8.4 Scour protection

Many walls situated in hostile environments are designed with some form of scour protection. Others have scour protection added when it is discovered that problems are occurring due to removal of toe ground. It is important to investigate the state of the toe of these walls and to detect any signs which would indicate loss of protection.

5.8.5 Dredging records

Although the ground at the toe of a wall may appear to be at the designed level it is possible that this may not have always been the case. In some instances there may have been dredging carried out in front of the wall, which has lowered the toe ground, and subsequent siltation has brought the ground back to its original level.

To detect the occurrence of a situation such as that described above, or to prevent it happening in the future, it is advisable to obtain records of past dredging activities and present dredging practices and to measure the level of the top of firm material below the silt (see Sections 5.2.5 and 5.8.1).

5.9 Review of functional wall loadings

Functional loadings on a wall will need to be identified to analyse wall behaviour. These loadings may be for past events, current practice or future functional requirements. They may be used as follows:

1. to identify reasons for past behaviour. An example of this is when a wall has partially failed, or shown signs of distress, and the loading condition at the time of the failure is known;
2. to establish current safety factors. The current wall loadings may be used to show the factors of safety for different types of failure mode. This may then be correlated with the results of a past failure, as in the previous item;
3. to establish factors of safety under proposed future loading conditions.

Loads may be categorized into horizontal and vertical. Examples of these would be:

horizontal: vessels berthing, bollard pull, fender forces;
vertical: surface loads, surcharges, crane loads, stacked material.

5.10 Constraints to implementation of maintenance and rehabilitation programme

5.10.1 Function of the structure

Many waterfront walls are in continuous use as piers, locks, bridge abutments, etc. It is thus important that any investigation of a wall should include a detailed study of its function to identify the constraints which its usage puts on any maintenance or rehabilitation programme. This study should include:

1. a review of long-term usage to identify periods of least or nil use, seasonal variations, etc;
2. a review of short-term usage to establish the usage pattern, whether the use is continual or intermittent, whether 24 hours per day or daytime only;
3. a review of the actual mode of usage to establish whether and how the usage would affect maintenance works.

The information gained from this study will be of particular importance in the programming and drafting of contract documents for maintenance or rehabilitation works. It may be necessary to specify night work only, work when the wall is not in use or work during the winter season. Commercial organizations, with high utilizations of their walls, may need to close down certain operations, or transfer them to other locations, during periods of maintenance.

5.10.2 Statutory responsibilities and regulations

When rehabilitation and particularly upgrading work is being considered, there are a number of points which should be investigated relating to the local authority and the responsibilities of the organization having the work carried out.

1. Does the work require consents and, if so, is there likely to be a problem gaining them (see box)?

> **An illustration of consents needed for certain types of work**
>
> On a Main River, consents for construction would be required from the National Rivers Authority, or equivalent, and, for flood improvement works, the Environmental Procedure specified in Statutory Instrument No.1217 must be followed. For coastal works, approval would be required from:
>
> MAFF (The Food and Environment Protection Act 1985, for consent to place material in the sea)
>
> Dept of Transport, Marine Division (to check that navigation is not impaired)
>
> Crown Estates Commissioners (who control much of the seashore)
>
> In addition, the Nature Conservancy Council and the Countryside Commission would also be involved at the planning approval stage.

2. If the wall belongs to a third party does the carrying out of rehabilitation work by another organization commit this organization to carrying out all future maintenance? For instance, if a water authority heightens a river wall to give improved flood protection, does the water authority then have to assume responsibility for the wall?
3. If the wall is part of a listed structure then any work carried out to the wall will need to be approved. It is likely that in such circumstances a very close check will be kept on the types of materials used and the visual effect of the finished work.

5.10.3 Effect on the local environment

Whenever work of a rehabilitation or upgrading nature is carried out there is an effect on the locality. If the structure is enhanced aesthetically this will affect local amenity values. Local property will become more desirable and thus more valuable.

If there are a number of developments being carried out at the same time it is necessary to look at the whole development area to ascertain where benefits are accruing. An analysis of this type may assist in providing the economic justification for carrying out maintenance work on a wall which, if viewed in isolation, might not warrant the expenditure.

Neighbouring developments will also bring some problems. Modern construction methods and the insertion of new services may interfere with parts of the wall's original construction. Also, the adjacent developments may bring increased loadings to the wall by virtue of increased traffic levels and changed drainage patterns, and may interfere with access to the wall. When new development is being permitted close to a wall there should be a requirement for the provision of suitable access to the wall and, if possible, a working apron should be left beside the top of the wall suitable for the use of light plant.

5.10.4 Access to site

Many waterfront walls are relatively inaccessible to construction plant and it is important that access to the wall is carefully checked prior to letting of maintenance and rehabilitation contracts. Such items as proximity to other buildings, limits on road and access bridges, apron loadings, and width and height restrictions should all be investigated.

Chapter six

Structural appraisal of a wall

Symbols used in this chapter

K_a Active soil pressure
K_o Earth pressure at rest
ϕ Internal angle of friction of the soil
ϕ' Effective angle of shearing resistance of the soil
δ Coefficient of friction between the backfill and the wall

6.1 Introduction

Waterfront walls are notoriously difficult to assess, both in terms of their stability and in their rate of deterioration. In many cases apparent abnormalities may be relatively harmless while the factors which are actually crucial to the continuing performance of the wall may go undetected. The reasons for this are that most of the wall is hidden under the soil or water and that a single inspection may well not reveal the rate of deterioration.

It is necessary at the outset of a structural appraisal to establish that the information at hand is sufficiently comprehensive to determine which modes of failure or deterioration are likely, and which are unlikely. A clear picture of the condition of the wall should be available to act as a baseline against which further deterioration can be assessed.

The procedure for carrying out an evaluation is shown in Figure 6.1 and is as follows:

1. assemble all the available data (see Chapter 5);
2. list all the apparent symptoms of degradation and changes in the condition of the wall;
3. use the diagnostic approach illustrated below to identify those causes which may be resulting in the effects (symptoms) listed;
4. if necessary, carry out stability evaluation for the various possible modes of failure;
5. divide the wall defects into critical and non-critical;
6. on the basis of the function of the wall and its environment establish which risks to its stability and deterioration are foreseeable and determine the probability of these risks occurring;
7. determine whether the identified risks may be controlled;
8. forecast the effects of continuing deterioration on the wall's future performance.

6.2 Assembly of available data

Before attempting to carry out an evaluation of the stability of a waterfront wall it is essential that the maximum amount of data has been assembled relating to the wall and its surroundings. In particular, detailed information concerning the history of the wall, its original design, construction, and subsequent

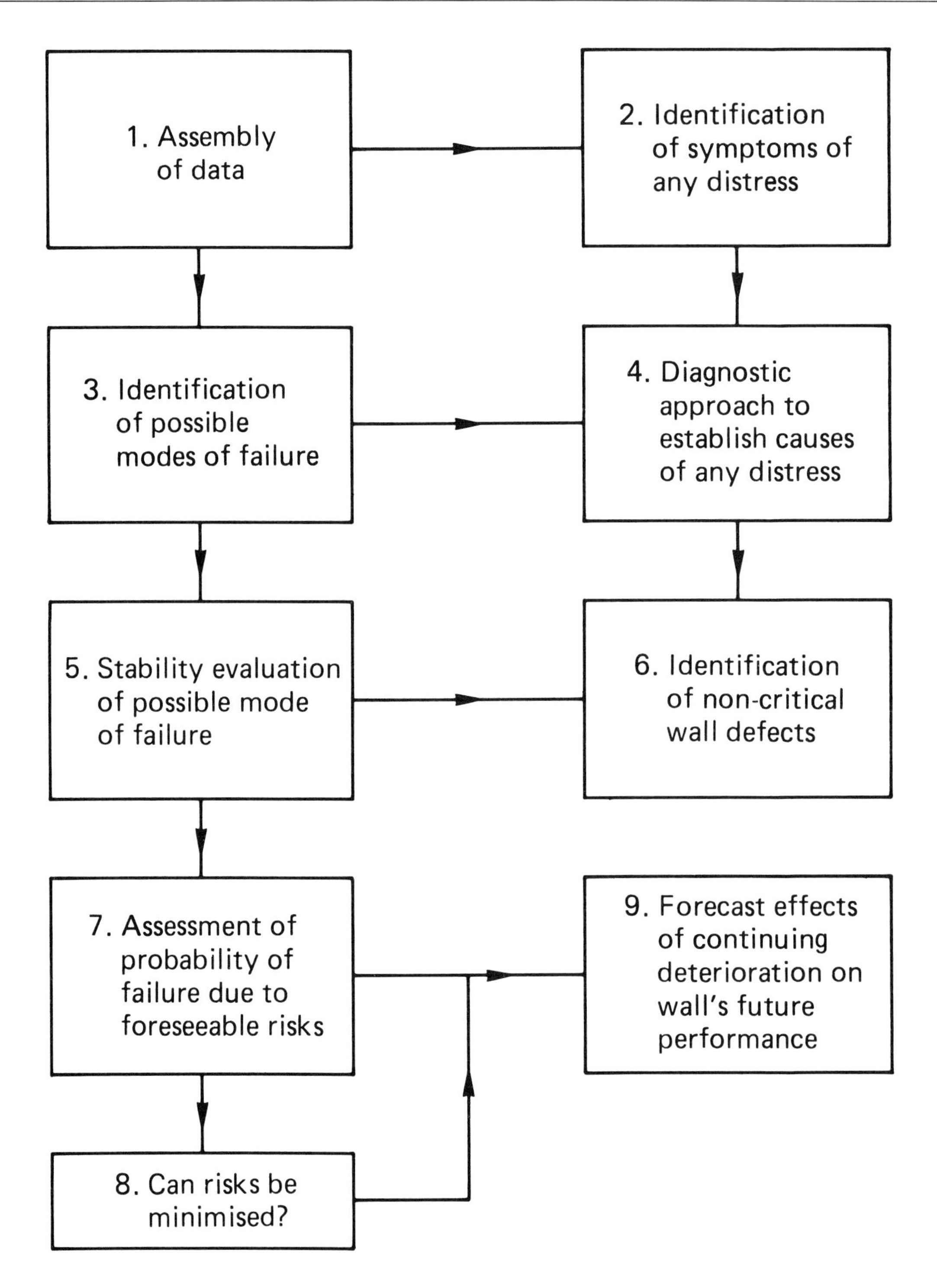

Figure 6.1 Flow chart showing typical stability evaluation procedure
Source: Livesey Henderson.

performance is of great benefit in the evaluation process. Money spent on detailed archival research may be of considerably greater use in the evaluation process than the equivalent sum spent on conventional site investigation. This is particularly true if the historical search is carried out an an early stage in the investigation.

Information may be classified as follows:

1. historical information relating to the design, construction, repair, maintenance and performance of the wall;
2. current and recent information on the behaviour of the wall as presently used, such as might be revealed by any routine inspection or monitoring;
3. information relating to the environment in which the wall is situated, including hydraulic regime, soils data, etc;
4. information relating to the present level of service, including a definition in terms of the size and frequency of occurrence of loads imposed on the wall;
5. detailed investigation results including an assessment of their reliability and an estimate of how representative they are of the length of wall to which they apply;
6. information relating to the apparent degradation of the wall such as deterioration of materials, loss of materials, temporary and permanent movement of the structure, movement of the surrounding ground, etc.

It is common to find conflicting information at this stage. For this reason it is important that the reliability of the data should be assessed as carefully as possible prior to the main stability evaluation.

6.3 Diagnostic approach to evaluation

The evaluation of a wall's condition calls for a process of deduction which is partly intuitive and partly logical. Where distress is apparent, and the cause is similarly obvious, the evaluation process may be straightforward; although the stability evaluation may be complicated by the existence of non-standard forms of construction.

In many cases, however, although there may be a number of distinct signs of distress, the cause and the mode of failure which has been initiated may be unclear. Often the most obvious signs of distress prove to be localized degradations of the wall or adjustments to wall shape which occurred in the past and which do not necessarily signify a decline in the wall's overall integrity or stability.

It is suggested therefore that a diagnostic approach should be used to identify the causes of any wall distress. This approach will not only help to identify the underlying causes of the real distress but, almost as importantly, will also allow the 'non-critical defects' to be logged for future reference. 'Non-critical defects' are those which do not affect the serviceability of the wall and will not lead to progressive degradation.

In simple cases the diagnostic approach is based on a list of identified defects set against possible causes, as given in Table 6.1. For more complex cases a matrix is suggested in Figure 6.2 relating to the following:

1. evidence of failure;
2. type of failure;
3. possible causes of problem.

The matrix is used by marking the items of 'evidence' and identifying which type of failure corresponds to the combination of items of evidence. At first not all items of evidence may have been identified and the wall may need to be rechecked. At the bottom of the matrix is a list of 'possible causes of problem' – usually only *one* of which could have been the actual cause.

Occasionally a wall may have two causes of distress, e.g. the cutting of a tie rod may only reduce a factor of safety, with the wall suffering distress later when excessive loads are placed on

Table 6.1 Diagnostic approach to wall failure. Simple cause and effect examples

Ship damage	Cracking and break-up of cope Cracking and impact damage just above and below waterline due to ship's belting and other projections Low-level impacts by bulbous bows
Abrasion by wave-driven gravel	Loss of wall thickness
Decay of wall material	Freeze/thaw cycles Alternate wetting and drying Wave erosion Wave impact Vessel wash erosion Acid water attack Erosion by water passing through wall Propeller or bow thruster scour Plant root damage Poor initial quality of stone, brick, concrete or mortar Alkali/aggregate reaction in concrete

the backfilling. The 'possible causes of problem' are further expanded and explained in Table 6.2

Where there are cracks in the wall, they should be examined for evidence of their cause. Particular points to note include the following:

1. the crack widths;
2. whether a crack is wider at one end than the other;
3. the pattern of cracks;
4. the depth of the cracks;
5. the position of cracks in relation to restraints in movement such as exterior or interior corners or changes in wall section;
6. the displacement forward of one side in relation to the other and whether the displacement is uniform along the crack;
7. seasonal or tidal variations.

Figure 6.3 suggests some interpretation of cracks but is not exhaustive. A typical example of where a crack can be used to determine how a wall has behaved is shown in Figure 6.4, which shows the entrance of the Bowling Lock on the Forth and Clyde canal. Since the crack is of uniform width from top to bottom, and one side of the crack stands proud of the other side (as evidenced by the sunlight striking the inner face), it is possible to deduce that the whole of the wall to the left of the crack has moved forward.

The diagnostic procedure may need to be carried out more than once for a particular length of wall if there are different types of defect to be considered. It may also have to be done in conjunction with a stability analysis as described below (see Section 6.7).

6.4 Identification of non-critical defects

Section 6.3 gives one method of identifying non-critical defects. However, there are many other ways in which similar information comes to light. It is particularly important to list all the non-critical defects which are discovered since this may well save a considerable amount of time and money at some later date when another evaluation is being carried out.

Because of the low priority given to maintenance work in the past and the fact that staff engaged in this work will change over a period of time, it is quite common to find that non-critical defects are the cause of much time-wasting and alarm. This in itself is more

EVIDENCE OF FAILURE:	Horizontal sliding	Overturning	Foundation failure	Deep slip	Overturning of top of wall	Loss of fill through wall	Consolidation of backfill	Crushing of wall mortar	Rotation about tie rod
Bulging in cope line	X	X		X	X			X	
Horizontal movement	X								
Vertical cracks	X					X			
Settlement of backfill	X	X			X	X	X		
Forward rotation of wall		X							
Vertical or sloping crack with greater displacement at top		X		X					
Settlement of cope			X	X				X	
Backward rotation of wall				X					X
Cracks in backfill parallel to wall		X		X					
Fence/lamp posts out of line or out of vertical				X					
Forward rotation of top of wall					X			X	
Vertical or sloping crack in top part of wall					X			X	
POSSIBLE CAUSE OF PROBLEM:	**TYPE OF FAILURE:**								
Loss of strut to toe of wall	X								X
Erosion or decay of mortar								X	
Crack in wall						X			
Failure of tie rod	X	X			X			X	
Decay of timber pile or grillage		X	X						
Softening of soil/rock under foundation		X	X						
Scour		X	X	X					X
Excessive vertical loading on wall			X					X	
Excessive bollard pull	X	X	X		X			X	
Draw-down of water in front of wall	X	X	X	X	X			X	X
Raising of water level behind wall	X	X	X	X	X			X	X
High superimposed load on backfill	X	X	X	X	X		X	X	X

Figure 6.2 Matrix relating evidence, types of failure and possible causes of problem
Source: Livesey Henderson.

Table 6.2 Detailed 'possible causes of problem' given in matrix shown in Figure 6.2

1. Erosion or decay of mortar:
   - freeze/thaw
   - wave erosion
   - vessel wash erosion
   - acid water attack
   - erosion by water passing through wall
   - propeller or bow thruster scour
   - plant root damage
   - poor initial quality
2. Crack in wall:
   - caused by other type of failure
3. Failure of tie rod:
   - corrosion
   - damage by traffic
   - accidental or deliberate cutting
   - accidental excavation in front of anchorage
   - increase in lateral pressure on wall due to surcharge in backfill
4. Decay of timber piles:
   - exposure to atmosphere
   - changes in water level
5. Softening of soil/rocks under foundation:
   - long-term weathering
   - effect of exposure
6. Scour:
   - waves
   - currents
   - littoral drift
   - overdredging
   - deliberate removal of beach material
   - bow thruster scour
   - propeller scour
   - movement of vessels with small clearance
   - loss of scour protection
7. Excess vertical load on wall:
   - mobile cranes
   - Ro/Ro ramps
   - heavy unit load
8. Excess horizontal load on wall:
   - dredging wires
   - construction operations
   - high winds on moored vessel
   - misuse
9. Drawdown of water in front of wall:
   - failure of dock or lock gates
   - conversion from impounded to tidal or semi-tidal dock
   - leakage from canal
10. Raising of water behind wall:
    - broken water main
    - new soakaways
    - broken surface water drain

discharge from contractor's pumps
connection to other body of water
effect of cut-off walls
wave overtopping and failure of waterproofing layer on top of fill

11. High superimposed load in backfill:
    stacking bulk materials
    stacking of containers
    high-density materials
    heavy mobile cranes
    road traffic and vibration
    large unit loads

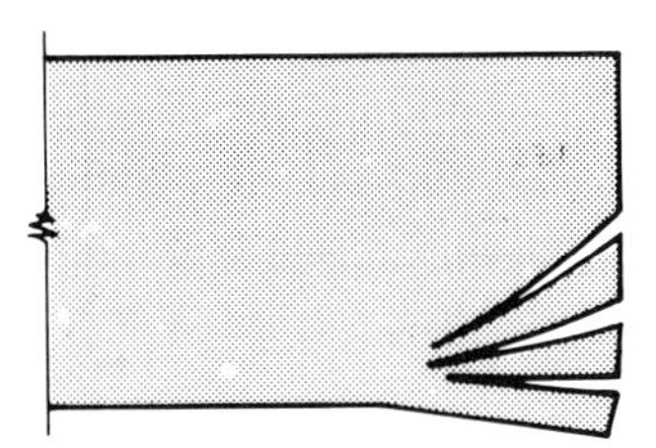

(a) Elevation of wall face: settlement of end of wall

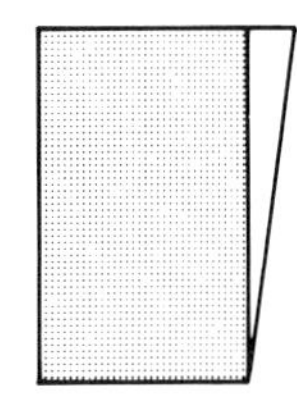

(d) Section through wall showing one part of wall rotating about toe

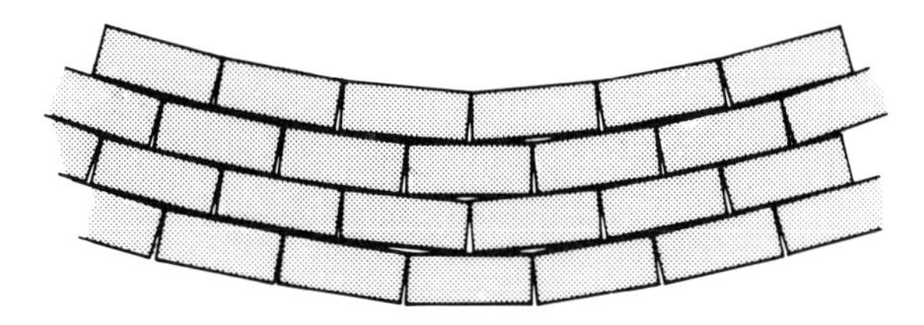

(b) Elevation of wall face: settlement in middle of wall

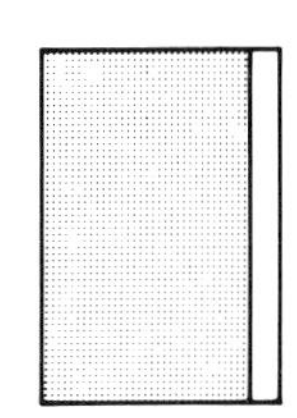

(e) Section through wall showing one part of wall sliding horizontally

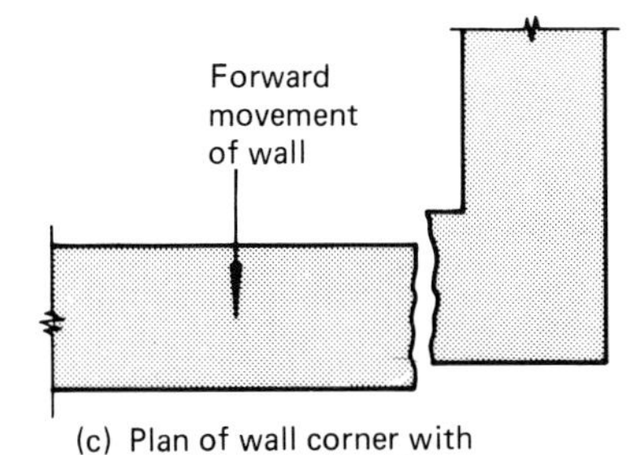

(c) Plan of wall corner with crack due to restraint at corner

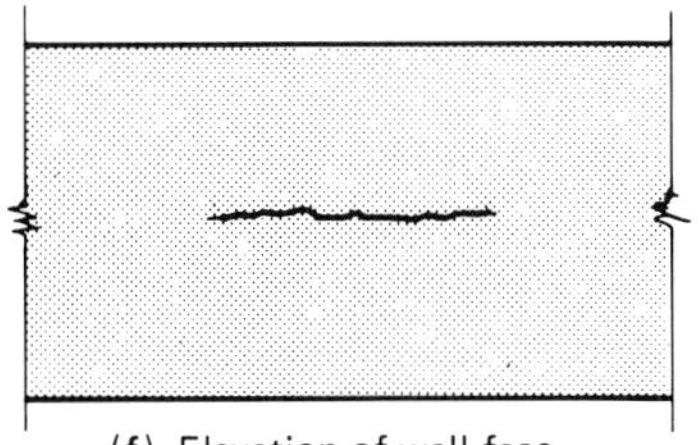

(f) Elevation of wall face showing crack due to corrosion of steel insert

Figure 6.3 Typical crack formations and their causes
Source: Livesey Henderson.

Figure 6.4 Bowling Lock, Forth and Clyde Canal. Vertical crack clearly indicating forward movement of whole section of wall
Source: British Waterways.

than enough justification for the small amount of effort and cost required for the proper logging of these defects.

It will be appreciated that in many cases a regular and comprehensive inspection programme will be all that is required to establish the presence of non-critical defects, since it will be possible to show the defects have not deteriorated between one inspection and the next (see Section 4.7).

6.5 Assessment of wall condition

The condition of a discrete portion of a wall – normally a short length – may be described by reference to the following three characteristics:

1. the appearance and significance of superficial defects;
2. the presence and significance of local defects;
3. the stability of the whole structure.

It is the relationship between these characteristics and the ability of the wall to function satisfactorily that defines its condition.

6.5.1 Wall appearance

There are two aspects of wall appearance that are important. If the function of the wall is primarily aesthetic, such as might be the case for one which forms part of a listed building, then any degradation of the appearance will contribute to a lowering of its level of service. Alternatively, if the wall performs a function related to its structural integrity then its appearance *per se* may be of little consequence unless the superficial defects will lead, in due course, to a local or general weakening of the structure. An assessment of wall appearance must therefore refer to the effect of the superficial defects on its function and to the potential for further degradation leading to weakening of the structure.

6.5.2 Local defects

Local defects may be caused by occasional severe localized overloading, regular overloading, or the effect of attrition due to natural agencies. It is thus important to establish which of these is the cause of the defect. Having established the cause of the defect the following questions should be addressed:

1. does the defect affect the strength of the whole structure or a major part of it?
2. what is the probability of the defect becoming worse in the course of time?
3. will the defect lead to progressive or sudden failure of the whole structure, or a major part of it?

On the basis of answers to the above questions the significance of the local defect may be evaluated and its contribution to a deterioration of the wall's level of service assessed.

Where a wall has moved, but its stability is not in question, the effect of the movement may be minimal. For instance, the movement of a canal bank may have little effect on the performance of the canal. However, if one of the lock walls moves then the effect could be very significant since the width in locks is critical.

6.5.3 Stability of whole structure

The assessment of the condition of a wall with respect to its overall stability is often a complex procedure and it is important to establish the correct framework for the assessment at the outset. There are essentially two approaches to stability evaluation:

1. when the standard of service is defined and the intention is to establish factors of safety relating to the various failure modes under normal and exceptional loading;
2. when the standard of service is to be defined by reference to the loads which the structure is capable of sustaining without undue risk of failure (back analysis).

More detailed guidance to the methods available for this evaluation is given in Section 6.7.

6.5.4 Wall condition codes

Where the details of a wall's condition have to be entered into a computerized database it will be necessary to devise some form of condition coding system. If this is not done it is not possible to retrieve statistical information from the database. The type of coding system will depend on the function of the wall but is likely to relate to both the appearance and the structural state of the wall (see Section 4.6.3). Wirral Borough Council, for instance, have a coding system and it is understood that many of the coastal authorities are introducing similar systems.

6.6 Using the wall condition assessment

The wall condition assessment may be used in a number of different ways.

1. It may be used to assess the rate at which the structure will deteriorate in the future and the effects of this deterioration on the stability and level of service of the wall.
2. It may be used to assess the serviceable life of the structure.
3. When related to the method of back analysis in Section 6.5.3 above it can be used to define appropriate levels of service.
4. It may be used as a guide to the level of maintenance or rehabilitation work required for the portion of wall assessed.
5. It may be used to determine the way in which normal loads might be controlled to reduce the risk of failure in cases where walls are close to their safety limits.
6. It may give pointers to the way in which local environmental regimes may be controlled to reduce the chances of exceptional loads being applied.

6.7 Stability analysis

6.7.1 Codes of practice

Where the stability of a wall is in doubt, or where a change in the loading on a wall is proposed, the wall should be analysed in accordance with the following codes:

1. BS 6349: Code of Practice for Maritime Structures:[54]
   (a) part 1: general criteria;
   (b) part 2: design of quay walls, jetties and dolphins (Section 5.3, 'Design of Gravity Walls' is particularly relevant);
   (c) part 7: design of breakwaters and training walls.
2. BS 8002: Code of Practice for Earth Retaining Structures (in preparation).[31]

In some cases, the analysis of wall stability carried out in accordance with the above codes can result in factors of safety which are very low, and occasionally even less than one, for walls which have been standing satisfactorily for many years. Great care should therefore be taken when basing decisions on the results of such analysis and it is essential that experience and engineering judgement be used in their interpretation.

In the above cases it is often necessary to re-examine the input data to the calculations to see if any of the values adopted are too pessimistic. Where any of the input data is based on assumptions or estimates, further efforts should be made to verify the information before contemplating wall-strengthening works. Various aspects of stability analysis are considered below.

6.7.2 Loading and strength parameters

The minimum uniformly distributed live load which should be considered on backfilling behind a wall (or on the hearting of a breakwater) is 5 kN/m^2, which corresponds to the crowd loading of pedestrians. Other vertical live loads should be based on the use to which the area is to be put. Advice on such loading is given in Clauses 44, 45 and 46 of BS 6349: Part 1 and Clause 3.3.3 of BS 8002.

Bollard pulls may be assessed in accordance with Clause 42 of BS 6349: Part 1 or on the basis of a more detailed investigation of the type of vessel and the local wind speeds. Where dredgers are to be used the loads imposed by the side and head wires should be investigated.

Dead loads should, whenever possible, be based on actual measurements of the densities of the various parts of the structures. Brick, stone and concrete have dry densities which may vary from 16 kN/m^3 to 26 kN/m^3 (i.e. effective submerged densities of 6 kN/m^3 to 16 kN/m^3) and all three materials may be used in a single structure. Some designers varied the densities deliberately to maximize the stability or minimize the bearing pressure under the foundations (e.g. see Figure 1.61, Port of Glasgow).

Soil properties should be based on recent investigations and an assessment by a geotechnical engineer. Separate parameters should be identified for the backfilling, the various layers in the original ground behind the backfill and the soil below and in front of the foundation. Effective stress parameters should be measured or assessed for cohesive materials.

Water levels behind the wall should preferably be based on measurements in boreholes to show how the water levels vary in relation to the water level in front of the wall, and on a review of likely changes which may affect the water level in the future. The full range of factors which may affect the water levels behind and in front of the wall need to be evaluated. These include the following:

1. Behind the wall:
   (a) groundwater levels;
   (b) effect of rivers and flooding;
   (c) connections now and in the future to other sources of water;
   (d) surface water drainage arrangements;
   (e) drain connections through the wall and effect of drainage layers behind the wall;
   (f) risk and effect of overtopping by waves.
2. In front of the wall:
   (a) normal and river-in-flood levels;
   (b) range of levels in impounded docks system;
   (c) extreme tide levels including meteorological effects;
   (d) canal operating levels;
   (e) lowest water levels used in maintenance.

Wave loading should be based on Clause 39.4.2 (revision AMD 5488 dated December 1986) of BS 6349: Part 1 or Chapter 5 of *Technical Standards for Port and Harbour Facilities in Japan* (published in 1980 by the Bureau of Ports and Harbours, Ministry of Transport, Tokyo).[55] A list of 'normal loading conditions' and 'extreme loading conditions' should be prepared in accordance with the recommendations given in Clause 3.2.3 of Part 2 of BS 6349.

It may occasionally be necessary to assess the strength of the wall materials, particularly where new loads are to be applied to the wall. Ideally, a composite block should be cut out of the wall for testing. Alternatively, cores may be tested.

6.7.3 Modes of failure

The principal modes of possible failure of a gravity wall as a whole are:

1. deep slip;
2. overturning;
3. foundation failure;
4. sliding.

Overturning and sliding can also affect parts of the wall, as opposed to the whole cross-section. Each of these four modes is considered separately below.

6.7.4 Deep slip

'Deep slip' refers to instability involving shear failure of the earth mass on which a wall is built. Circumstances where waterfront walls are particularly at risk from a deep slip include the following:

1. where the wall is founded on clay or on a soil overlying clay or other weak strata;
2. where the soil behind the wall slopes up from it, as occurs with some coast-protection walls;
3. where the ground in front of the toe of the wall slopes downwards;

4. where high pore-water pressures can exist below the wall.

The stability of the slope should be analysed using an appropriate computer program to find the most critical slip surface (BS 8002 and BS 6031[56]) and the lowest factor of safety, which is expressed as:

$$\frac{\text{Sum of restoring moments}}{\text{Sum of disturbing moments}}$$

(see Clause 6.4.2 of BS 6031).

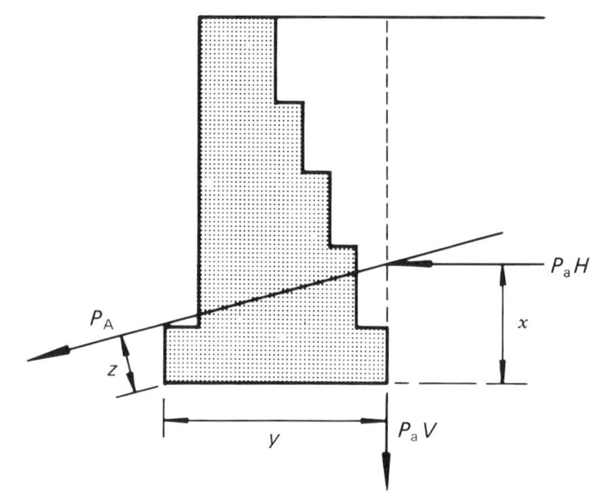

Overturning moment = $P_az = P_aHx - P_aVy$

Net sliding force = $P_aH - P_aVx$

Figure 6.5 Illustration of the active pressure components of force acting on a retaining wall for considerations of overturning
Source: Livesey Henderson.

6.7.5 Overturning

'Overturning' means failure of the wall by rotation about the toe of the wall. The factor of safety against overturning is expressed as:

$$\frac{\text{Moment of restoring forces about toe}}{\text{Moment of disturbing forces about toe}}$$

Computer programs can be used for the analysis of overturning (and also for sliding and bearing pressure failures on the same program). They are particularly useful for situations where it is necessary to carry out a sensitivity analysis or to consider various loading cases.

The following points should be noted with regard to overturning calculations for old waterfront walls:

1. the soil pressure acting against the back of the wall should usually be calculated as the 'active' pressure because even the most massive of walls moves sufficiently for soil pressure to reduce to the active value. Higher pressures should be allowed for walls with ties and for swelling or overconsolidated clays and where soil is to be compacted behind the wall;
2. account should be taken of the shape of any excavation out of the original ground before backfilling was placed and of the high internal angle of friction of designed rubble backfills;
3. effective stress parameters should be used for backfilling;
4. the angle of wall friction should be carefully assessed allowing for the effect of a 'virtual back' for a wall with a stepped rear face (see Figure 35 of Part 2 of BS 6349) and for the effect of projecting wall bases, oversails (artificial projections to increase wall friction), counterforts and the roughness of unfaced rubble masonry. Variations in the angle of wall friction have a significant effect on wall stability (see Appendix A);
5. where the resultant of the active pressure is resolved into vertical and horizontal components, both components should be treated as 'disturbing' forces with the vertical component shown as negative (see Figure 6.5);

6. dry densities of the various materials in the wall should be used for calculations of the restoring moment of the wall; the effects of horizontal water pressure and of uplift should be calculated separately. This procedure makes it easier to allow for a non-uniform water pressure under the base of the wall and to carry out a succession of calculations with different water level assumptions;
7. horizontal concentrated loads such as bollard pulls should be distributed at 45° down through and along the wall unless the wall is constructed in separate sections of a smaller length than given by the 45° dispersion;
8. the effect of passive pressure in front of the toe is usually insignificant due to the small lever arm and the low pressure which can be mobilized without excessive movement;
9. it may be necessary to check the overturning at different levels above the base of the wall, particularly where it is proposed to apply heavy concentrated loads on filling behind the wall.

6.7.6 Sliding

The factor of safety against sliding is expressed as:

$$\frac{\text{Horizontal disturbing forces}}{\text{Horizontal resisting forces}}$$

The same points should be taken into account as mentioned in relation to overturning except for the following:

1. the base friction arising from the vertical component of the active pressure should be considered as a restoring force;
2. the angle of friction between the base and the ground should be evaluated on the basis of the detail of the base and foundation: serrations, upward slopes and concrete cast directly on to the ground may allow the angle of friction to be taken as the full ϕ of the soil under the foundation;
3. the amount of passive resistance available from any soil in front of the wall is likely to be small for reasons discussed in BS 8002. Where passive pressure is included in the resisting forces, it should be restricted to a value at which movement of the wall will be small;
4. the resistance to sliding may need to be checked at various levels in walls of varying thicknesses, particularly if some new localized horizontal loading is proposed.

6.7.7 Bearing pressure

The maximum bearing pressure under the wall should be calculated from the position, magnitude and direction of the resultant force derived from soil pressures, water pressures, imposed loads, and structure weight acting on the base of the wall. These figures can be derived from the data used in the overturning calculations. Bearing pressures under the heel of the wall due to wave or berthing loads are unlikely to be critical as such forces are usually resisted by the passive resistance of the backfilling or hearting. Permissible bearing pressures should be estimated in accordance with BS 8004, *Code of Practice for Foundations*.[57]

6.7.8 Counterforts

For walls with counterforts (see Figure 6.6) the following assumptions can be made for calculating the resistance to *overturning*:

1. active pressure (K_a) acts on the back of the wall between the counterforts and on the back of the counterforts and can be resolved as described in Section 6.7.5;
2. as the wall tends to rotate about the toe of

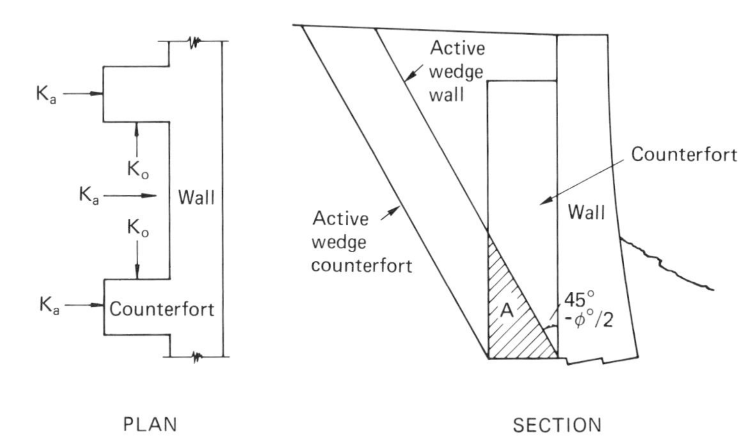

Figure 6.6 Plan and elevation of a counterfort application of earth pressures
Source: Livesey Henderson.

the wall, the counterforts rise and mobilize friction on their sides;

3. the soil pressure on the side of a counterfort is earth pressure at rest (K_o) because the soil is constrained by the counterforts;
4. vertical friction (K_o tan δ) on the whole of each face of a counterfort assists in resisting overturning provided that the counterforts are not too close together. In the latter case the weight of the soil between the counterforts would limit the amount of side friction which can be mobilized;
5. to allow for the varying section of the wall, a length equal to the spacing of the counterforts should be considered at one time, instead of the more usual one metre length.

In considering the effect of counterforts on the resistance of a wall to *sliding* the following can be assumed:

1. active pressure (K_a) acts on the backs of the wall and the counterforts, and can be resolved into horizontal and vertical components;
2. the lower part of the side of the counterfort, which is outside the active wedge for the wall without counterforts (area A in Figure 6.6), resists horizontal sliding directly by friction based on K_o and tan δ;
3. as the wall tends to move forwards, the active wedge behind the wall moves downwards and transfers a vertical friction load to the part of the side of the counterfort within the wall active wedge (the whole side except area A in Figure 6.6);
4. the vertical friction is a function of K_o and tan δ, and is transferred as an additional vertical load to the base of the counterfort where it increases the base frictional resistance.

The bearing pressure under a counterfort wall with a base slab of non-uniform width (see Figure 6.7) can be calculated by estimating the width of the wall foundation in compression, and calculating bearing stresses using the Z value for the T shape of the area in compression. A better estimate of the width in compression can then be made and the calculation repeated until the centroid of the pressure diagram coincides with the resultant of the disturbing and restoring forces. The tension at the top of the counterfort where it joins the main wall should be checked as failure at this point has been noted in some cases.

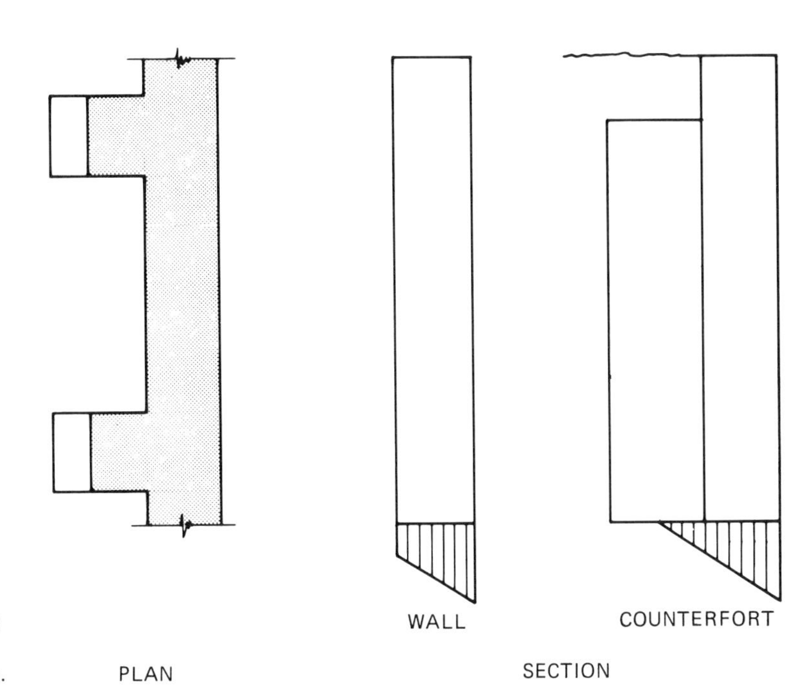

Figure 6.7 Plan and section of a counterfort showing bearing pressures
Source: Livesey Henderson.

6.7.9 Dewatering

Projects which involve dewatering of dock basins, or locks, require particular care to ensure that the likely water pressures behind and under the walls and under any floor are correctly estimated. Although most such structures were originally constructed in a dewatered site, the water pressures will be more onerous when the water is pumped out again many years later. This is because the water level in the soil may take a long time to respond to changes, particularly where clay exists under or behind a wall.

6.7.10 Factors of safety

Ideally, the following factors of safety should be achieved:

	Normal loading	*Extreme loading*
Deep slip	1.40	1.20
Overturning	2.00	1.50
Sliding	1.75	1.50
Foundation failure	Calculated allowable	Allowable + 25%

1. 'Normal' and 'Extreme' loadings have the meanings given in Clause 3.2.3 of Part 2 of BS 6349.
2. 'Resultant within middle third' and 'Factor of Safety against overturning of 2' are given as alternatives in BS 8002, Clause 3.2.1.2. The two methods produce very different results (see Appendix A).

Frequently these factors of safety are not achieved and it is then necessary to review the input data to the calculations. The following points should particularly be examined:

1. the soil parameters: are they realistic? Would it be more realistic to adopt mean strengths rather than a lower bound? Were laboratory tests based on the complete range of particle sizes present or only on the finer fraction? Would further investigation and testing produce a more reliable and possibly more favourable answer?
2. the loading on the backfilling: is it realistic?
3. the water levels at the back of the wall: would further field measurements, including checking how they respond to changes of water level at the front, be useful?
4. wall and base friction: sensitivity analyses should be carried out to check the effect of different assumptions.
5. wall geometry: are all the assumptions about the shape of the wall justified by the evidence?

A number of these issues are examined in more detail in Appendix A.

Chapter seven

Maintenance and rehabilitation methods

7.1 Introduction

The replies to questionnaires (see Appendix C) showed that the commonest faults in old waterfront walls are undermining of the wall and degradation of the wall materials. As a result, many of the repairs are, in principle, fairly simple, except for the complications introduced by working under water, in the tidal zone or exposed to waves. However, as discussed in Chapter 2, there are a wide variety of faults which have to be remedied and this section examines some of the most commonly used methods.

Many walls standing today would be considered underdesigned by modern standards. When faced with the task of repairing one of these walls the engineer is in a dilemma. He may either upgrade the design to modern standards, thereby probably involving the expenditure of unnecessary funds (since he will have to rely on the 'belt and braces' principle), or he can reinstate the wall to its original condition, and by doing so leave himself open to criticism and possible legal action if the wall should subsequently fail. There is thus a need for the engineer to draw the attention of those responsible for the wall to this conflict and to agree on the approach to be adopted.

Irrespective of the agreed type of overall design philosophy, there is still an onus on the engineer to produce designs which are cost-effective. It is surprising how a small mistake can have dramatic consequences (see box).

The reconstruction of the piers at West Bay, Dorset

In the early 1970s the piers at the entrance to West Bay in Dorset, which were then over 100 years old, were rehabilitated by driving sheet piles around the structures and filling the space between the piers and the piling with shingle and concrete. Figure 7.1 shows a cross-section of the works as designed. In a relatively short space of time the beach shingle had abraded away the bottom of the piles, thereby allowing the shingle fill to escape from below the concrete. The piers are now (1989) being reconstructed. This recent work would not have been necessary if the concrete filling had been taken down well below any anticipated beach level.

It is also surprising how much material can be removed from a masonry structure, for the purpose of repair or reconstruction, without the structure showing any apparent sign of distress. Figure 7.2 shows the cavities formed under a lock wall when the cill timbers are replaced. Similar work of a more comprehensive nature was carried out on the five-lock staircase of the Caledonian canal at Fort Augustus[58] and monitoring of the movement of the structure, using

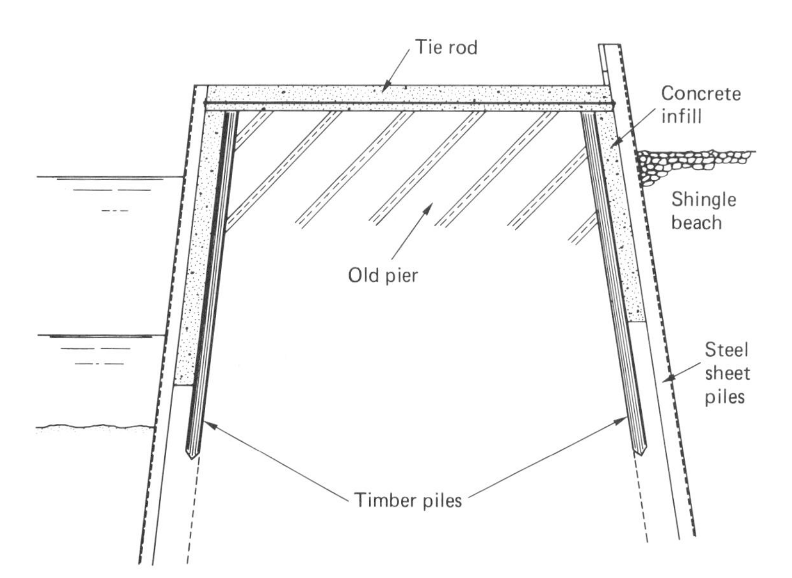

Figure 7.1 Cross-section of the 1970s design for the reconstruction of the East Pier at West Bay, Dorset
Source: Dobbie and Partners/West Dorset District Council.

Figure 7.2 Kytra Lock, Caledonian Canal, showing cavities left after removal of decayed timbers
Source: British Waterways.

one metre long vibrating wire strain gauges, showed no signs of distress during repairs.

7.2 Modifying loads on the structure

7.2.1 Redefining the standard of service

Where a wall has been overloaded or is at risk from some type of overloading, it may be possible to reduce loads by regulating the usage of areas affecting the wall. Typical possibilities include the following:

1. restricting stacking loads on the backfilled area behind the wall (see box). Particular attention should be paid to the densities of stacked materials. Tables 24 and 25 of BS6349: Part 1 give typical densities of bulk and stacked materials;
2. imposing partial restrictions on loading so that the loaded area is kept a certain distance from the face of the wall;
3. controlling the weight and position of equipment. It may be preferable to use heavier equipment further from the wall or to use lifting equipment on the ship or on rails on independent foundations. In some places operators of heavy mobile equipment have to obtain the port engineer's approval prior to use of the equipment near waterfront walls. An example of a request form, as used at Hull, is shown in Figure 7.3;
4. controlling the use of wall bollards by dredgers or other vessels;
5. controlling the water level in front of the wall either by restricting the times at which gates are opened in relation to the tide or by deciding to avoid dewatering of the area in front of the wall for repairs, by using other repair methods;
6. controlling the water level at the back of the wall. An example of this is shown in the repairs carried out at Chatham (see box).

Load restrictions, South Alfred Dock, Liverpool

The top half of the south wall in the Alfred Dock in Liverpool started to overturn due to excessive imposed loads. The wall had been weakened by the loss of face material and mortar over a period of time, coupled with rapid drawdown of water (as the dock is used as a half-tide dock). Movement of the wall appears to have been curtailed as the result of the enforcement of loading restrictions.

Repairs to the bullnose, Chatham Dockyard

Chatham Dockyard in Kent has a bullnose separating the entrances of the North and South locks. This is a mass concrete structure with granite cope and quoins, backfilled by miscellaneous material and founded on gravel overlying chalk. Investigation showed that the structure had a low factor of safety against sliding, due to the partial loss or absence of its toe, and that propeller wash and bollard pull from modern vessels were likely to induce collapse of an already highly cracked structure.

Repairs included the grouting of a granite-filled trench around the toe of the wall, the installation of tie bars in the back face of the wall (tensioned back to a steel frame in the centre of the bullnose) and improvements to the drainage paths. The last of these was achieved by drilling drain holes through the wall, installing a filter membrane and the placing of free-draining granular backfill.

APPLICATION FOR USE OF MOBILE CRANE

Date

I/We . . . . hereby make application

for permission to operate a mobile crane at . . . . Dock

in the loading/discharge of a Vessel named . . . . at

. . . . Berth on . . . . 19.

I/We set out below the particulars of the crane and the worst loadings which will be imposed on the quay in the operation of the crane for your consideration.

Manufacturers name . . . . Crane Type No./name . . . .

Crane dimensions

Pad centres

a	b	c	d	e

Pad sizes

	A	B	C	D	E	F
X x Y	x	x	x	x	x	x

Load to be lifted Tonnes

All up weight of crane tonnes

Loadings

Pad loadings with:

	A	B	C	D	E	F
Jib at (1)						
Jib at (2)						
Jib at (3)						
Jib at (4)						
Jib at (5)						

Engineer's comments:

Signed . . . .

Operations Officer's approval:

Signed . . . .

Figure 7.3 Form for application for use of a mobile crane, as used at Hull Docks
Source: Associated British Ports, Hull (ABP Hull).

1. reducing the depth of water available in front of the wall;
2. reducing the level of the ground behind the wall;
3. reducing the exposure of the wall to waves by other works such as beach raising.

In many instances such solutions will not be possible but in others it may be feasible to save considerable expenditure on civil engineering works which would otherwise be needed to strengthen the wall.

7.2.2 Reduction of soil load on back of wall

Where a wall has an inadequate factor of safety against overturning, sliding, etc, one solution is to reduce the loading on the back of the wall by modifying the backfill. Such works include the following:

1. excavating the fill behind the wall and replacing it by a reinforced earth structure (see box);

> **Use of reinforced earth at Gateshead**
>
> The Metropolitan Borough Council of Gateshead needed to improve the stability of a river wall on the south bank of the Tyne, as part of the preparations for the 1990 National Garden Festival. Cracking of the wall indicated forward movement and monitoring of the cracks showed that movement occurred throughout the tidal cycle. It was decided to construct a thrust-relief structure behind the existing wall to reduce the imposed loads and hence improve the stability.
>
> The thrust-relief structure was formed of earth reinforced by geogrids, manufactured from high-density polyethylene, and tied to a bagwork-facing unit (see Figure 7.4). The gap between the wall and the thrust-relief structure was filled with a coarse drainage medium.

2. replacing part or all of the backfill by material with a high internal angle of friction or with cement (Figure 7.5);
3. removing part of the filling behind the wall where the use of the area will permit this. For example, at Gateshead a quay wall had been damaged by a deep slip. It was no longer required to function as a quay but could continue as a retaining wall with a lower ground level behind it;
4. replacing part of the backfill with lightweight material, such as expanded polystyrene slabs, which have been used on road-embankment construction. Expanded polystyrene can only be used above the level of the groundwater because it is buoyant. Sufficient fill must be placed above the polystyrene to spread heavy point loads;
5. grouting up the backfill to reduce the active pressure on the wall (see Figure 7.6). It is very easy to show, in a design drawing, a neatly grouted zone of filling behind a wall. In practice it can be difficult to achieve and this method should only be used if it can be ensured that:

   (a) the material will be effectively grouted;
   (b) the grout will not spread to other areas;
   (c) drains will not be blocked by grout;
   (d) the groundwater level will not be adversely affected;
   (e) the liquid grout will not apply high pressures to the back of the wall;
   (f) the additional weight of grout in the soil will not excessively increase pressure on the wall below the grouted zone.

There have been a number of cases where grouted backfills have been subsequently excavated and little evidence of the grout has been found. One reason for this is that the material

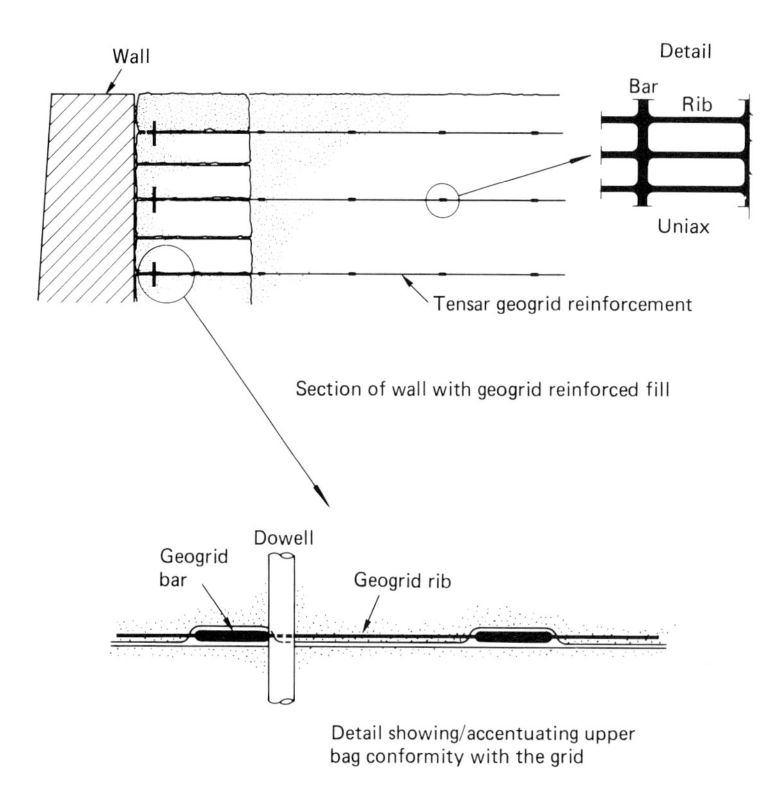

Figure 7.4 Method of using a reinforced earth structure to reduce the pressure on the back of a wall
Source: Netlon.

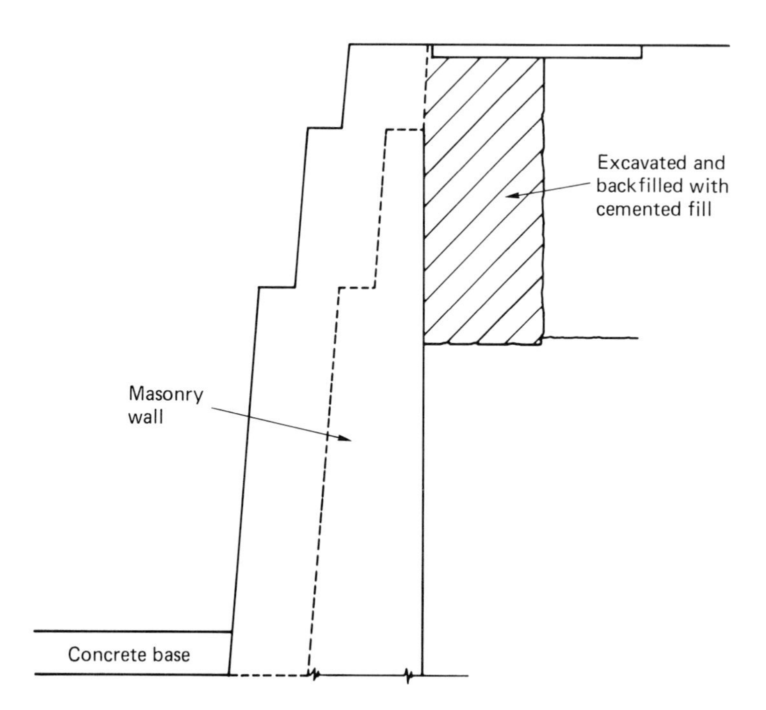

Figure 7.5 Example of soil being replaced with cemented material, Milford Haven
Source: Wallace Evans and Partners.

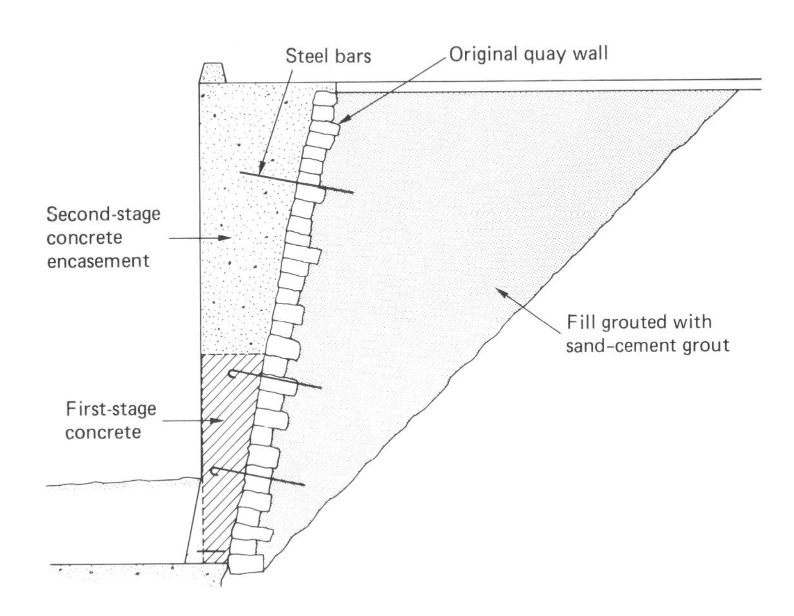

Figure 7.6 Grouting of fill behind a quay wall in Appledore
Source: Wallace Evans and Partners.

of some backfills is non-uniform and the grout takes the path of least resistance through any coarse material and leaves the fine material ungrouted. Trial pits should be excavated to check the suitability of the material for grouting. It is also prudent to carry out trial grouting and to check the result by excavation. Sowden[8] gives further details of the grout treatment of fills.

7.2.3 Pressure-relieving slabs

Where the main cause of wall distress is the surcharge applied to the backfilling, one solution is to carry the loads on a slab which is not supported on the filling. Possible methods are the following:

1. a concrete surface slab supported on piles behind the wall. The slab should extend sufficiently far back from the wall to remove any loading from the active wedge behind the wall or any part of a potentially unstable soil mass. Bored piles should normally be used because the installation of driven piles would apply additional horizontal pressure to the wall;
2. where the wall had adequate bearing capacity, it may be possible to support one end of the relieving slab on the wall (see Figure 7.7);
3. further relief can be given to the back of the wall by removing part of the backfill and constructing a piled relieving slab at a lower level.

7.3 Remedial works to the wall toe

7.3.1 Quay wall toe erosion repairs

Repairs to the toes of quay and similar walls usually have to be carried out under water. Where the toe of a wall has been partially undermined or eroded the following methods of repair can be considered:

1. grout-filled bags: the void should be cleaned out and then measured so that grout bags of the correct size can be made. Figure 7.8 shows a cross-section of the North Deepwater Quay at Cork, Eire, where the wall has been partially undermined. Pressure grouting into a purpose-made fabriform

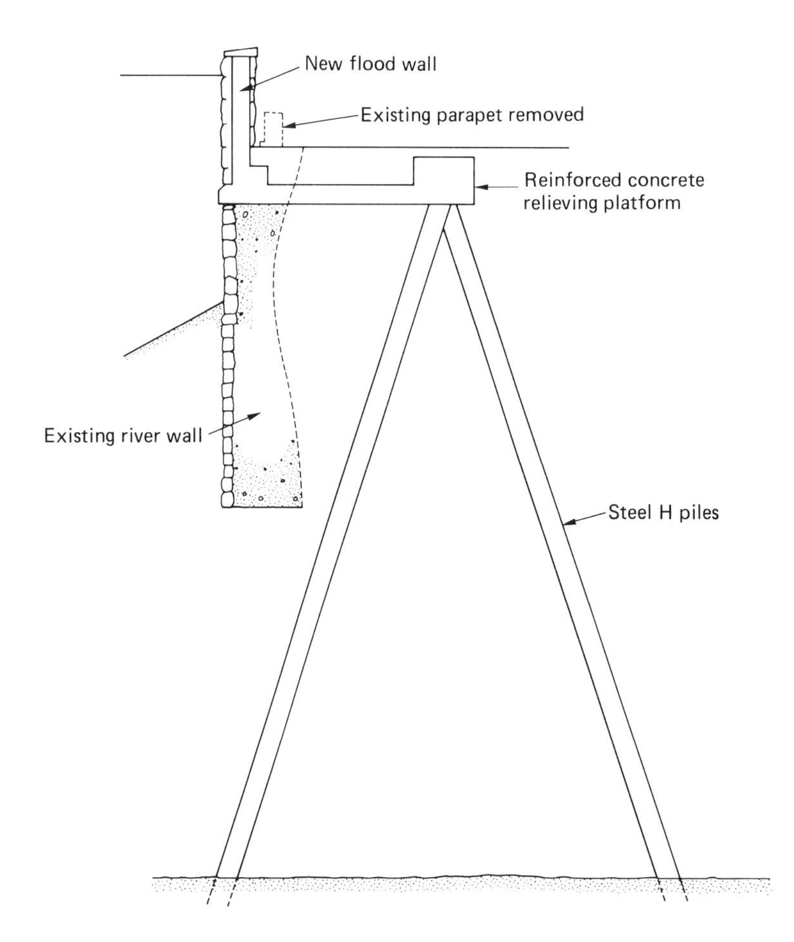

Figure 7.7 Use of a pressure-relieving slab for a river wall at Barnstaple
Source: MAFF.

bag was used to fill the void;

2. tremie concrete: the void should be cleaned out and appropriate permanent formwork devised to suit the particular situation. Examples of such formwork are precast concrete blocks and concrete bagwork. The shape of the void for this method has to be accessible for the placing of tremie concrete and capable of being totally filled without leaving voids (see box);
3. grouted aggregate: in this method large size aggregate is placed in the void and grout pipes are inserted so that grout can be placed from the bottom upwards.
4. non-tremie underwater concrete: recently developed admixtures allow concrete to be

Repairs to St Mary's Quay, Isles of Scilly

The outer half of St Mary's Quay, in the Isles of Scilly, suffered a number of problems including settlement, leaching of fines from the core, local bulging of the quay face, erosion of the face (see Figure 2.4) and undermining of the foundations; the last being exacerbated by the scour from the bow thrust of a new ferry.

Repairs have included grouting of the hearting from the deck, stabilizing bulging areas by grouting reinforcing bars into vertical holes drilled down the face, and the filling of erosion pockets and undermining by the use of bagwork and tremied concrete.

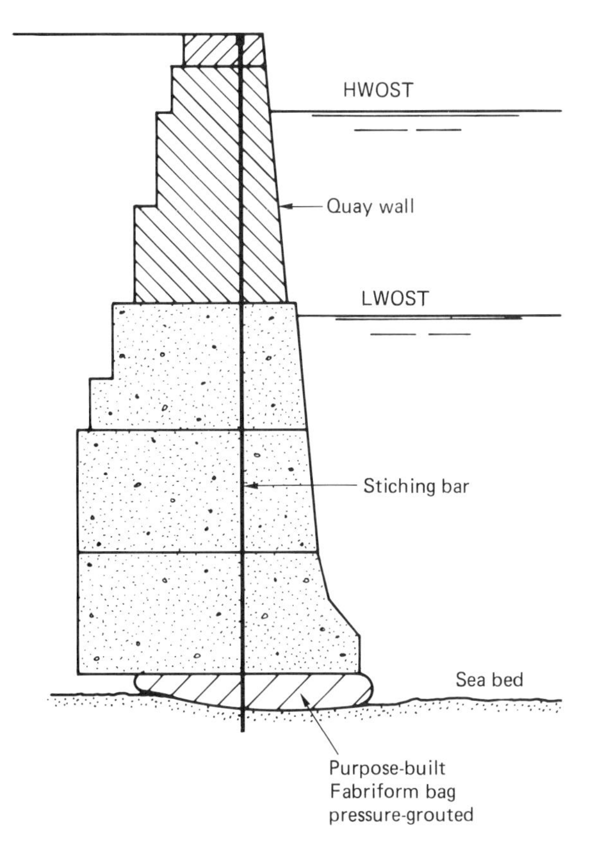

Figure 7.8 Pressure grouting into a pre-formed bag, North Deepwater Quay, Cork, Eire
Source: Cork Harbour Commissioners.

placed underwater without the use of tremie pipes (see Figure 7.9).

Where the bed has been eroded in front of the toe of the wall, and the erosion is likely to continue, a layer of rip-rap or a grouted mattress may be used as anti-scour protection. This can, however, interfere with the depth of water at the berth.

7.3.2 Coast protection sea wall toe repair

Before considering how repairs should be carried out to the toe of a sea wall, the possibility of changing the conditions which caused the erosion should first be examined. Such changes could include:

1. improvement to the littoral drift regime such as halting the removal of beach material updrift of the site of the sea wall;
2. the installation of groynes;
3. the protection of the beach from wave attack (e.g. by the use of an offshore breakwater);[59]
4. beach nourishment.

If the erosion is caused by wave reflection from the wall, changes to the shape of the wall or the placing of a rubble mound in front of it should be considered, such as proposed for Blue Anchor, Somerset (see box). Examples of this technique may also be found at Aberystwyth, Lancaster and Holderness.

Repairs to sea wall at Blue Anchor, Somerset

The sea wall at Blue Anchor, Somerset, which protects the B3191 coast road, was built around the turn of the century. Since its construction a number of schemes have been introduced to try to reduce scour and reduction of beach level at the toe of the wall. These included shingle traps and the present stone-pitched bitumen grouted apron in front of the wall. Considerable erosion has taken place at the toe of this apron resulting in the removal of beach material.

The rehabilitation works proposed, which were tested in a deep random wave flume, are illustrated in Figure 7.10, and include a double layer of 3·6 tonne local limestone rock. A trial section has been placed on-site and has already validated the proposed construction method.

Figure 7.9 Placing concrete underwater without tremie pipes by means of a suitable admixture
Source: Conference *Ports '86*, page 439, figure 6.

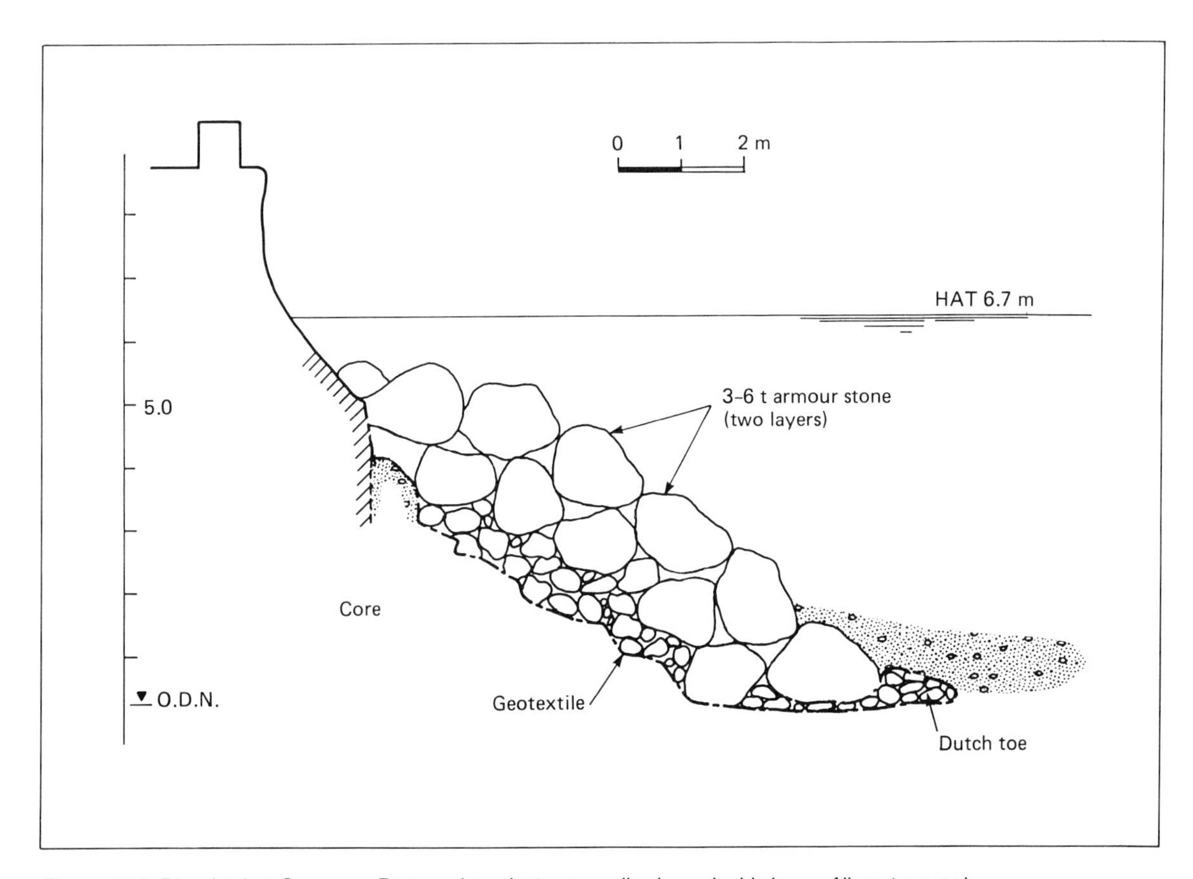

Figure 7.10 Blue Anchor, Somerset. Proposed repairs to sea wall using a double layer of limestone rock
Source: Hydraulics Research, *The use of rock in coastal structures*, 1986.

Repairs to the toe of a sea wall can usually be carried out in dry conditions, as in the majority of such walls the toe is exposed at low tide. *In situ* concrete stepped aprons of various shapes with some type of cut-off – sometimes of steel sheet piling – are often used to protect the toe. Where the cause of the toe erosion is not controlled, further trouble can be expected in the future and there are many instances of successive repairs and extensions to the aprons of sea walls (see Figure 1.102). Repairs to the toe itself can be made with concrete using an accelerator so that the repair can resist the effect of the rising tide.

Additional information relating to the design and repair of sea wall toes is given in reference 30.

7.3.3 Breakwater toe repair

For breakwater walls there is usually no opportunity to modify the sea conditions which cause damage to the toe of the wall, since the breakwater is, by its nature, exposed to wave attack. Where the breakwater consists of a vertical wall on top of a rubble mound, repairs to the rubble mound should be preceded by a check that the size, density, shape, slope, profile and thickness of the armouring are properly designed to resist the wave attack.

In some cases the toes of gravity wall breakwaters are permanently underwater and can be repaired by methods similar to those used for quay walls (Section 7.3.1) although the causes of the damage are not the same as for sea walls. In other cases the lower part of the breakwater wall is exposed at low tide – even if only for a short period on favourable tides. Repairs to the toe itself (and to the lower part of the breakwater wall) can be made with concrete as for sea walls. As an example, Holyhead breakwater wall has been successfully repaired by Sealink Ltd by using shotcrete (see Section 7.5.10) with an accelerator added at the nozzle so that a setting time of 20 minutes is achieved. This technique allowed the maximum use to be made of the available four-hour tidal window.

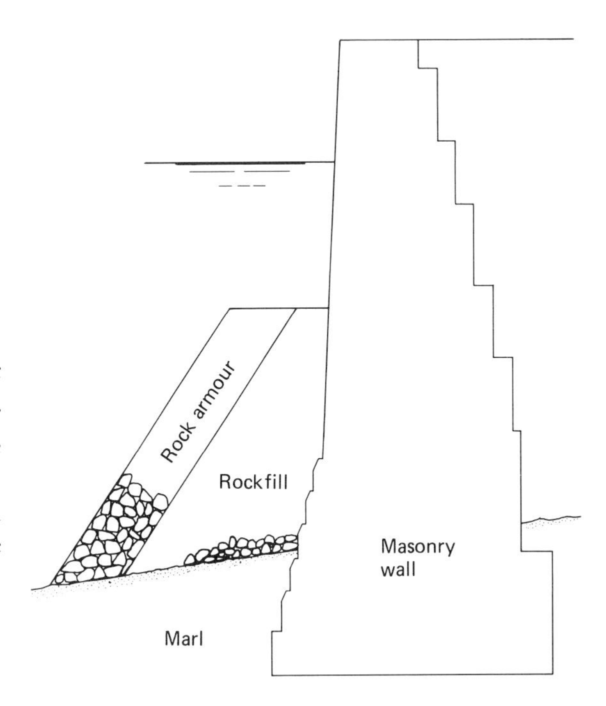

Figure 7.11 Typical cross-section of the river wall at the Huskisson Dock, Liverpool, showing the rock slope in front of the wall
Source: Mersey Docks and Harbour Company.

7.4 Increasing wall stability

7.4.1 Rock placed in front of wall

Where the soil mass under, behind and in front of the wall has an inadequate factor of safety against a deep slip failure, it may be possible to improve the factor of safety by placing rock on the toe as a counterweight, but this is only feasible where the water depth in front of the wall can be reduced. This solution has been proposed for stabilizing a dock wall in the Cumberland Basin in Bristol.

A rock slope placed in front of the wall is also effective in stabilizing a wall which is failing by sliding horizontally. An example of this technique being used at the Huskisson Dock, Liverpool, is shown in Figures 7.11 and 7.12.

Figure 7.12 Wall stabilization at the Huskisson Dock, Liverpool, illustrated in Figure 7.11
Source: Mersey Docks and Harbour Company/Land and Marine Contractors Ltd.

Reference should also be made to the report *The use of rock in coastal structures* published by Hydraulics Research Ltd.[60]

7.4.2 Ground and rock anchors

Ground and rock anchors can be used to increase the resistance of the wall to sliding and overturning. The anchors have to be designed so that no part of the wall is put into tension. To achieve this the centroid of the anchor forces has to pass through the wall at a relatively small angle to the vertical, ensuring that the resultant of all forces acting on the wall remains within the middle third of the width, at all levels of the wall. (An example of this, at the Albert River Wall, Liverpool is shown in Figure 7.13.) Such anchors increase the vertical loading on the foundation and, where the soil is insufficiently strong, compression piles under the wall may be required (see box).

An example of a different solution devised to preserve a Grade 2 listed lock wall at the Old City Canal in the Isle of Dogs in London is shown in Figure 7.16. The foundation of the wall was deemed unable to take any increase in load and the vertical component of the anchor force has been transferred to a vertical pile.

Rock anchors have also been used in the deepening of a dock in India (see box, page 207).

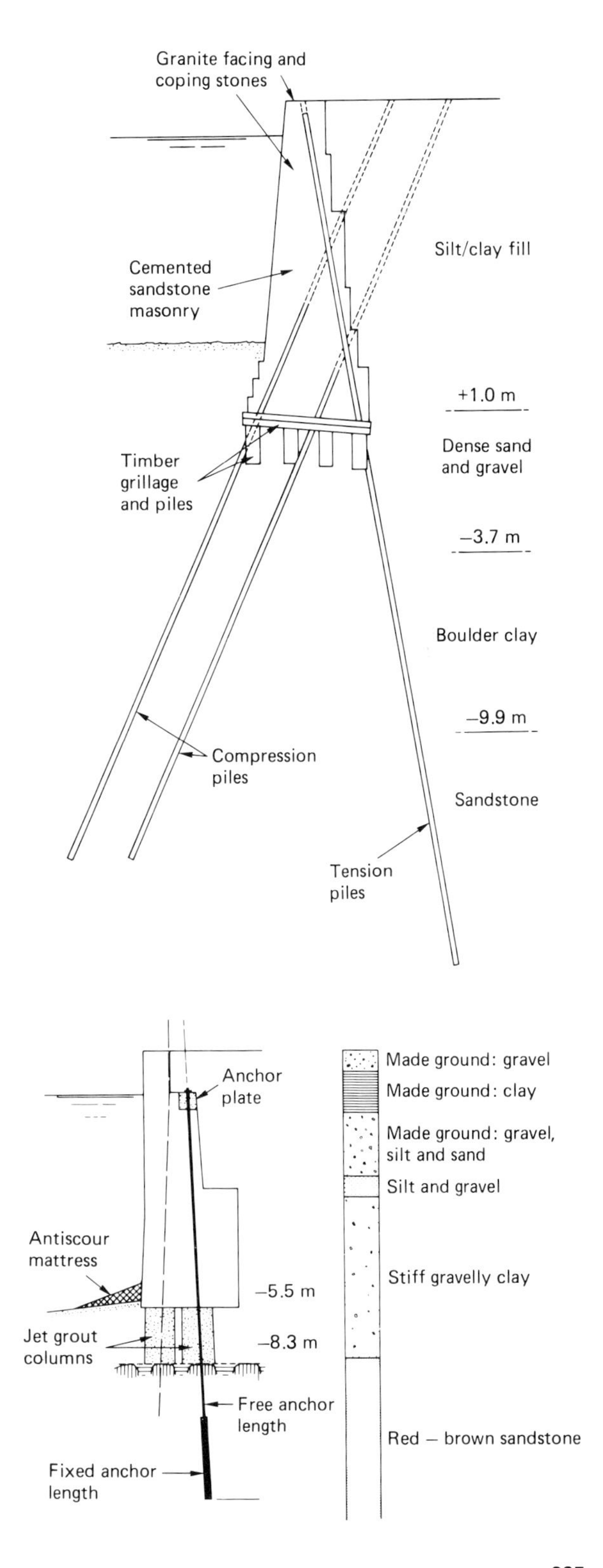

Figure 7.13 Typical section of the Albert river wall, Liverpool, showing strengthening works
Source: Jones, B.I., *The Restoration of the Historic Docks of Liverpool*.

Compression piles in use in two repairs at Liverpool

There are two examples of the use of compression piles for wall repair work in Liverpool. These are in the east wall of the Brocklebank Dock and the south wall of the Canning Half-tide Dock. In the Brocklebank Dock a row of columns was installed under the rear of the wall (see Figure 7.14) using a jet-grouting method. After installation of this row each column was redrilled to take a rock anchor of 20 m length which was stressed to 1.25 times its working load and then de-stressed to the working load. Subsequently the row of columns under the front face was installed. These columns formed a contiguous interlocking impermeable support underneath the wall adjacent to the dock.

In the case of the Canning Dock partial collapse of the wall had occurred which necessitated the installation of a temporary rubble embankment for support (see Figure 7.15). Two rows of near-vertical compression piles were then installed by using the mini-pile system and drilling through the wall. Finally a row of tension piles was placed at an angle of 35° to the horizontal.

Figure 7.14 Typical cross-section of the dock wall, Brocklebank Dock, Liverpool, showing rock anchorage and jet-grouted columns
Source: Civil Engineering, Jan/Feb 1987, p.51.

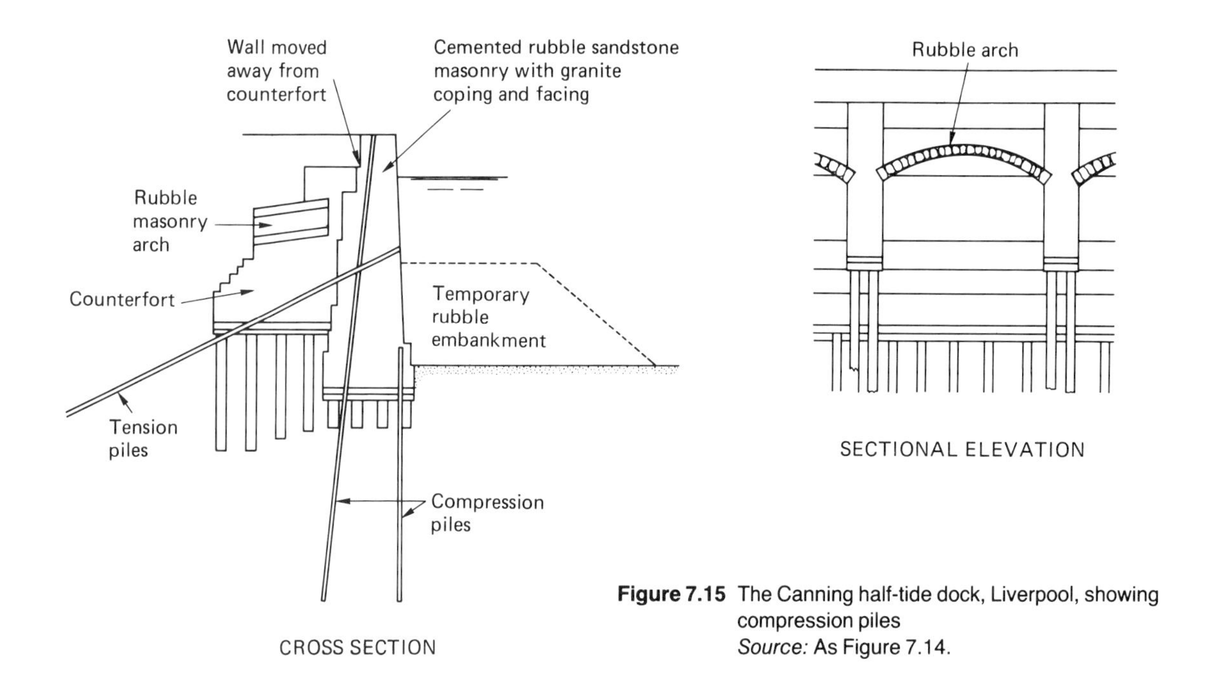

Figure 7.15 The Canning half-tide dock, Liverpool, showing compression piles
Source: As Figure 7.14.

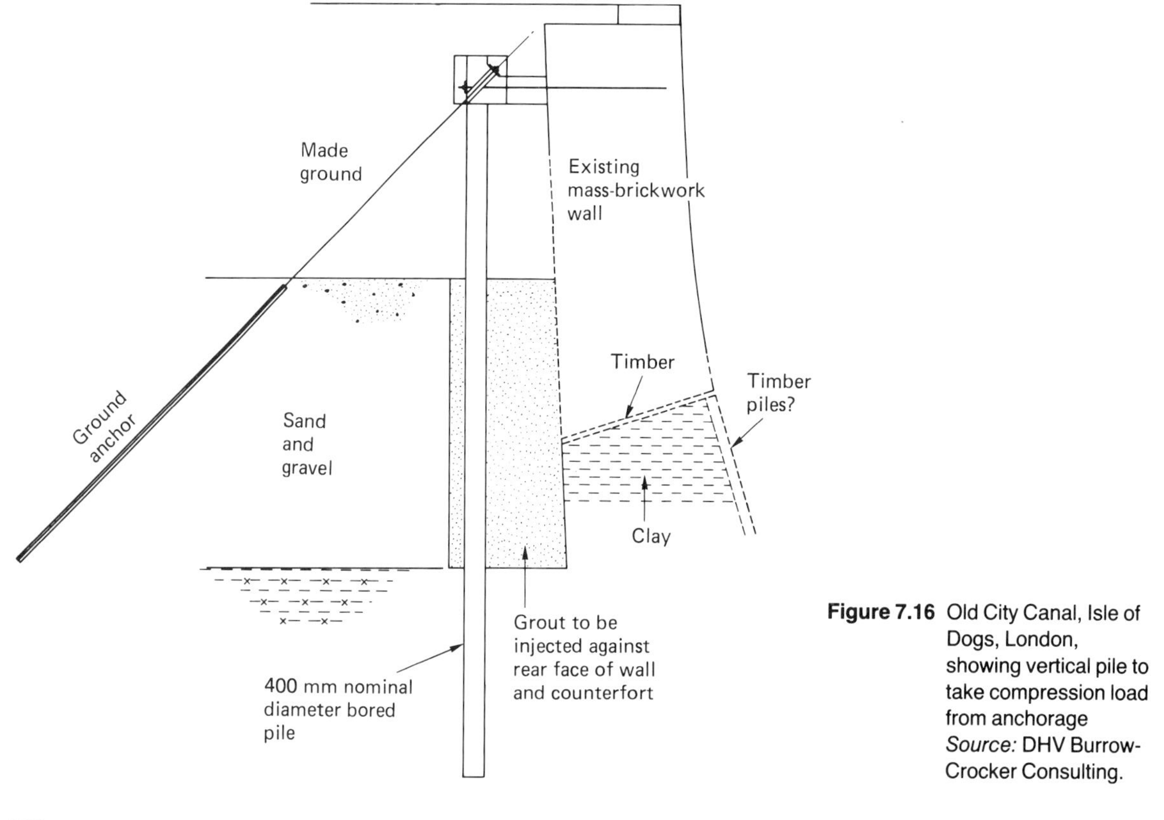

Figure 7.16 Old City Canal, Isle of Dogs, London, showing vertical pile to take compression load from anchorage
Source: DHV Burrow-Crocker Consulting.

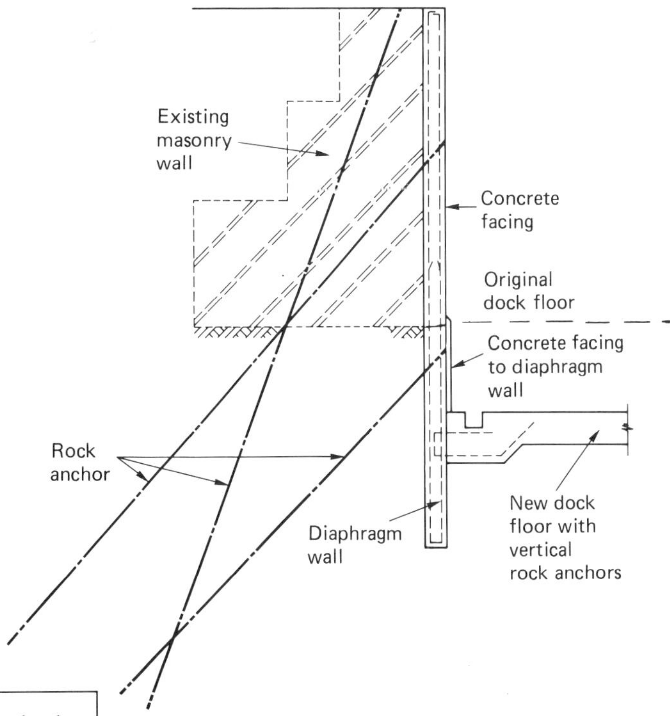

Figure 7.17 Rock anchors being used for a dock-deepening project in India
Source: Consulting Engineering Services (India) Ltd.

Use of rock anchors for deepening a dock in India

A wet dock in India is being deepened and converted into a drydock. Part of the dock wall consists of an old masonry wall founded on weak rock overlying stronger rock. Figure 7.17 shows how rock anchors and diaphragm walling are being used to convert the wall to its new function. The following sequence of construction illustrates the importance of correct phasing of this type of work:

install top anchors;
dewater dock to existing level of bottom of dock;
construct diaphragm wall panels (excavation by mechanical rock cutters);
install bottom anchors;
construct RC facing to existing wall;
install middle anchors;
excavate dock floor to new level, construct new dock floor and facing to diaphragm

In this case the sequence of the work was particularly important.

The effect of any horizontal restraints (such as wall corners) to movement of the wall caused by inclined anchors should be assessed. The diameter of holes through the wall for anchors should be kept as small as possible to avoid weakening the wall. Anchors should be designed and installed in accordance with BS 8081.[61]

7.4.3 Piling

Piling through the wall to improve the bearing strength of the wall can be done by jet grouting (see Figure 7.14) or small-diameter drilled piles (as small as 90 mm diameter and reinforced with a single bar) (see Figure 7.18).

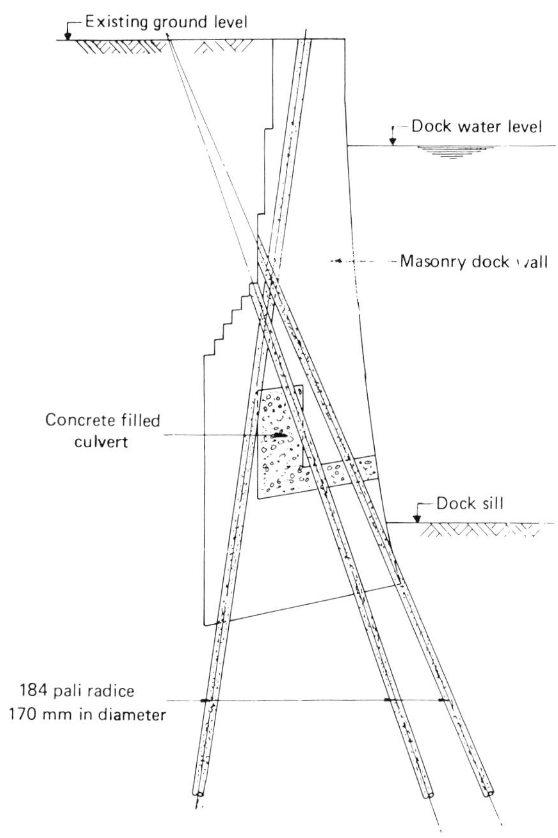

Figure 7.18 Illustration showing the use of small-diameter drilled piles (Pali Radice piles)
Source: Fondedile Foundations Limited.

In the Mazagon Dock in Bombay,[62] the foundation of a wall was strengthened by underpinning within a dewatered cofferdam by excavating one metre 'hit and miss' sections and replacing clay and weathered rock by concrete bag work and *in situ* concrete. In most cases in the UK dewatering within a cofferdam is not now a financially viable option.

7.5 Repair of the wall structure

7.5.1 Patching of concrete walls above water

Where a concrete wall has been damaged above water, the damaged area should be cleaned and all loose material removed by high-pressure water jets. Pumps with delivery pressures of over 1000 bar are available but the pressure has to be limited to avoid removing sound material. Where the shape of the hole is likely to lead to a poor bond between the new and the old concrete, it may be necessary to consider using stainless steel dowel bars drilled and grouted into the wall.

Formwork should be fixed over the damaged area with space left to place and vibrate the concrete from the top. In the intertidal zone, the use of accelerators will assist in preventing damage from the rising tide. Alternative methods of repairing concrete are given in Section 7.5.10, Allen[63] and Perkins.[64]

7.5.2 Patching of walls below water

Damaged parts of a wall which are always below water should be cleaned with a high-pressure water jet and repaired with concrete, irrespective of the original material of construction, since the repair will never be seen. The jets for use under water must have a balancing jet to avoid putting a high reactive force on the diver. Formwork should be used as described in Section 7.5.1. Concrete with an additive to prevent segregation under water should be pumped in at one end until it emerges at the other or, in the case of larger holes, concrete without an additive can be placed by tremie. (See also Sections 7.8 and 7.9.)

7.5.3 Repointing

Where the mortar in a brick or masonry wall is missing or crumbling away, the joints should be cleaned out with a jet of water which has sufficient pressure to remove loose or defective material without damaging sound mortar. Care should be taken not to displace individual bricks or stones. Where all the original mortar is weak, hand raking or mechanical removal of loose material may be preferable to avoid unnecessary damage to the wall. Where the resulting voids are shallow, the joints should be filled with pressure pointing.

In some types of pressure pointing, OPC, fly ash, sand, and a non-ionic wetting agent are mixed, then passed via a pressure pot to a nozzle. Mortar is forced out of the nozzle by compressed air. Narrow joints up to 100 mm deep can be filled, as can wider joints up to 300 mm deep in rubble masonry.

Where the joints are deeper than 300 mm, injection pointing should be used with nozzle sizes appropriate to the width of the joint. Where the voids in the joint are wide and of moderate depth but the appearance of the wall is not important, spray concrete (see Section 7.5.10) may be more economic.

In the intertidal zone, accelerators should be used in the mortar to achieve sets of between 20 seconds and 30 minutes according to the particular situation. The accelerator is added at the nozzle. Setting times of between 1/20th and 1/50th of a second can be achieved where the mortar is being placed on a wall in running water.

When extensive repointing is contemplated, consideration should be given to the possibility that the reduction in permeability of the wall may cause problems by preventing free drainage of water from the back of the wall. If this is likely to be the case, weep pipes should be inserted before repointing.

7.5.4 Grouting of wall structure

Grouting of a wall is usually carried out to improve its integrity i.e. to bond together all the individual stones or bricks of the wall so that it acts as one coherent mass. Grouting may also be done to reduce the permeability of the wall and to fill voids in the foundation. The technique is not new. Grouting has been in use for over 150 years. There are records of the Wilson bridge in Tours being grouted in 1835[65] (see Figure 7.19).

One of the more common applications of grouting is to solidify the loose filling in breakwaters such as the one shown in Figure 7.20. In this type of structure, the outer coursed skin walls were originally designed to retain the rubble filling and resist the wave forces. The outside walls can have a low height/width ratio because of the high internal angle of friction of the rubble filling. Although some examples of this type of breakwater wall have survived for a hundred years or more with only minor maintenance, the structures are vulnerable to progressive failure once the outer wall is breached, as the internal filling has no resistance to wave action.

Grouting has the attraction of solidifying the loose filling and, in theory, bonding the whole structure together. A higher-risk solution is to repair the outer wall by replacing missing stones, repairing the pointing, sealing the top surface of the breakwater to prevent internal water pressures caused by overtopping waves, and protecting the toe of the wall. The decision on whether to grout the filling or not depends on the history of the severity and frequency of the damage. Another factor to consider is whether making the structure rigid, or less porous, is desirable. A reduction in porosity may affect the breakwater's ability to dissipate wave energy.

If grouting is to be used, trial pits should be excavated in the filling to estimate the volume of voids to be filled and to provide information for determining the details of the grouting process. The outer walls will have to be made grout proof.

It is most important that the specification for the work should require that the grouting be carried out in defined stages so that the pressure of grout against the back of the skin wall is limited to a safe value. The applied grout pressure should also be limited to a safe value. Arrangements should be made to check how far the grout has spread.

It is advantageous to carry out grouting trials in advance of a contract to grout the whole wall. The effectiveness of the grouting can then be checked as well as the amount of grout used and the cost.

Grouting is also carried out on masonry and concrete coast protection and quay walls, bridge

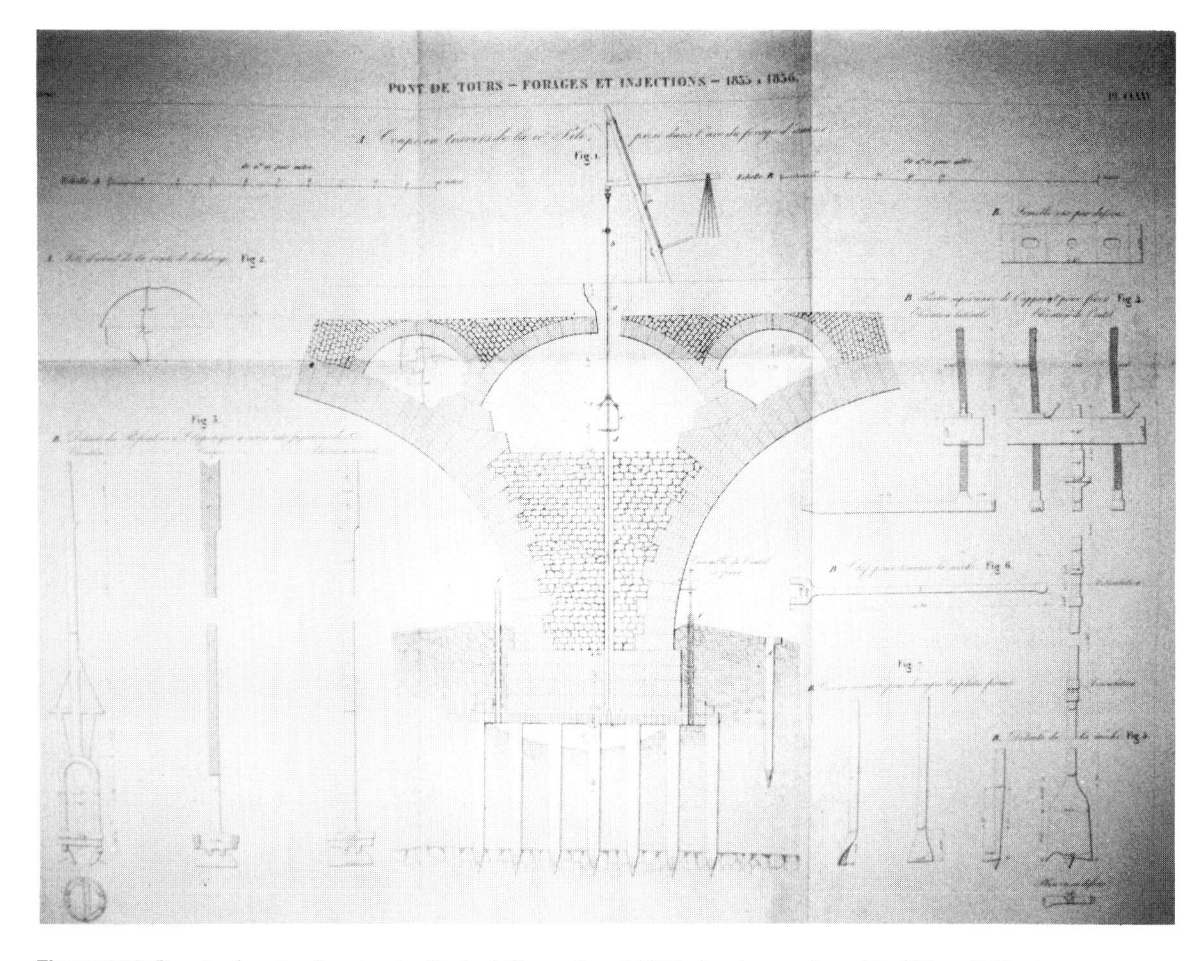

Figure 7.19 Drawing from the Annales des Ponts et Chaussées of 1835, showing grouting of the Wilson Bridge in Tours, France
Source: Le Franc, M. and Viriogeux, M. 'Symposium on civil engineering management . . . and An example of an ancient bridge rehabilitation', IABSE Symposium, Washington 1982.

abutments and piers; to fill up cracks and voids within the coursed masonry structure. Where the face of the wall is accessible it has been reported that grouting into horizontal holes is more effective than using vertical holes. As it is usually impossible to seal the rear face of quay and coast-protection walls, the possibility and any effects of leakage of grout from the rear of the wall should be considered. In some cases grouting of masonry has been found to be ineffective. This may be due to a low void content or to an inappropriate type of grout.

Points which require consideration before grouting of a structure is carried out include the following:

1. the spacing and penetration of grout holes;
2. the need to use staged grouting;
3. the sizes of fissure to be filled;
4. the specification of the grout;
5. the means of escape for the air or water within the voids to be filled by grout;
6. the grout pressure to be used;
7. the need for intermediate grout holes if

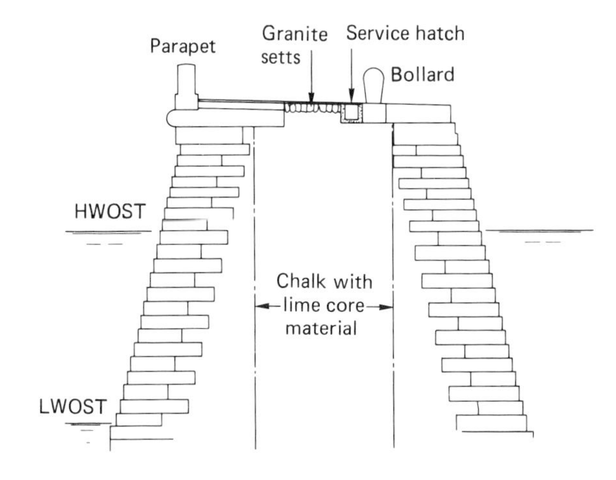

Figure 7.20 Ramsgate breakwater 1750–92
Source: W.S. Atkins & Partners.

grout fails to appear at the next grout hole down the wall;

8. action to be taken if the wall continues to accept grout without any reduction in flow (e.g. use of special grouts);
9. whether vacuum grouting should be used to increase penetration;
10. whether water leakage needs to be controlled by water-reactive grouts;
11. methods of monitoring the grouting operation, i.e. grout take up at each place, location and depth of grout holes, observation of grout spread, etc.

As there are no codes of practice for remedial grouting, advice should be sought from specialist contractors or consultants, who will be able to assist in the selection of suitable techniques and types of grout. For instance, grout for structural void filling will have a much higher 28-day compressive strength (typically 14 N/mm^2) than one designed for ground stabilization (around 2.2 N/mm^2). See Sowden[8] (Chapter 16) for grouting techniques.

It would appear that further research could be beneficial to record the current state of the art in grouting of structures and to codify the procedures to be adopted.

7.5.5 Cracks and joint sealers

As stated above, where walls are grouted it is first necessary to seal cracks in the face of the wall. Above water or in the intertidal zone the normal method would be to use pressure or injection pointing. Below water one method is to caulk the cracks, e.g. with hemp, although underwater pressure pointing is also possible. In areas below water, where joints between blocks of concrete or masonry require sealing, propri-

Figure 7.21 A Gloucester Docks wall, showing the curve in plan from movement of the structure
Source: P.F.B. Tatham.

etary joint sealers which swell when immersed in water may be considered.

Where isolated large cracks have occurred due to differential movement of two parts of a wall, rigid sealing of the crack may be ineffective. Such cracks sometimes occur near corners in dock walls: away from a corner walls can rotate slightly to allow the backfill pressure to reduce to the 'active' level, but near a corner each wall is prevented from moving by the restraint provided by the other.

Where the factor of safety against overturning is high, the relative movement will be small and may be accommodated by the normal flexibility of a masonry wall. In such a case a curve in plan in the wall will be observed (see Figure 7.21). Where the factor of safety is lower, the movement will be larger and will exceed the capacity of the masonry wall to flex so that a crack forms. If the crack is large enough, filling will be washed out from behind the wall and settlement of the surface behind the wall will occur.

Repair of such cracks is difficult. Where the wall movement has ceased, it may be possible to seal the front face of the crack and then drill and grout it. Alternatively, it may be possible to excavate the fill from behind the wall and replace it with coarse rubble which is too large to pass through the crack.

7.5.6 Masonry bonding, stitching, dowelling and wedging

Conventional methods of calculating sliding, overturning and the distribution of concentrated loads such as wave forces, bollard pulls and ship impacts assume that a wall acts as one structure. A number of different methods have been used in the past to ensure that the individual stones in a masonry wall act together as one coherent mass. Some of the methods used, such as mortar joint, wedging, timber and iron cramps, may deteriorate and fail.

Mortar joints close to the exposed face can be repaired, but wedges and cramps used to hold a wall together cannot be replaced when they deteriorate. It is then necessary to hold the wall together using a different method, such as injection pointing of the outer skin and grouting of the inside. If these are not feasible, the wall has to be refaced or reconstructed. These methods are described elsewhere in this chapter.

A further alternative is to tie the wall together with stainless steel rods, either drilled and grouted within the body of the wall or passing right through and anchored at both sides. Figure 7.22 shows the repair of the North wall and Figure 7.23 the High wall both at Lyme Regis, and Figure 7.24, the breakwater at Weymouth, where dowelling techniques have been used.

7.5.7 Replacement of brick and stone

Where individual bricks or stones have decayed, or fallen out of the above-water or intertidal face of a waterfront wall, the cavity can be cleaned out by chiselling and high-pressure water, and a new stone or brick put back in the wall. Injection pointing may help to complete the repair as it is otherwise difficult to fill all the surrounding space with mortar.

An example of the use of the replacement technique may be found in Swansea, where maintenance work has been carried out on the sea defences. Here, loosened stone blocks, 1.00 m by 0.65 m by 0.65 m in size, were bonded back into the wall with rapid-curing epoxy adhesives, and then back-grouted and repointed.

Where substantial areas of facing brickwork or stonework have fallen out (see Figure 7.25) the situation is more difficult. In many walls the facing is of different construction from the body of the wall. (Figure 7.26 shows the cross-section of such a wall.) The body of the wall may first require repair by grouting or injection pointing. If the new facing has no adequate key, stainless steel anchorage bars can be drilled and grouted into the wall. The face of the wall can then be concreted or, if the original appearance of the wall is important, the brickwork or stonework

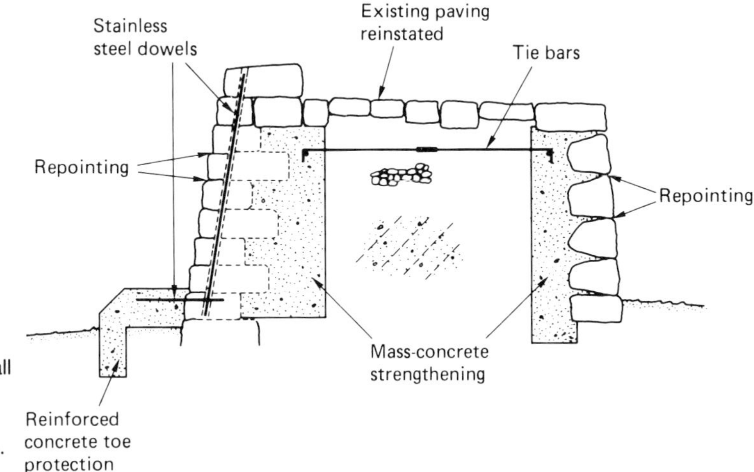

Figure 7.22 Section through the North Wall at Lyme Regis, showing ties used for strengthening
Source: Dobbie and Partners.

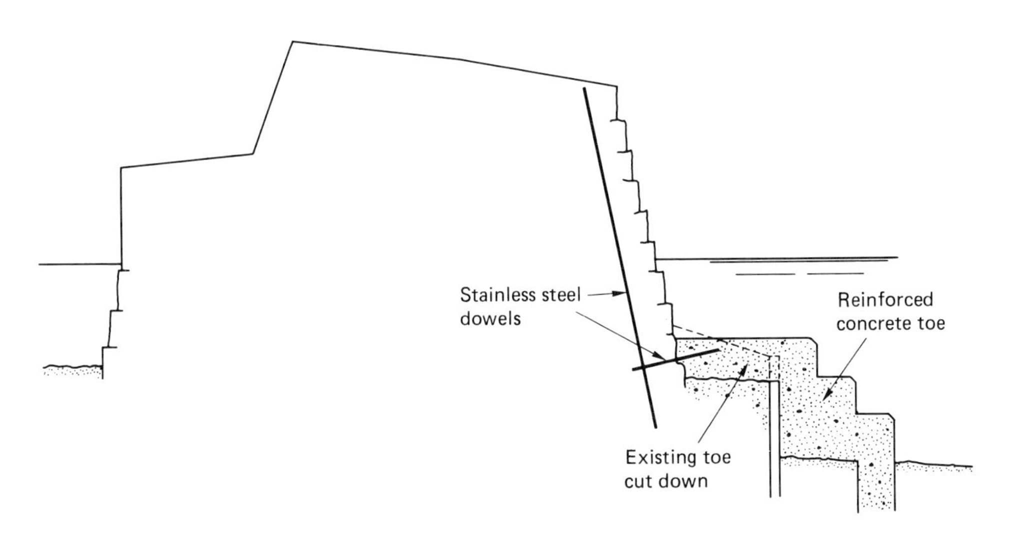

Figure 7.23 Section through the High Wall at Lyme Regis, Dorset, showing the use of stainless steel dowels
Source: Dobbie and Partners.

can be reconstructed – possibly with a concrete backing.

7.5.8 Additional skin of brickwork

Where a wall or a breakwater has suffered considerable degradation to its facing, and where the appearance of the wall is important, a complete new skin of brickwork may be an appropriate solution. Figure 7.27 shows a section through the British Rail breakwater at Starcross, Devon, which has been refaced with a new skin of brickwork tied to the existing core by steel anchors.

7.5.9 Protection of the top surface of wall or backfill

Protection of the top surface of a wall is a particular requirement for breakwaters, exposed

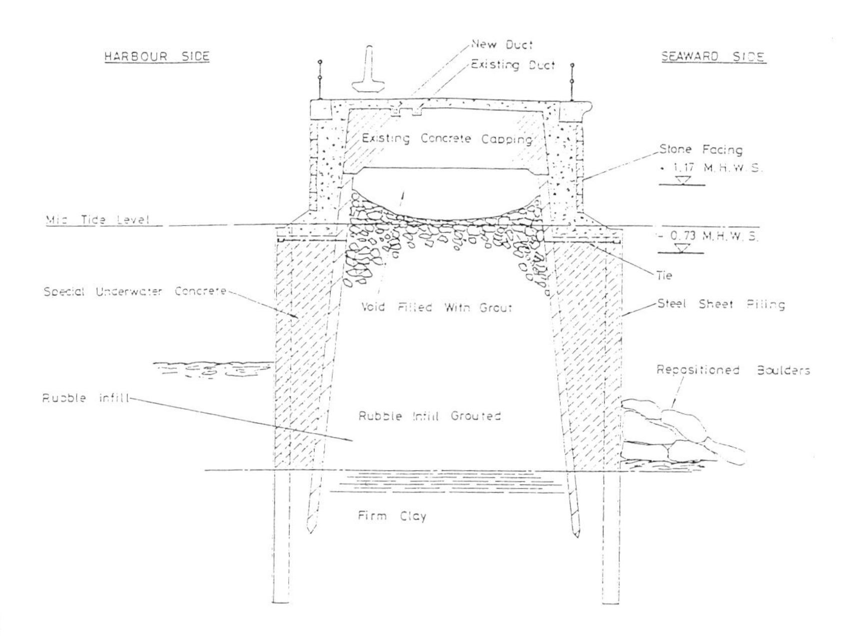
HARBOUR SIDE
New Duct
Existing Duct
SEAWARD SIDE
Existing Concrete Capping
Stone Facing
1.17 M.H.W.S.
Mid Tide Level
0.73 M.H.W.S.
Tie
Special Underwater Concrete
Void Filled With Grout
Steel Sheet Piling
Repositioned Boulders
Rubble infill
Rubble Infill Grouted
Firm Clay

(a)

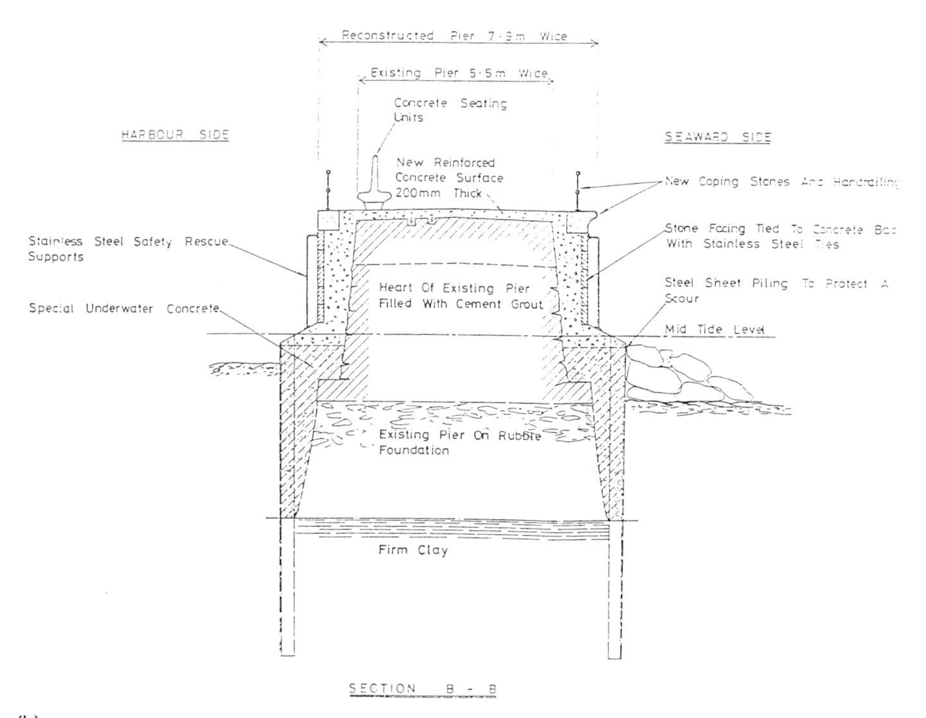
Reconstructed Pier 7·9m Wide
Existing Pier 5·5m Wide
Concrete Seating Units
HARBOUR SIDE
SEAWARD SIDE
New Reinforced Concrete Surface 200mm Thick
New Coping Stones And Handrailing
Stainless Steel Safety Rescue Supports
Stone Facing Tied To Concrete Bac
With Stainless Steel Ties
Special Underwater Concrete
Heart Of Existing Pier Filled With Cement Grout
Steel Sheet Piling To Protect A
Scour
Mid Tide Level
Existing Pier On Rubble Foundation
Firm Clay
SECTION B - B

(b)

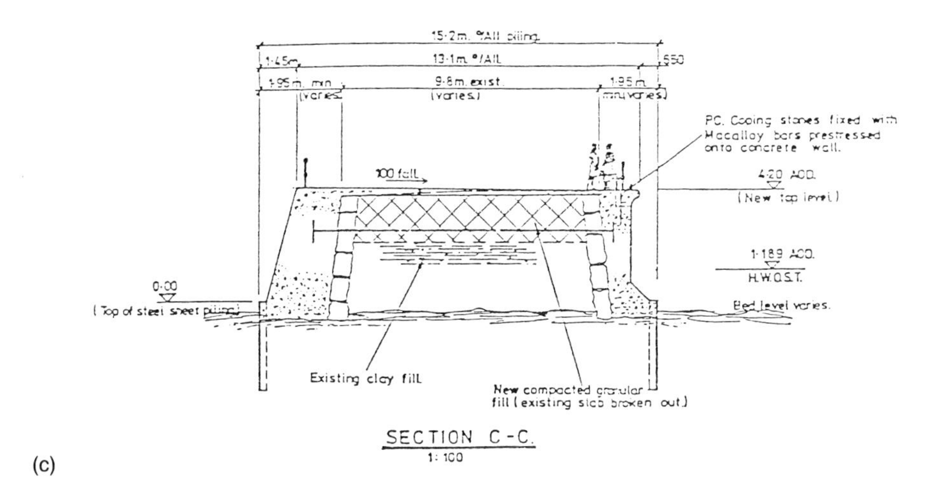

(c)

Figure 7.24 Sections through the reconstructed stone pier at Weymouth (a) Seaward end; (b) mid region; and (c) shore end
Source: DHV Burrow-Crocker Consulting.

Figure 7.25 Photograph of the south side of the Alfred Dock, Birkenhead, showing an area of missing brickwork *Source:* Mersey Docks and Harbour Company.

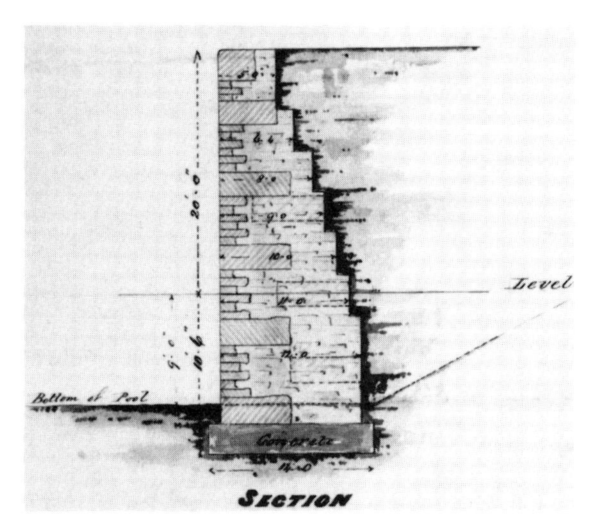

Figure 7.26 Typical section of a wall, showing the face to be of different material from the mass of the structure. The Great Float, Wallasey *Source:* Mersey Docks and Harbour Company.

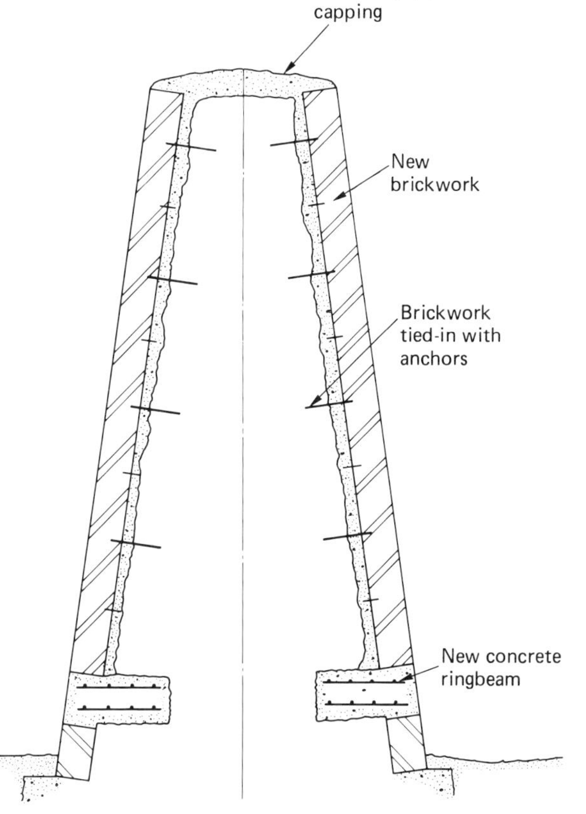

Figure 7.27 Section through the Starcross breakwater, Devon, showing the new skin of added brickwork *Source:* British Rail.

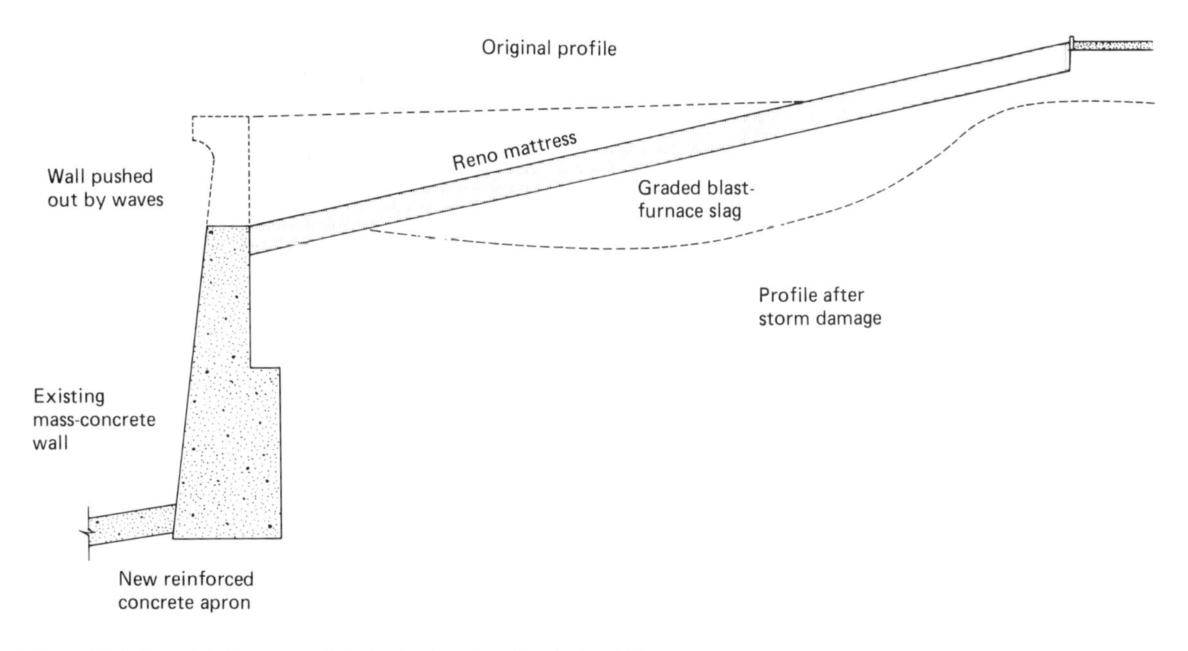

Figure 7.28 Repair to the sea wall at Harrington, showing the backfill protected by a revetment
Source: Allerdale District Council.

piers and other structures subject to overtopping by waves. The surface of the backfill to coastal defences also needs to be protected to prevent erosion of the backfill. Protective surfaces should be designed to resist overtopping by the maximum forecast wave and to be sufficiently waterproof to prevent the build-up of water pressure within any permeable hearting or filling, which then increases the outward load on the wall forming the seaward surface of the structure.

The amount of overtopping will be affected by the profile of the wall and of the seabed or beach seaward of the wall. In the past the surfaces subjected to wave overtopping have been protected by stone pitching – this may still be appropriate in cases of listed structures and where the appearance is important. In other cases concrete, asphalt or a designed revetment should be considered.

Figure 7.28 shows an example of a sea wall where the backfill has been protected by a revetment after failure of the top of the wall due to overtopping.

7.5.10 Sprayed concrete

'Sprayed concrete' is the process whereby a cementatious mix is propelled into place by a high-pressure air stream. In the maintenance and rehabilitation of old waterfront walls sprayed concrete is used for complete refacing, the filling of voids and the repointing of wide mortar joints. The finished material is a high-strength concrete which will form an integral part of the wall. Lime can be added to the mixture to give masonry walls some ability to accommodate movement. Examples of typical uses are shown in Figures 7.29 to 7.32. The first two illustrations show refacing work being carried out to the sea wall at Colwyn Bay, North Wales, and the second two show repointing and void filling operations on a sea wall at Dundee, Scotland.

These figures illustrate the relatively simple access arrangements which can be used for repairing a wall with sprayed concrete. Figure 7.32 also illustrates the extent of marine growth which often has to be removed by high-pressure

Figure 7.29 The spraying of concrete on to the sea wall at Colwyn Bay, North Wales
Source: Tarmac Structural Repairs Ltd.

Figure 7.30 The sea wall at Colwyn Bay after the application of sprayed concrete
Source: Tarmac Structural Repairs Ltd.

Figure 7.31 Filling voids with sprayed concrete, Dundee sea wall
Source: Tarmac Structural Repairs Ltd.

water jetting before repair can begin. It is important that the access arrangements used should be approved by the Health and Safety Executive or other similar appropriate regulatory body.

The term 'Gunite' is used for sprayed concrete where the aggregate is smaller than 10 mm and 'Shotcrete' where the aggregate is 10 mm or larger. There are also dry and wet processes requiring different techniques and equipment. In the dry process the dry mix is conveyed by air to the nozzle where water is added. In the wet process, wet concrete is pumped to the nozzle where high-velocity air is used to propel the mix into position.

The dry process is more suitable where small volumes of concrete are to be replaced or where access is difficult, as the plant may be positioned up to 400 m from the point of application. The wet process is more suitable where

Figure 7.32 Repointing sea wall with sprayed concrete, Dundee, Scotland
Source: Tarmac Structural Repairs Ltd.

large volumes of concrete are to be placed. A specialist contractor should be consulted on the most suitable process for any particular situation.

The quality of sprayed concrete depends on the following:

1. correct specification: advice is given in Concrete Society Publication (1979) *Specification for Sprayed Concrete*;[66]
2. thorough preparation and cleaning of the surface to which the sprayed concrete is to be applied;
3. correct application: select only reputable contractors, e.g. those who are either members of the Sprayed Concrete Association, who vet all members prior to admission, or other firms who can demonstrate their competence. There is also a move to test and certify 'nozzlemen' in the same way as welders, in view of their importance in achieving quality in the finished product.

Sprayed concrete can be reinforced with stainless steel or polypropylene fibres added to the mix, or steel mesh can be fixed to the wall before spraying the concrete. The latter should be used wherever the thickness of the sprayed concrete is greater than 25 mm. Sprayed concrete can be used for intertidal work if an accelerator is added to the mix. Further information can be obtained from the Sprayed Concrete Association, London, who publish a booklet on the use of the material.

7.5.11 Ferrocement

'Ferrocement' is a cement-rich mortar reinforced with layers of small-diameter steel mesh. It has been used for many years for making boats and other thin sections by a process of plastering and building the material up in thin layers. More recently it has been employed as a thin liner to rehabilitate sewers and other structures, including sea walls. Its use for sewer lining was approved by the Water Industry Certification Scheme in 1989. In the form used for sea walls the material is placed by a Gunite process, in layers, with a mesh fixed to the wall.

7.6 Replacement of wall by a new structure

Where the existing wall is beyond repair, the structure can be replaced, completely refaced or reconstructed. If ground conditions are suitable and the appearance of the wall is not particularly important, one solution is to drive sheet piles in front of the wall and use ground or rock anchors to secure the top of the sheet piling, such as has been carried out in Portsmouth and Axmouth (see Figures 7.33 and 7.34). The space between the sheet piles and the original wall is usually filled with concrete to reduce corrosion of the sheet piles and transmit the anchor force from the sheet pile to the original wall. One disadvantage is that the life of the sheet piles will usually be far less than the life of the original wall due to corrosion of the steel. Figure 7.35 illustrates the further repair which is needed when the sheet pile corrodes. The particular advantage of sheet piles is that they can be installed from the surface even when most of the wall to be protected or replaced is permanently below water level.

Where the wall is exposed for at least part of the tide, there are many possible alternatives. Figures 7.24 (a), (b) and (c) show three different sections along the length of the reconstructed Stone Pier at Weymouth with variations to suit the height and type of the original structure.

Where the appearance of the wall is important, but its structure is beyond repair, the outside face of the wall can be retained or simulated by a replica, but with the soil loads being carried by a new wall behind (see Figures 7.36 and 7.37, Crail in Fife). There are obviously many different ways of constructing a hidden new wall.

Figures 7.38 to 7.44 show the variety of

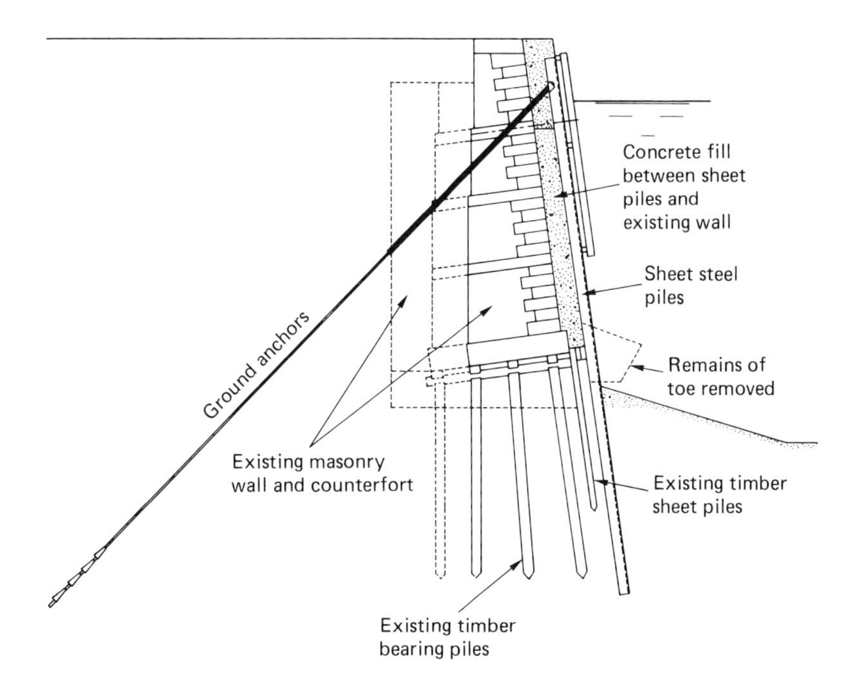

Figure 7.33 Tied sheet piling being used to strengthen a wall at Portsmouth
Source: T.F. Burns and Partners.

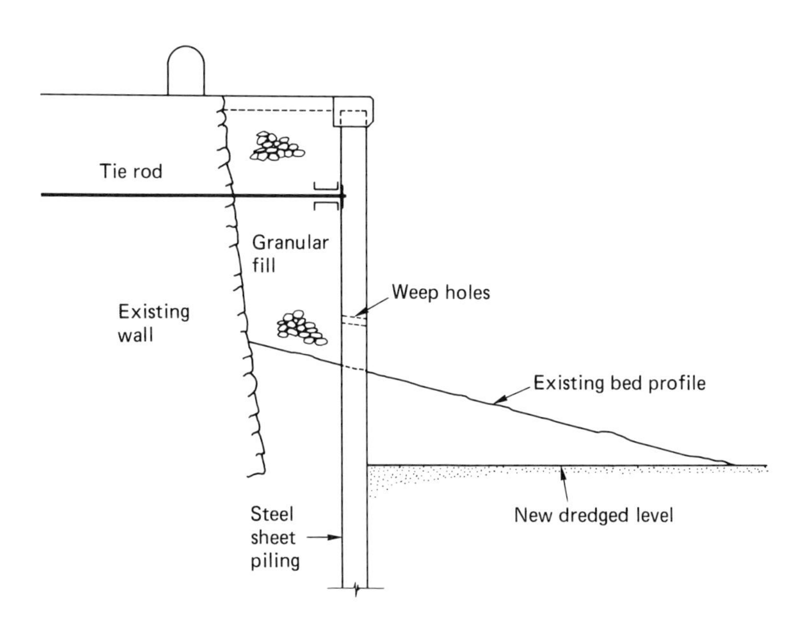

Figure 7.34 Tied sheet piling being used to strengthen a wall at Axmouth
Source: East Devon County Council/Le:vis & Duvivier.

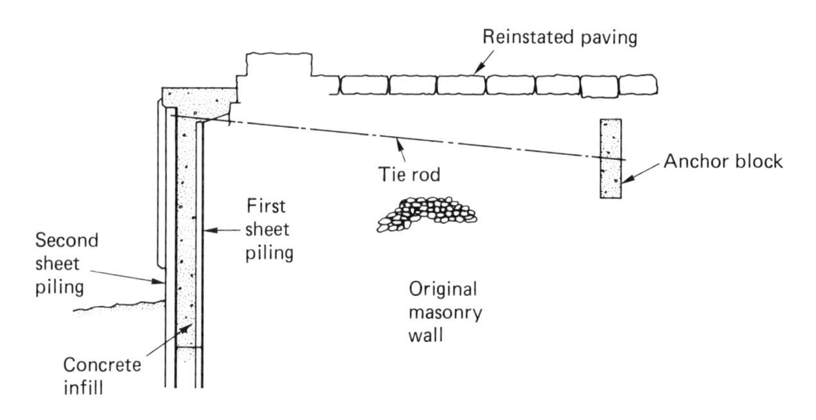

Figure 7.35 Sheet piling being used to repair a previous sheet piled repair, North Wall Roundhead, Lyme Regis, Dorset
Source: Dobbie and Partners.

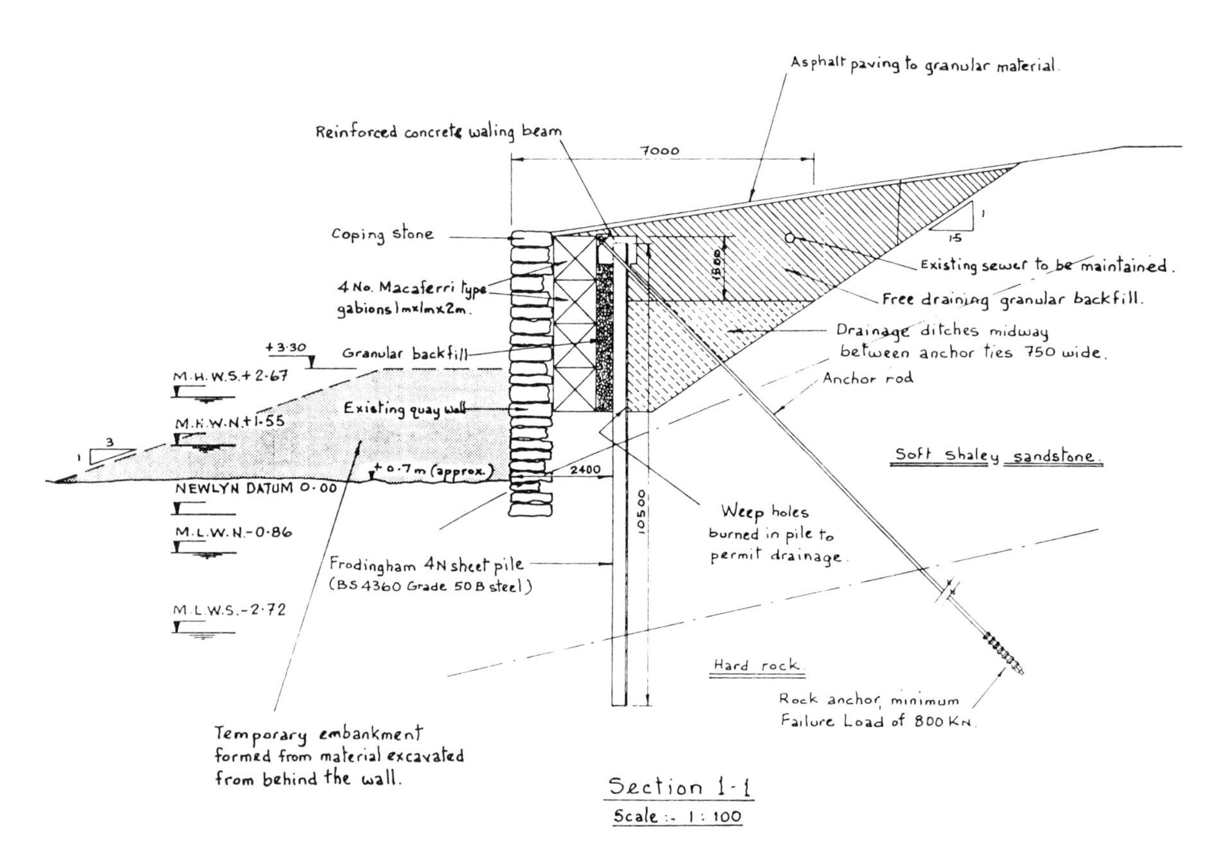

Figure 7.36 New, hidden wall behind a reconstructed wall at Crail, Fife
Source: Wallace Evans and Partners.

Figure 7.37 The reconstructed wall at Crail shown in Figure 7.36
Source: P.F.B. Tatham.

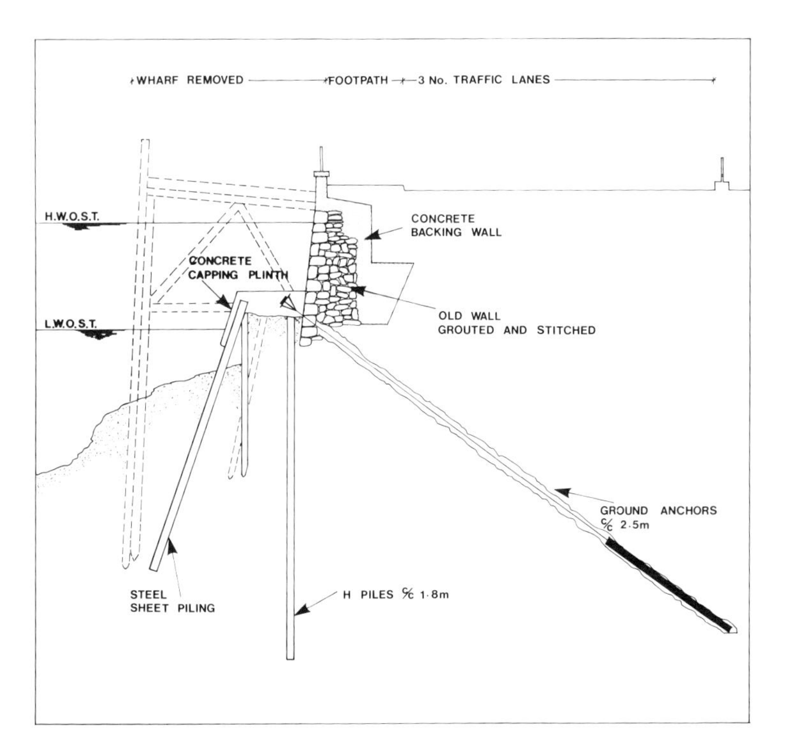

Figure 7.38 Remedial works for Anderson's Quay, Cork *Source:* Mr Liam Fitzgerald, City Engineer, Cork Corporation (client) and Mr Peter Langford and Mr Seamus Mulherin, Ove Arup & Partners Ireland, Consulting Engineers, authors of *Cork City Quay Walls*, 1982.

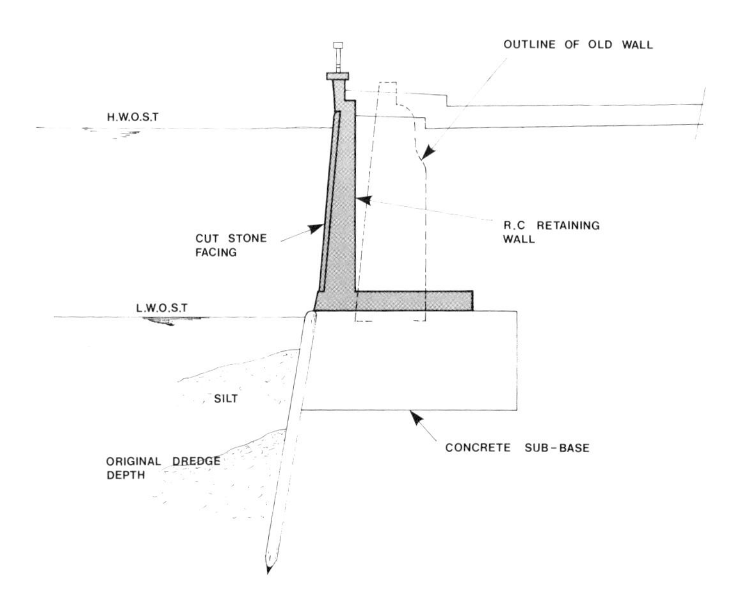

Figure 7.39 Reconstruction of McSwiney Quay, Cork *Source:* See Figure 7.38.

Figure 7.40 Wandesford Quay, Cork, showing collapsed portion
Source: See Figure 7.38.

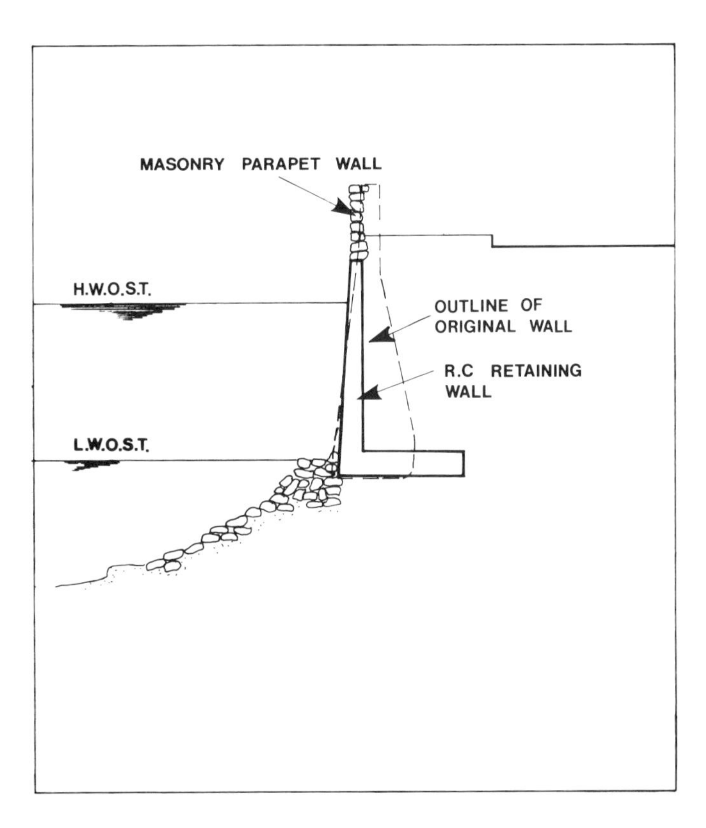

Figure 7.41 Reconstruction of Wandesford Quay, Cork
Source: See Figure 7.38.

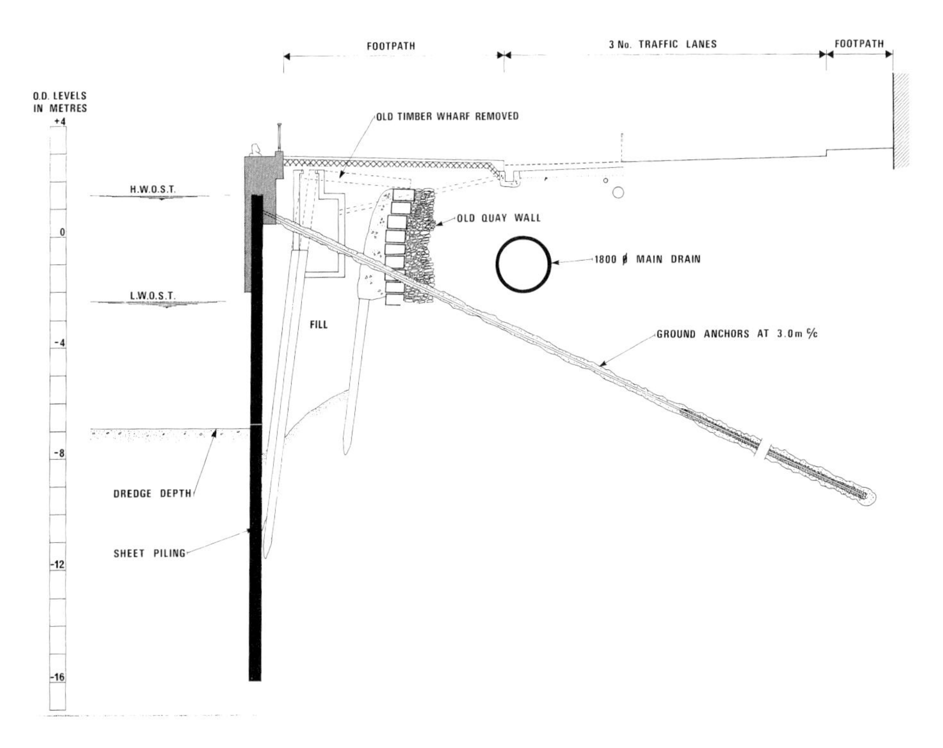

Figure 7.42 Phases 1 and 2 reconstruction works at Penrose Quay, Cork
Source: Mr Liam Fitzgerald, City Engineer, Cork Corporation (client) and Mr Peter Langford and Mr Seamus Mulherin, Ove Arup & Partners Ireland, Consulting Engineers, authors of *Cork City Quay Walls*, 1982.

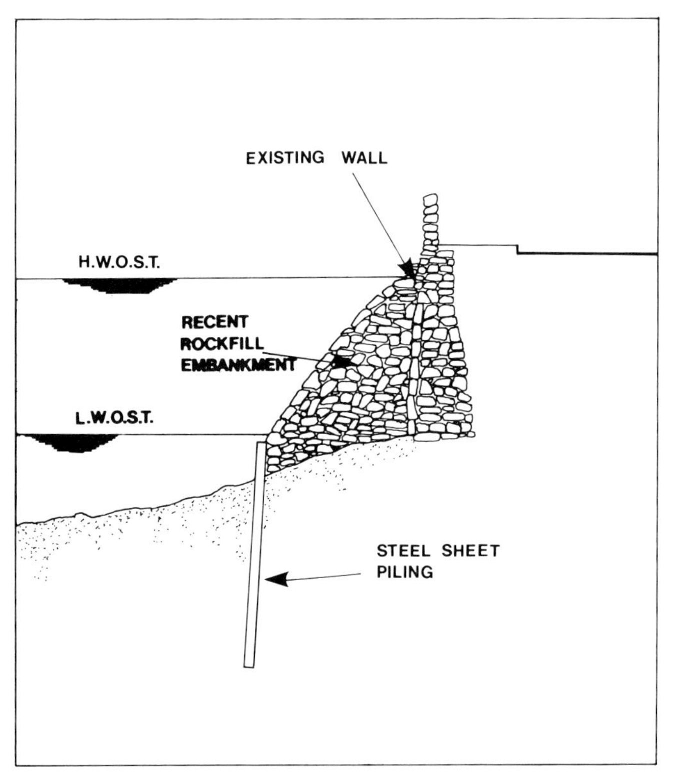

Figure 7.43 Remedial works at Glanmire Road, Cork
Source: See Figure 7.42.

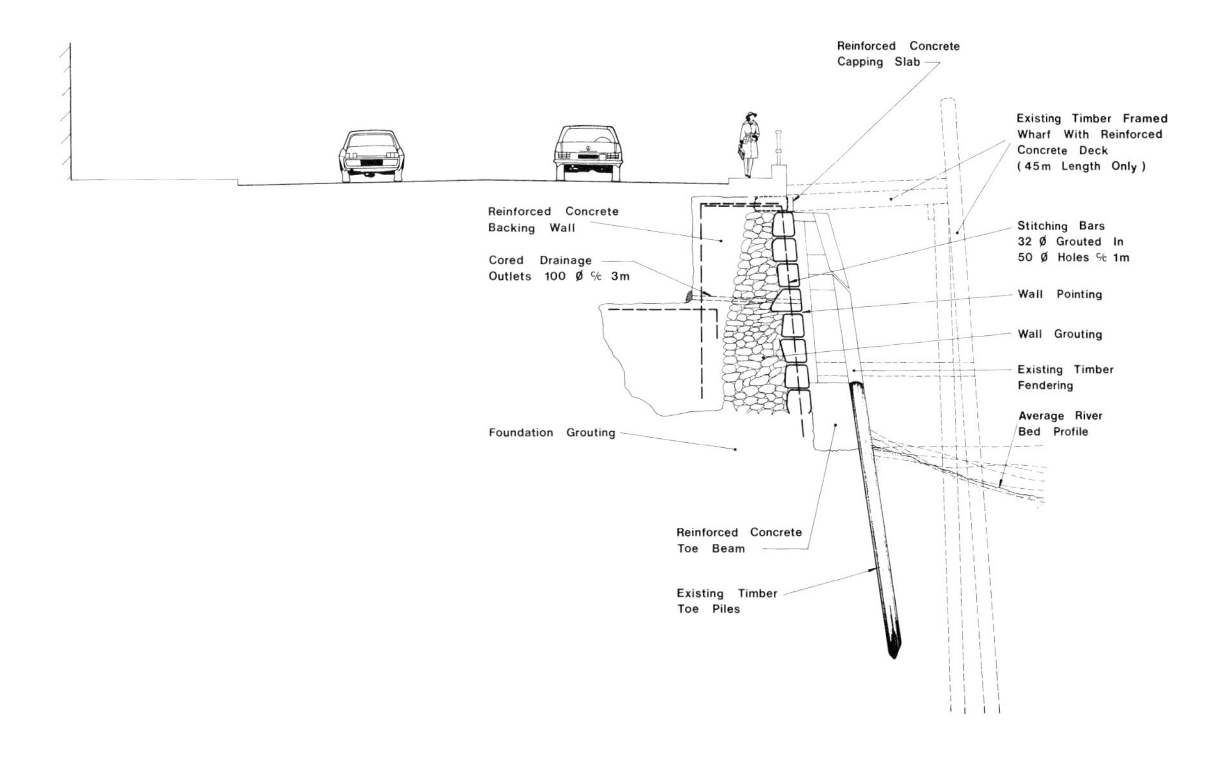

Figure 7.44 Remedial works at George's Quay, Cork
Source: Mr Liam Fitzgerald, City Engineer, Cork Corporation (client) and Mr Peter Langford and Mr Seamus Mulherin, Ove Arup & Partners Ireland, Consulting Engineers, authors of *Cork City Quay Walls*, 1982.

solutions adopted for the quay and river walls at Cork and Figures 7.45 and 7.46 illustrate two of the designs adopted for the harbour wall at Saundersfoot.

7.7 Modification of wall to suit new purpose

One of the commonest modifications required is to increase the height of the wall for flood defence purposes. Three examples of this are shown in Figures 7.47 to 7.49. A more difficult modification is lowering of the ground in front of the wall – typically to provide a greater depth of water for a ship berth at a quay. Anchored sheet piling is again a suitable solution. Where sheet piling is not possible or aesthetically desirable, some type of underpinning is required, which will involve drilling piles through the wall and anchoring to resist the increased overturning moment, as described above.

7.8 Techniques for working under water

All work carried out under water is slower and more costly than that above water and is also usually of lower quality. Therefore every opportunity to make repairs in the dry should be taken, even though the available working periods may be quite short. Such opportunities may arise from tidal working, making use of low river levels, lowering the level in an impounded dock, or by constructing a cofferdam and

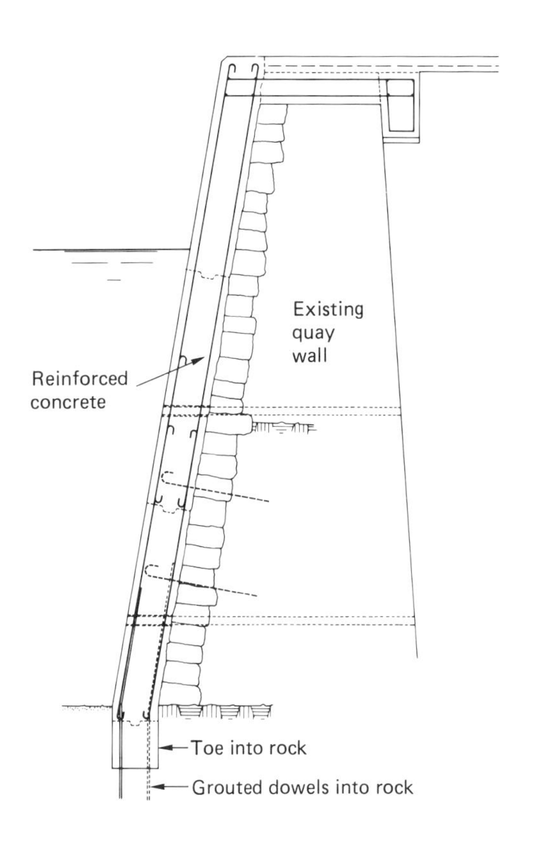

Figure 7.45 Encasement of harbour wall at Saundersfoot, Dyfed
Source: Saundersfoot Harbour Commissioners.

dewatering the area of repair.

Any method involving dewatering the area in front of a wall should be preceded by a check that the stability of the wall is not endangered (see Figure 2.7), as the water in front of the wall may be providing a vital component of the forces supporting the wall.

One method, which allows dewatering of small areas of wall for repair work to be carried out, is the limpet cofferdam: this is a small open-topped caisson which is shaped to fit against the front face of a wall. The limpet dam is lowered into the water, placed against the wall and pumped out by a high-capacity pump to effect an initial seal. The outside water pressure then forces the dam and its rubber seals against the wall and holds it in place (Figure 7.50).

Unlike a normal cofferdam, a limpet cofferdam does not affect the forces acting on the

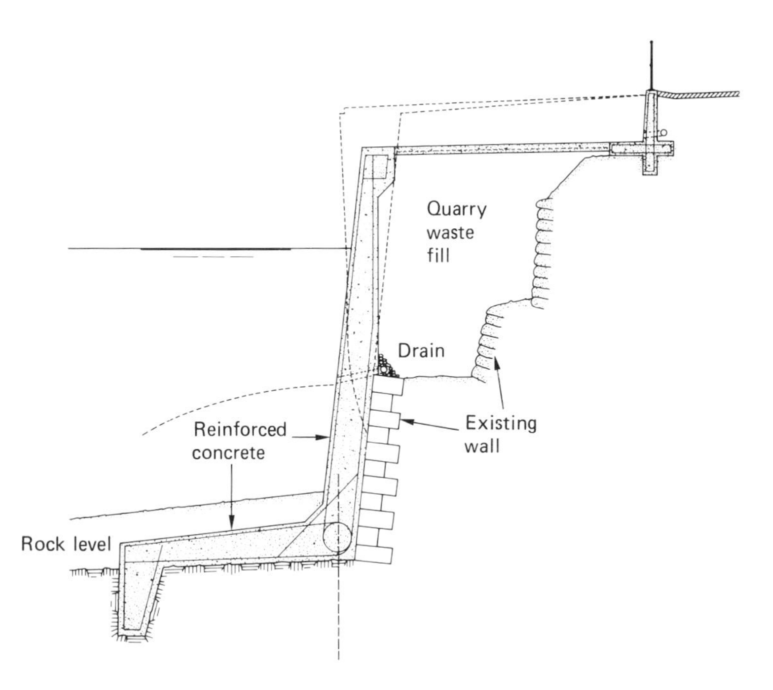

Figure 7.46 Cross-section of quay wall and retaining wall, Saundersfoot, Dyfed
Source: Saundersfoot Harbour Commissioners.

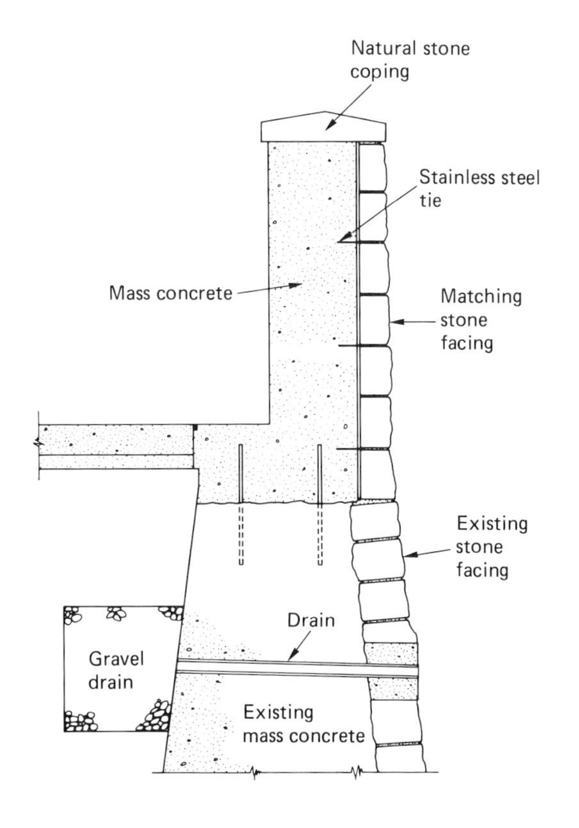

Figure 7.47 Design for increasing the height of the wall of the left bank of the Monnow, Monmouth
Source: Welsh Water/Sir William Halcrow & Partners.

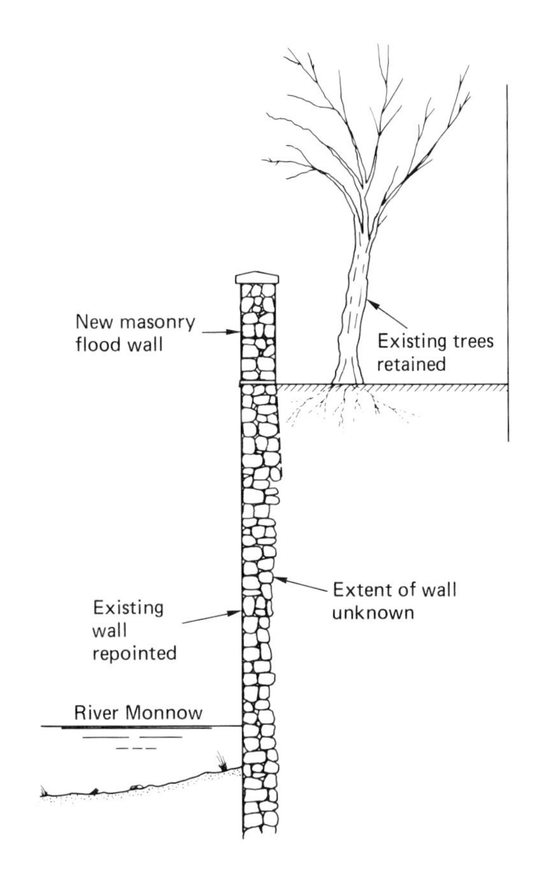

Figure 7.48 Design for increasing the height of the wall of the right bank of the Monnow, Monmouth
Source: Welsh Water/Sir William Halcrow & Partners.

wall, apart from a redistribution of the hydrostatic forces locally to the perimeter and a local outward pressure on the wall within the limpet dam. The latter may make the use of a limpet dam inappropriate where the wall is of composite construction with a thin facing, which could be blown off by water pressure behind the face.

Access to a limpet dam for workers, materials and equipment is via the open top. All work on the wall can then be done in the dry, even though the access and working conditions will be somewhat restricted. No personnel should be allowed within the limpet while shipping is moving in the vicinity as accidental breaking of the seal to the wall will lead to immediate flooding. Different limpet dams will be required for each different wall profile and will not be effective if there are appreciable leakage paths through cracks in the wall.

Other methods for dewatering small areas such as open-topped caissons, diving bells (with or without airlocks) and underwater habitats are for work on the sea bed and are not normally suitable for work on the vertical face of a wall.

Most repair work on parts of walls which are always below water will have to be done by qualified divers. This should be planned having regard to the difficulties with which divers have

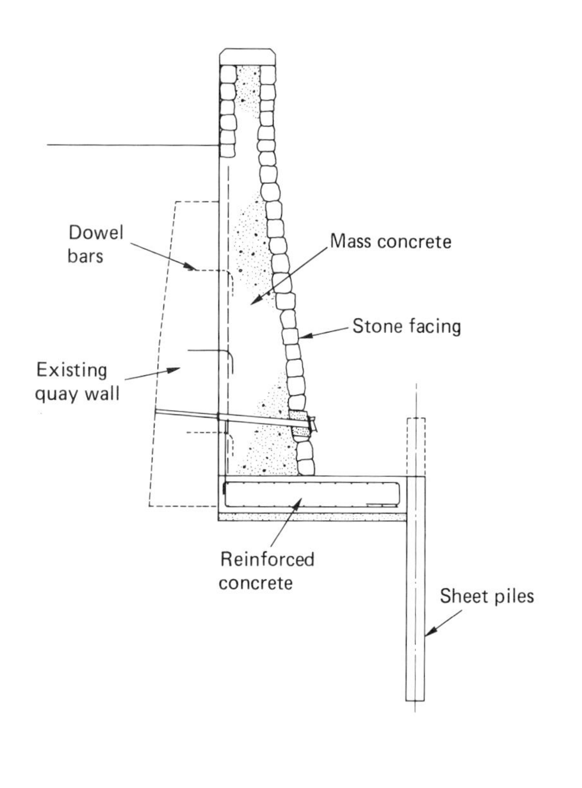

Figure 7.49 Design for increasing the height of Baker's Quay, Barnstaple, Devon
Source: MAFF.

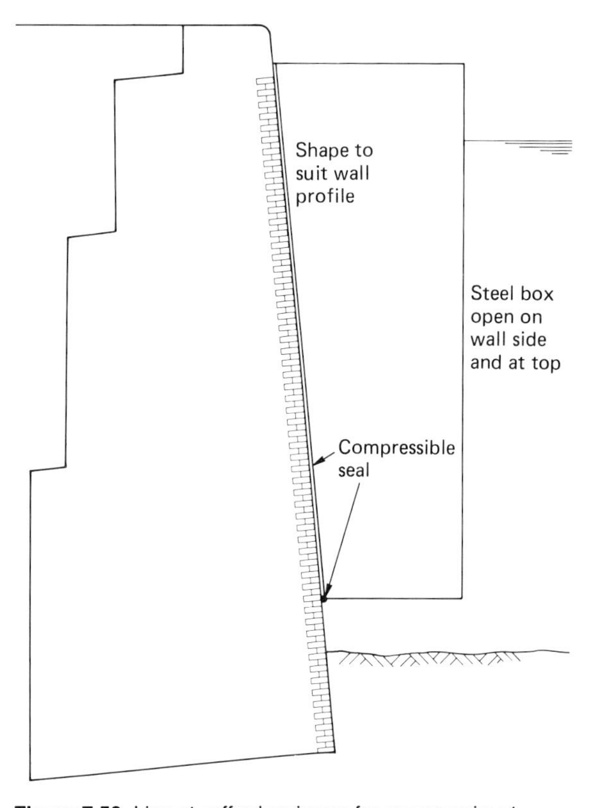

Figure 7.50 Limpet cofferdam in use for quay repairs at Harwich
Source: Livesey Henderson.

to contend, such as the following:

1. poor visibility (in some docks the visibility is virtually nil);
2. the inability to apply great force to the work in hand due to the diver's low negative buoyancy;
3. the problems caused by boats and other craft, waves or currents in some locations.

For general information on diving, reference should be made to the Underwater Engineering Group publications *The principles of safe diving practice*[67] and *Handbook of underwater tools*.[68]

The details of the underwater work should be as simple as possible. A design for use under water may be quite different from one appropriate for use above. An example of this is the technique for filling a hole under the toe of a wall. Above water, a suitable technique might be making and fixing formwork and filling with concrete. Below water one of the techniques discussed in Section 7.3.1 is probably more appropriate.

7.9 Selection of materials

Section 7 of BS 6349: Part 1 (Code of practice for maritime structures: general criteria) covers the general requirements for materials to be used in maritime construction. Additional points are given below.

Concrete: BS 6349 covers the requirements for underwater and other types of concrete used in marine works except for where concrete is made with an admixture which allows it to be placed without the use of a tremie or similar device. A range of materials is now available for use under water so that mortar patching, grouting, concreting and pressure pointing can all be done using appropriate proprietary materials. These materials are polymer modified cement-based systems. It is important to use the appropriate material for the particular situation.

The requirements in BS 6349: Part 1 for concrete placed in the intertidal zone and for concrete to resist abrasion should be noted.

Amendment No. 4 of BS 6349: Part 1 contains important changes to the recommendations for making durable concrete for use in maritime conditions.

Brickwork: bricks must be frost resistant, and the mortar in intertidal work must be protected by a quick-setting water-proofing compound to prevent damage from the rising tide as described in Clause 65 of BS 6349: Part 1.

Stone masonry: where stone masonry has to be reconstructed rather than replaced by concrete, it is usually desirable to clean off and re-use the stone from the existing wall. If replacement stone is needed, stone from the original quarries should be obtained wherever possible. The selection of new stone should be based on reference 60.

Mortar: old waterfront walls of stone masonry, brickwork and block were usually constructed without movement joints. Mortar that can creep and yield should therefore be specified as described in Chapter 5 of Sowden,[8] which deals comprehensively with mortar as it was used in the past and how it should be specified now for repairs made to existing structures.

Sulphate-resisting Portland cement (SRPC) is necessary where leachates and sulphates are present in the bricks, blocks or stone, or in the groundwater.

Factors of safety and selection of key design parameters

A.1 Introduction

The purpose of this appendix is to examine the significance of some of the parameters used in checking the stability of old waterfront walls and to demonstrate their effect on factors of safety.

A.2 Active pressure coefficients and wall friction

The Coulomb equation for the coefficient of active pressure (Ka) in cohesionless soil is as follows:

$$K_a = \cos\delta \left(\frac{\operatorname{cosec} \alpha \sin(\alpha - \phi')}{\sqrt{[\sin(\alpha + \delta)]} + \sqrt{\left(\frac{[\sin(\phi' + \delta)\sin(\phi' - \beta)}{\sin(\alpha - \beta)]}\right)}} \right)^2 \quad \text{(A1)}$$

where the symbols are as follows (and as Figure A1):

δ = the coefficient of friction between the backfill and the wall

φ′ = the effective angle of shearing resistance of the soil

Where the backfilling is level so that β = 0°, and where the back of the wall is vertical (or with vertical steps) so that α = 90°, the expression for K_a simplifies to:

$$K_a = \cos\delta \left(\frac{\cos\phi'}{\sqrt{\cos\delta} + \sqrt{[\sin(\phi' + \delta)\sin\phi']}} \right)^2 \quad \text{(A2)}$$

and when δ = 0

$$K_a = \tan^2\left(45 - \frac{\phi}{2}\right) \quad \text{(A3)}$$

The active soil pressure normal to the back of the wall, P_{an}, is given by the expression:

$$P_{an'} = P_{v'} \,.\, K_a \quad \text{(A4)}$$

where $P_{v'}$ is the effective vertical pressure in the backfill.

In Figure A2 K_a has been plotted against φ′ for three values of δ. This figure shows that the differences in K_a for the three values of δ shown are small. However, the wall friction has a greater effect on the factors of safety against sliding and overturning than is apparent from Figure A2. This is demonstrated by the example given below.

A.3 Example of wall without water pressure

Figure A3 shows a 6 m high wall without water in front or behind. The wall is shown as being rectangular in order to simplify the example. The point being made applies equally to a wall with a stepped back. Figure A4 shows

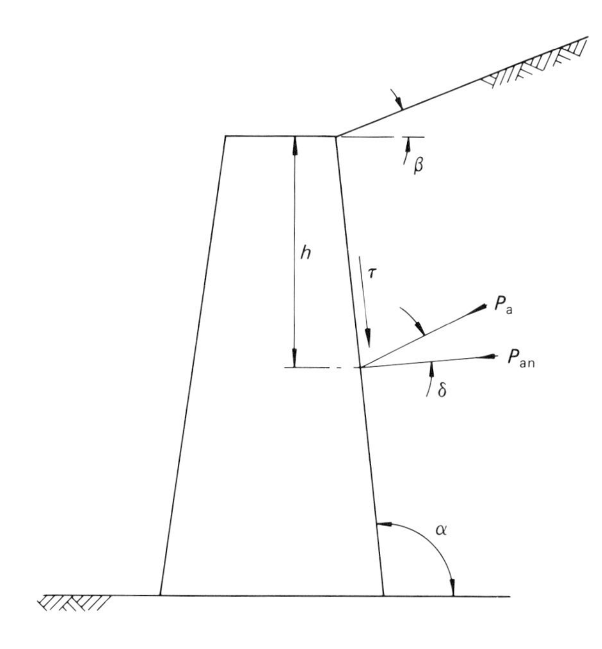

the approximate net horizontal forces applied to the wall by backfilling with various values of ϕ' and δ.

Where $\delta = 0$, no vertical force is applied to the wall. Where δ is greater than zero, a vertical shear force ($P_{an}\tan\delta$) is applied to the back of the wall and increases the resistance of the wall to sliding by increasing the bottom friction under the wall. This component has been subtracted from the active pressure to find the net horizontal force applied to the wall in order to demonstrate the effect of varying wall friction.

For a ϕ' of 25° it can be seen that the disturbing forces are equal to the resistance of

Figure A1 Symbols used in the Coulomb equation for the coefficient of active pressure in cohesionless soil
Source: Draft BS 8002.

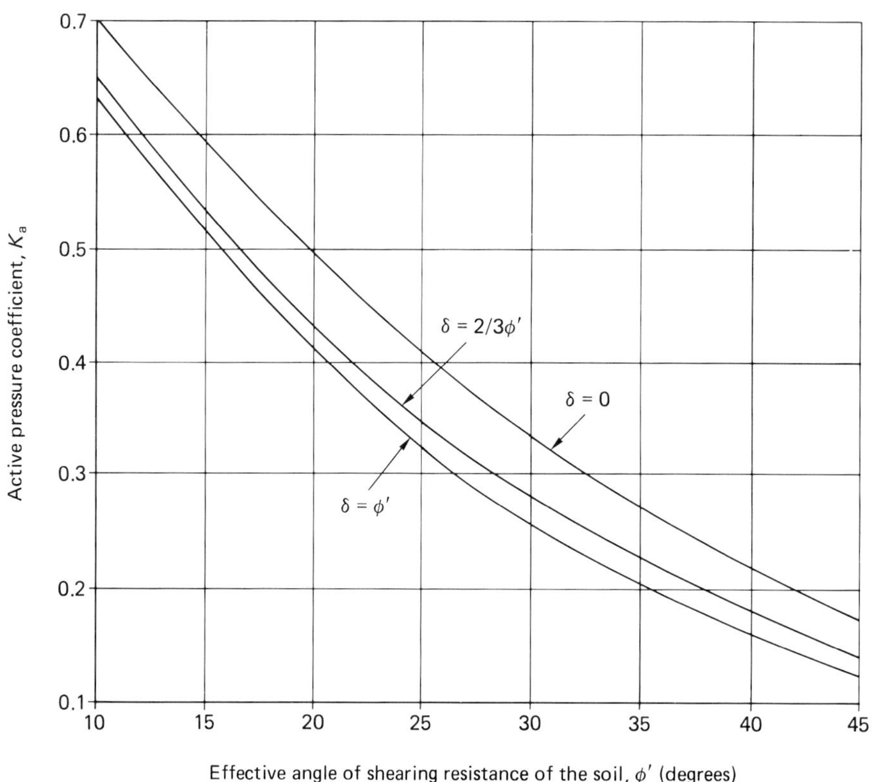

Figure A2 Plot of K_a, against ø′ for three values of δ
Source: Livesey Henderson.

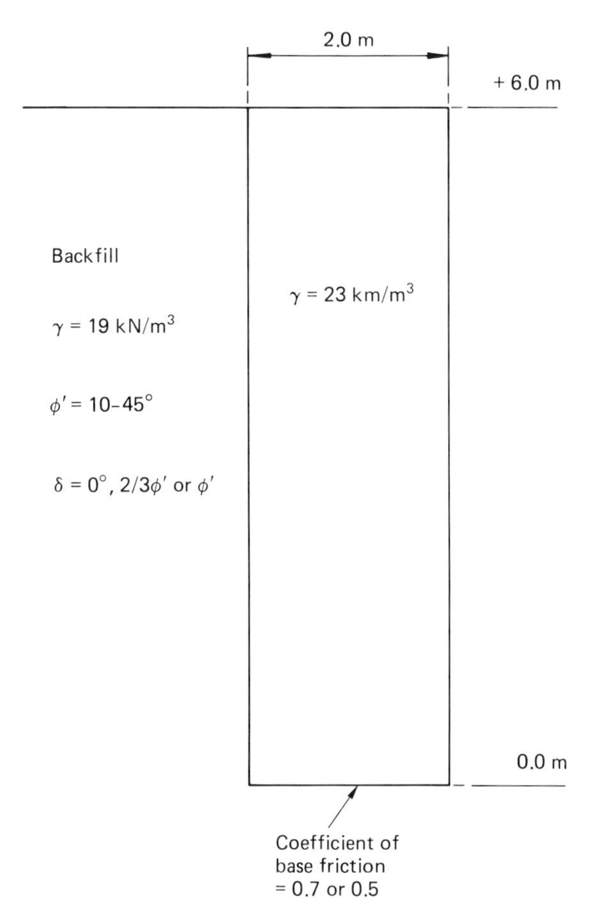

Figure A3 Example of a 6 m high wall without water
Source: Livesey Henderson.

the wall to sliding when the coefficient of wall base friction (μ) is 0.50 (i.e. Factor of Safety = 1.0). If $\delta = \phi'$, then the ratio of wall resistance to net disturbing forces would be 1.84.

Figure A5 shows the overturning and restoring moments for the same wall. Again for the example $\phi' = 25°$, when $\delta = 0$, the disturbing moment is equal to the restoring moment, but when $\delta = \phi'$ (i.e. the normal value for a stepped wall with a 'virtual' back) the restoring moment is 2.3 times the disturbing moment. (When $\phi' = 45°$ and $\delta = \phi'$, the disturbing moment is nil because the resultant of the active soil pressure passes through the toe of the wall.)

Both Figures A4 and A5 demonstrate the importance of determining the correct values of ϕ' and δ, and of being conservative in the choice of δ when in doubt. Figurc A4 also shows that the wall base friction must be correctly determined.

Where concrete has been cast directly on to the ground, where the base is of rough masonry or where stepped or upward sloping bases have been used, the coefficient of base friction will usually be tan ϕ', where ϕ' is the effective angle of shearing resistance of the soil under the base.

A.4 Example of wall with water pressure

Figure A6 shows an example of a wall with water at a higher level in the backfilling than in front of the wall. The water pressure diagram under the base shows the water pressure falling uniformly from the high level at the back to the lower level at the front. Where the wall was founded on clay and the fall of the level in the front had been relatively rapid, it would be prudent to assume that the water pressure under the base was a uniform value corresponding to that at the rear of the wall.

Figure A7 compares the horizontal disturbing and restoring forces for the walls shown in Figures A3 and A6 based on $\delta = ⅔\phi'$. The water affects the wall in two ways: it reduces the effective weight of the wall, and hence its resistance to sliding, and it also increases the horizontal disturbing forces in the backfilling. Figure A8 shows the same comparison for the overturning moments, which are also subject to the dual reductions in stability discussed above.

This example shows the great importance of measuring or estimating the most disadvantageous water levels correctly.

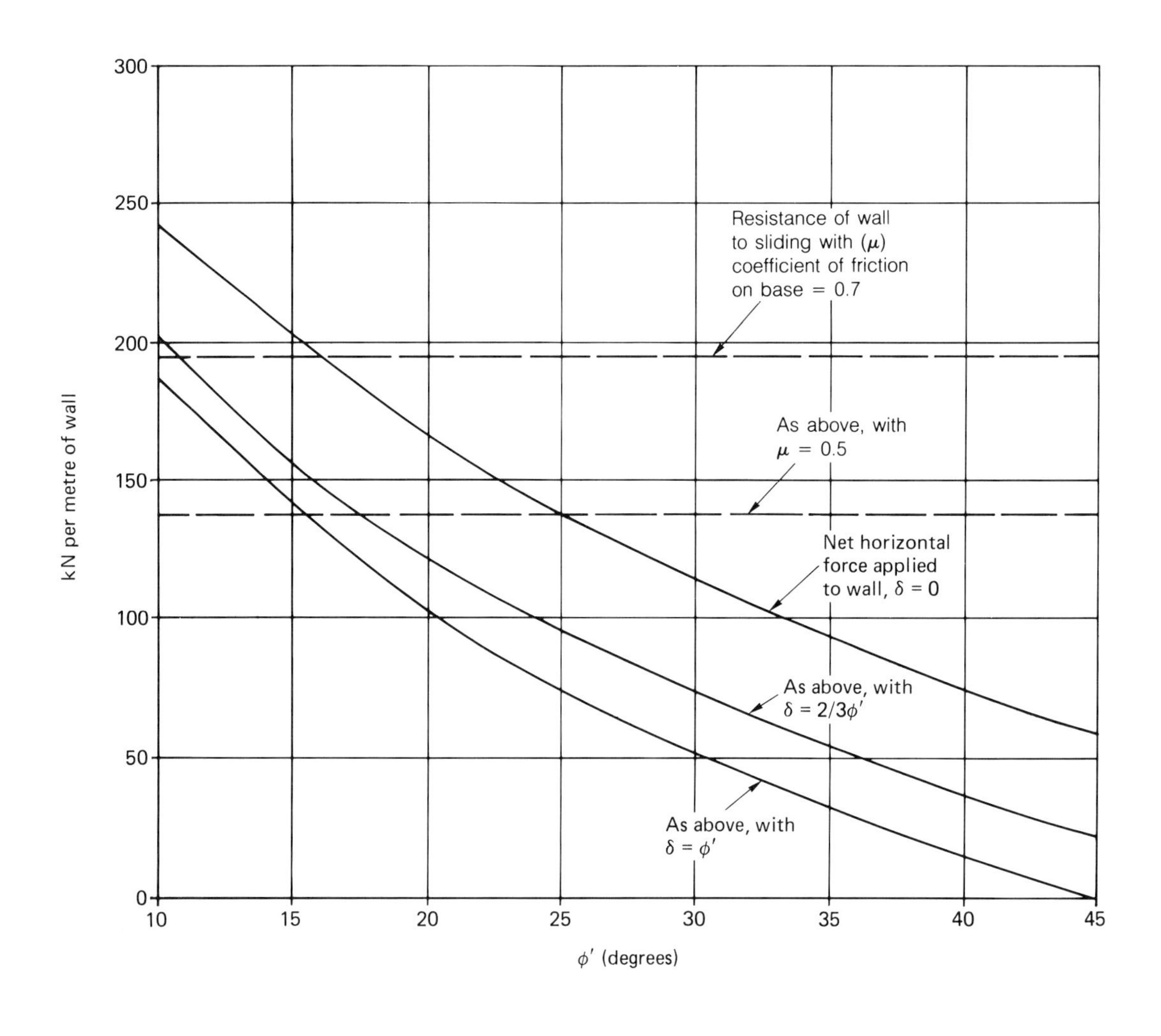

Figure A4 Horizontal forces on the wall shown in Figure A3 for various values of ø′ and δ
Source: Livesey Henderson.

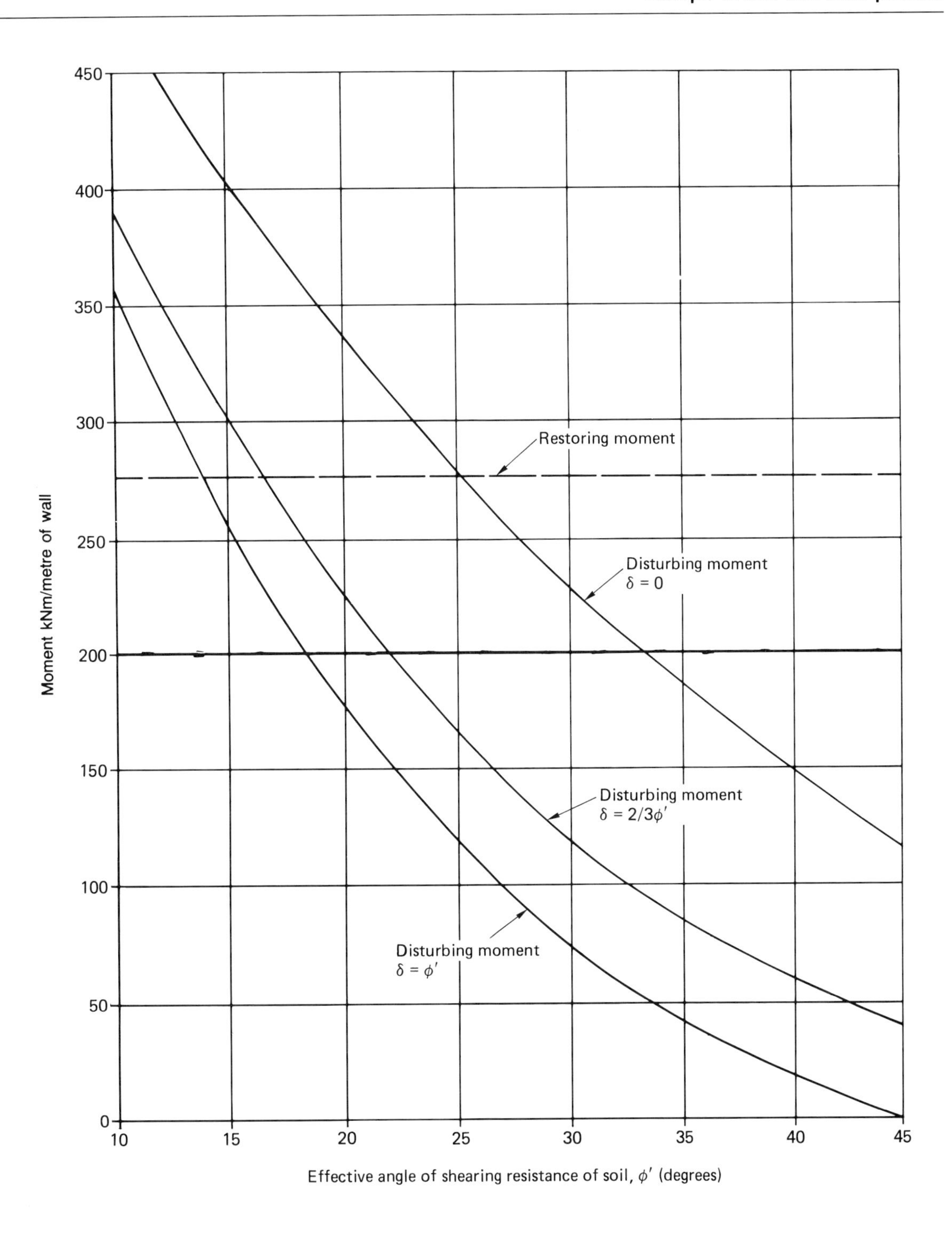

Figure A5 Overturning and restoring moments for the wall shown in Figure A3
Source: Livesey Henderson.

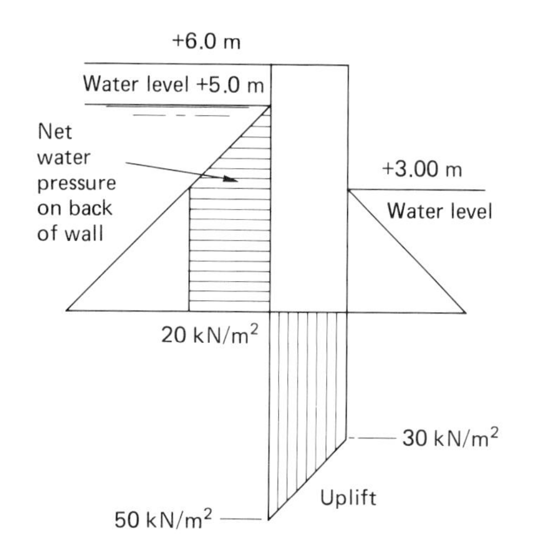

Figure A6 Example of a 6 m high wall with water at a higher level in the backfilling than the front
Source: Livesey Henderson.

Figure A7 Comparison of the horizontal disturbing and restoring forces for the walls shown in Figures A3, A6 and based on $\delta = 2/3\phi'$
Source: Livesey Henderson.

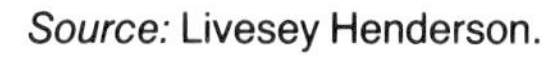

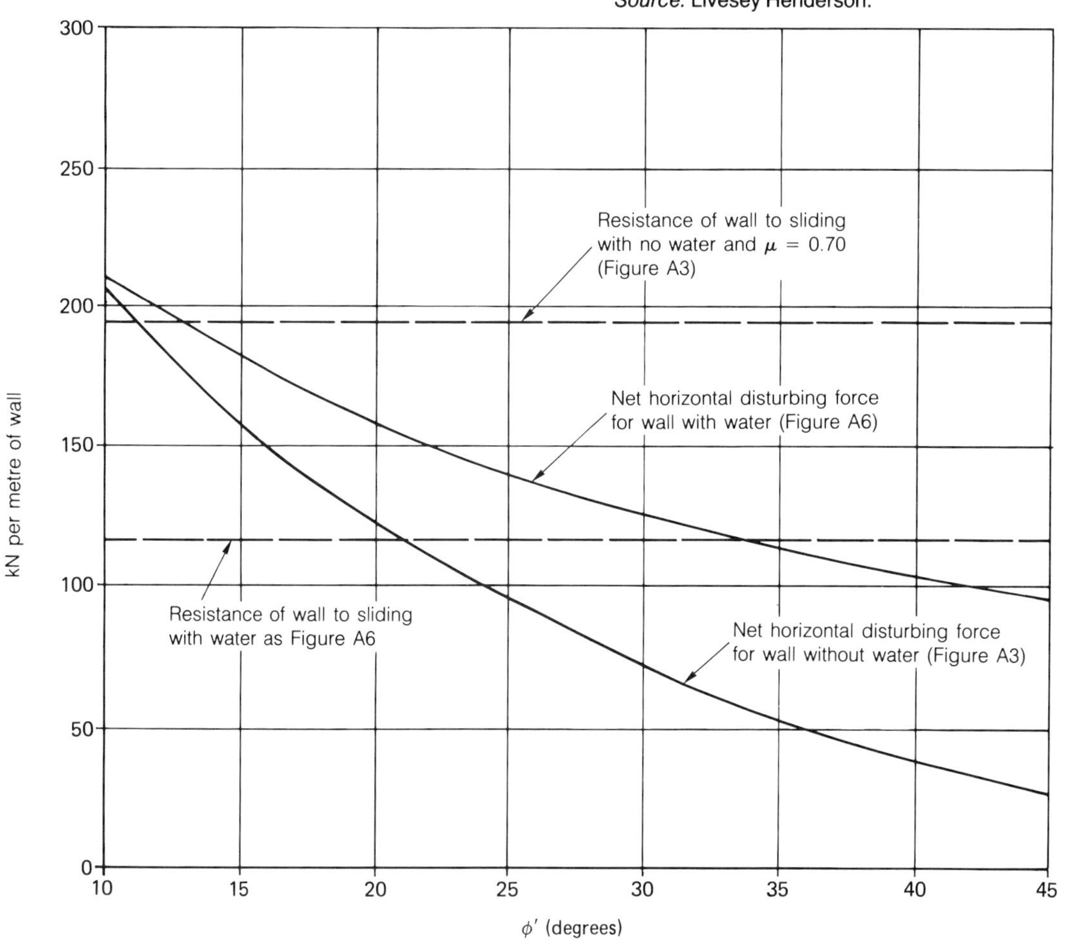

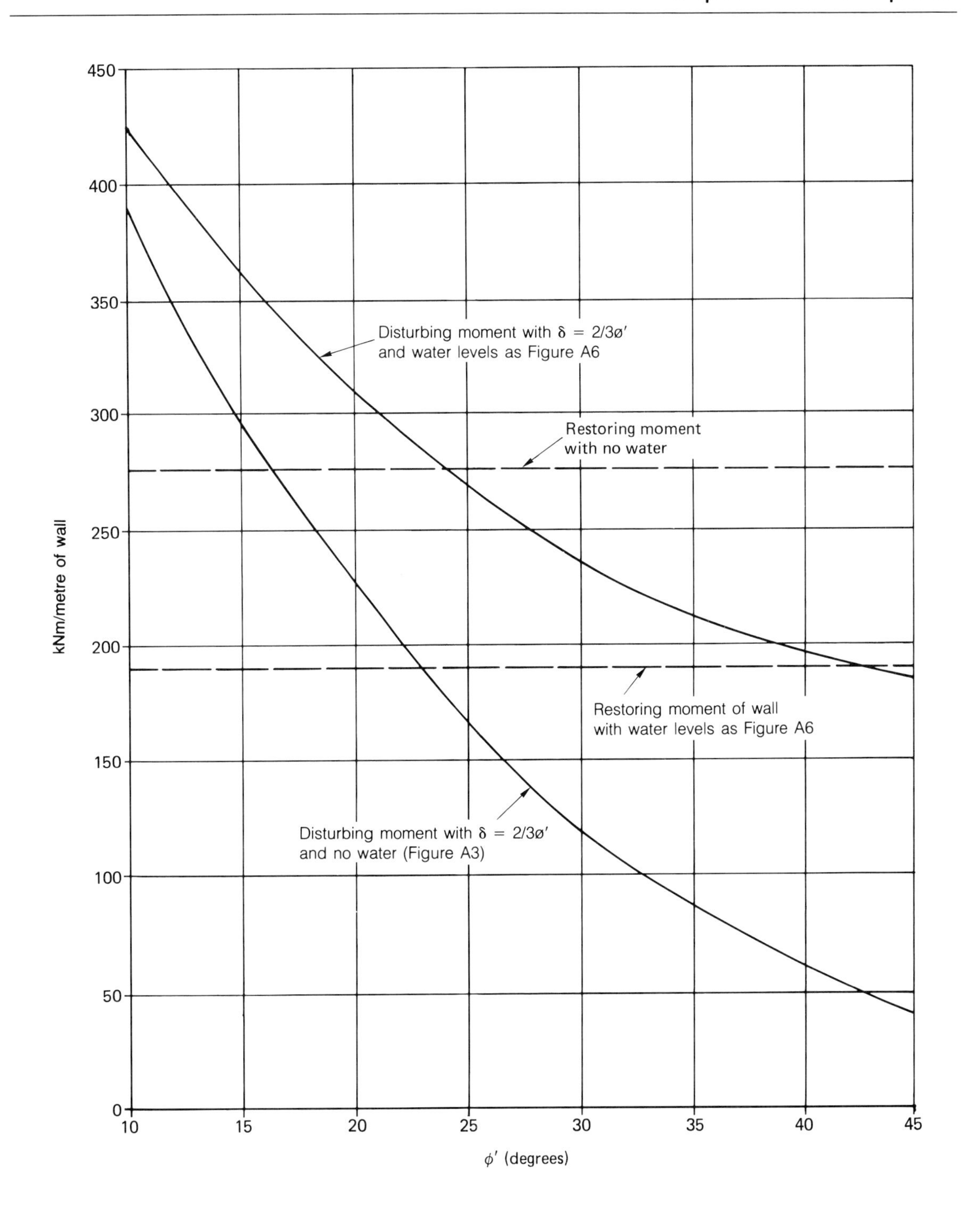

Figure A8 Comparison of overturning moments for the walls shown in Figures A3 and A6
Source: Livesey Henderson.

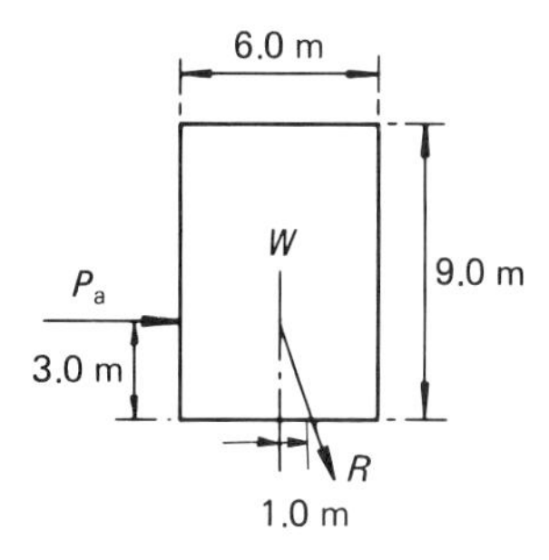

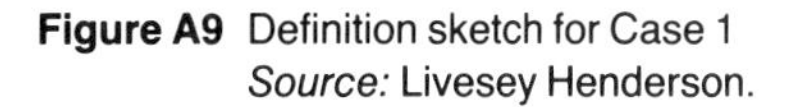

Figure A9 Definition sketch for Case 1
Source: Livesey Henderson.

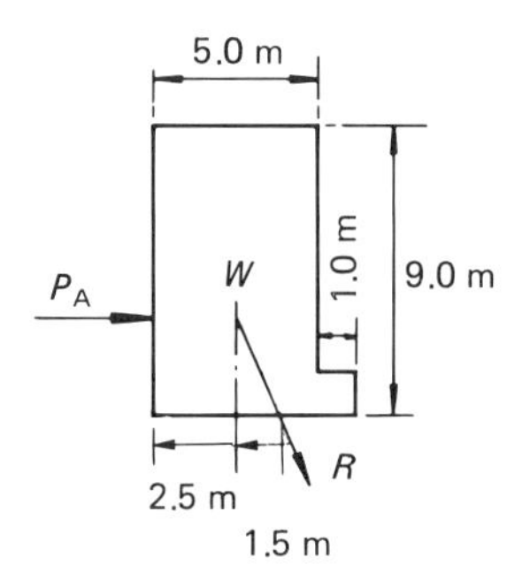

Figure A10 Definition sketch for Case 2
Source: Livesey Henderson.

A.5 Factors of safety against overturning

The draft for BS 8002[31] states, with regard to failure of gravity walls by overturning:

> There are two methods of calculation. The first is to design the wall so that the resultant thrust lies within the middle third of its base; this will ensure that the maximum edge pressure, imposed on the ground by the foundation, never exceeds twice the average pressure due to dead weight of the wall itself plus any surcharge resting directly above it. The second method is to design the wall so that the ratio of resisting to overturning moment is not less than 2 and then calculate the maximum edge pressure under the foundation. This should not exceed the allowable bearing pressure of the soil.

The two methods usually give different answers, but the ratio of one to the other depends on the shape of the wall. This is illustrated by the following examples.

Case 1 (see Figure A9)

Data:

1. wall of rectangular section, 9 m high by 6 m wide
2. resultant of active pressure (P_A) acts horizontally at ⅓ height of wall (3 m above base)
3. weight of wall (W) acts at centre of wall resultant (R) passes through edge of middle third of base (1 m from centre of wall)

P_A, W and R form a triangle of forces so that:

$$W = 3P_A \tag{A5}$$

The ratio of restoring moment to disturbing moment is given by:

$$\frac{W.\,3}{P_A.\,3} \tag{A6}$$

Substituting equation (A5) in this gives a factor of safety against overturning of 3.

Case 2 (See Figure A10)

Data:

1. wall section as shown in Figure A10
2. resultant of active pressure (P_A) acts horizontally at ⅓ height of wall (3 m above base)
3. weight of wall (W) acts at centre of main

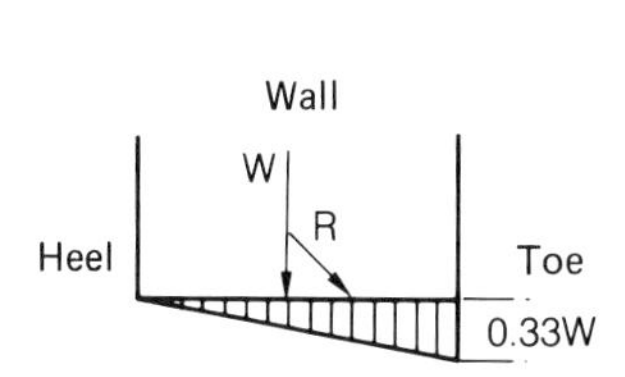

Figure A11 Bearing pressure diagram for Case 1. Factor of safety against overturning = 3
Source: Livesey Henderson.

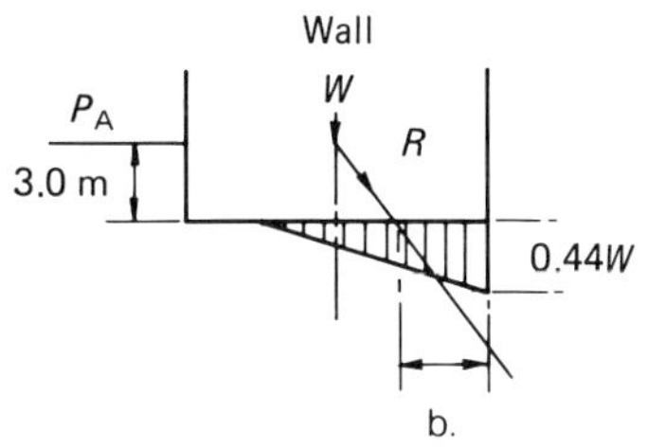

Figure A12 Bearing pressure diagram for factor of safety against overturning = 2
Source: Livesey Henderson.

body of wall (2.5 m from back face)

4. resultant (R) passes through edge of middle third of base (1 m from centre of base and 1.5 m from line of action of W)

P_A, W and R form a triangle of forces so that:

$$\frac{W}{P_A} = \frac{3}{1.5} \text{ or } W = 2P_A \qquad \text{(A7)}$$

The ratio of restoring moment to disturbing moment is given by:

$$\frac{W.\ 3.5}{P_A.\ 3}$$

Substituting equation (A7) in this gives a factor of safety against overturning of 2.33.

Thus in the two simplified cases above both have resultants passing through the edge of the middle third of the base but the factors of safety against overturning are quite different.

An examination of the examples of wall shown in Chapter 1 shows that, for some of the wall shapes, checking that the resultant passes through the middle third of the base is likely to give low factors of safety against overturning. This particularly applies in cases where the wall rests on hard material and the heel of the base has been cut away. In that case, if the width of the remainder of the base is too small, relying on the resultant being within the middle third could be unsafe.

A.6 Bearing pressure

Where the resultant is just within the middle third, the bearing pressure distribution under a wall will be triangular, with zero at the heel and twice the average pressure at the toe. For Case 1 in Section A5 above the pressure distribution would be as follows (see Figure A11):

$$\text{Max. pressure} = \frac{W}{B} \cdot \left\{\frac{1 + 6e}{B}\right\} \qquad \text{(A8)}$$

where:

W = vertical load on base
e = eccentricity of the resultant
B = width of base

This gives a maximum bearing pressure of $0.33W$.

If, in the above case, P_A was greater so that the factor of safety was reduced from 3 to 2 the pressure distribution would have been calculated as follows (see Figure A12):

To make factor of safety against overturning = 2, $W = 2.P_A$

Ratio of perpendicular to base in triangle of forces = 2.

$$\text{Maximum pressure} = \frac{2\,.\,W}{3b} = 0.44W$$

Thus, with a factor of safety of 2 against overturning instead of 3, the maximum bearing pressure has increased by 33% and there is tension at the heel. In most cases, therefore, it is necessary to check both the ratio of the restoring to disturbing moments and the position of the resultant to find the maximum and minimum bearing pressures. It is desirable that there be no tension within gravity walls which do not have reinforcement.

Application of geophysical methods to the investigation of old waterfront walls

B.1 Introduction

One of the many developing fields of science which has proved of recent benefit to the engineer is the use of geophysics in non-destructive investigation. To date, it has been applied to old waterfront walls in only a limited number of cases and the results have been mixed. There is, however, reason to believe that methods may be refined in the future to provide the investigative engineer with a more useful array of techniques at his disposal. This appendix gives a brief review of the geophysical techniques available, their applicability to the investigation of old waterfront walls and the progress which has been made in developing them for this use.

Two of the major difficulties for the engineer faced with an investigation of an old waterfront wall are the determination of wall shape and the assessment of the internal state of the structure. Whilst the submerged face of a wall may be detected by means of marine investigation techniques, that part of the wall hidden by the ground is less easy to configure. The internal state of the wall is also difficult to assess since it may be extremely variable and there is an economical limit to the amount of direct access or sampling that can be carried out.

B.2 Geophysical techniques

For the purposes of this review the geophysical techniques may be broken down into the following categories:

1. **seismic**; where a vibration or shock wave is emitted from a source and the behaviour of the transmitted pulse is monitored as it passes through, and is reflected off or refracted along, the various carrying media within the structure;
2. **electrical**; where a direct electrical current is passed through the structure and variations in resistance are measured as a function of depth;
3. **magnetic**; where the Earth's magnetic field is measured and interpretation of local conditions are made from any anomalies detected in relation to a local base station;
4. **gravitational**; where the Earth's gravitational field is measured and interpretation of local conditions are made from any anomalies detected;
5. **electromagnetic**; where continuous electromagnetic waves are transmitted from a known source and the resultant electromagnetic field is measured after the waves have passed through the structure. The resultant field indicates the nature and disposition of the various media present;
6. **radar**; where pulsed electromagnetic waves are reflected at the boundaries of the media of the structure through which they pass. A contrast in electrical properties is necessary to produce a measurable reflected pulse;

7. **well logging**; where instrumentation is introduced into the body of the structure through an already installed borehole.

These techniques may be used by deploying equipment on the ground or wall surface, or in some instances, from within boreholes inside or alongside the structure.

Data collected from a geophysical survey represents only the variation of a particular geophysical parameter over the survey area. It requires interpretation to be related to the structure being investigated. Interpretation is only possible and reliable when some of the measurements recorded can be correlated with a known portion of the structure or can be related to a previous example. Some other form of site investigation, such as trial pits or boreholes, is thus a prerequisite for satisfactory interpretation.

The value of a geophysical survey depends to a great extent on the clarity of the geophysical signal being measured and the extent to which this signal has been modified by the various media through which it has passed. The following factors are of particular importance:

1. penetration, which is affected by the rate at which the signal decays in each medium through which it passes;
2. resolution, which is a function of signal frequency and measurement related to position and thickness of medium;
3. signal-to-noise ratio, which is a function of the geophysical and man-made environment;
4. contrast in physical properties, which determines the clarity and configuration of the measured signal as recorded on trace.

Old waterfront walls are not ideal subjects for geophysical investigation. The internal fabric of the wall structures is frequently riddled with discontinuities, is variable in composition and saturated with either fresh or saline water. All these characteristics contribute to difficulties in the application of geophysical methods. In addition, these methods are normally used to investigate subjects on a macro scale, where depths of 30 to 1000 m are common. For waterfront wall investigations it is the material from 0 to 30 m which is of most interest to the engineer.

B.3 Surface methods

B.3.1 Seismic methods

Seismic exploration is based on the generation of seismic waves on the ground surface and the measurement of the time taken by the waves to travel from the source through the ground to a series of geophones, which detect the ground motion. These are usually deployed along a straight line from the source. The seismic energy can be generated by a number of means including a hammer and plate and falling weight.

Apart from the seismic energy travelling directly through the ground to the geophone array, two other main paths are possible:

1. the reflected or head wave travels horizontally along the interface between two media;
2. the reflected wave travels vertically downwards and is reflected back to the surface by the interfaces between the various media.

The travel time of the seismic energy through the ground is controlled by the physical properties of the media and the attitudes of the various strata beneath the geophone array. Simple interpretation procedures are normally adopted in civil engineering applications but in the oil industry all the data is recorded on magnetic tape for subsequent computer processing. In this way the received signals can be enhanced in relation to the environmental noise levels and the information displayed such that rapid geological interpretation can be carried out.

Seismic refraction

The seismic refraction method is widely used in civil engineering to determine depth of bedrock, since at relatively shallow depths head wave pulses are not masked by other disturbances. Its main disadvantage is that velocity inversions in the substrata can give rise to incorrect interpretation.

This method has been used extensively to investigate the internal structure of large dams and the *in situ* state of the foundation rock, as well as the effects of grouting. It has also been employed to identify cavities and fracture zones in a rock mass.

There are at present no recorded instances of this technique being used for the investigation of old waterfront walls. It could, however, possibly be developed to assist in the identification of voids and fractured zones in such structures.

Seismic reflection

The seismic reflection method does not suffer from the velocity inversion problem associated with refraction but is limited at shallow depths due to the masking of relatively weak reflected signals by the remaining effects of the direct wave. Although the reflection technique is widely used in geophysical work its use in shallow depths has been severely limited. There are no recorded instances of it being used to investigate waterfront walls, where it might assist in the definition of hidden wall shapes. To be of any value, the method would need to be adapted by the introduction of high-frequency seismic sources, multiple-array receivers and computer-assisted interpretation.

Seismic resonance

Seismic resonance, triggered by seismic waves, has been detected in the form of waves emanating from underground voids. The resonance phenomenon is not well understood and attempts to use it for the detection of voids have been of limited value, particularly where the void has been filled with water and/or other fill material. There are no practical results of this technique available related to waterfront walls although some research is in progress.

B.3.2 Electrical method

Electrical soundings are suited to horizontal stratified media, since the spatial distribution of the direct or alternating electrical current in the ground, and hence the depth of investigation, depends on the configuration of the conduction and the spacing of the electrodes. When using a standard Wenner array the depth of investigation increases with the current electrode spacing and this gives rise to a pseudo-electric section which can be related to the geological structure beneath the survey line. It is possible to postulate a geological model from which an electrical section can be computed to fit the experimental data.

An alternative approach is to move the electrodes over the survey area to delineate lateral variations using a constant electrode array. In this way it is possible to locate discontinuities, such as geological boundaries, dykes, in-filled sinkholes, or major drainage zones.

Resistivity surveys are used to detect jointing and fissuring in the ground and the presence of cavities and mineworkings. Problems are encountered when the natural background variations in resistivity in the ground are large enough to mask the value of the local variations due to a sub-surface feature. The method would be highly sensitive to the ingress of sea water to the internal portions of a wall and the washout of material. Changes in the degree of compaction of a wall's core would change the porosity of the structure and hence its resistivity. General degradation of the structure by weathering and/or fracturing would also change its resistivity.

This method should thus be useful for detecting structural or material defects within a wall and particularly when the defects are associated with the ingress of sea water. Very little work has been carried out in this field but some encouraging results have been obtained from one case history, where good penetration of the electrical current was reported.

B.3.3 Magnetic method

In a magnetic survey variations in the Earth's magnetic field are measured. The main requirements for achieving an accurate magnetic map of the area of interest is that the diurnal variation in the Earth's magnetic field should be monitored. This is usually carried out by means of a recording magnetometer base station so that the actual magnetic readings obtained during the survey can be reduced to a common datum.

To be of any value, normally magnetic rocks must be present in the survey area but possible magnetic anomalies on an engineering site can be man-made targets, such as brick-lined shafts or the remains of ancient dwellings. The method has been used extensively for the detection of abandoned mine shafts and adits, where no records of their position exist, and magnetometer surveys are often carried out over water to assist in the detection of ferrous debris, particularly relating to wrecks and munitions.

No examples of the use of the magnetic method in the assessment of waterfront structures has been found. However, it could be of value in determining the position of known, or suspected, iron and steel features buried within or behind a wall, such as tie rods, sheet piles, etc.

B.3.4 Gravity method

Gravity surveying involves the measurement of variations in the Earth's gravitational field. After the necessary corrections to the field data, gravity anomaly maps are made which represent variations in the mass distribution in rock mass below the survey area. Until recently little use had been made of the method in engineering investigation except for the detection of large masses over extensive areas. With the availability of more sensitive gravity meters, which are capable of measuring changes in the gravity field to 5 gal, the method could be used to locate large voids within a marine structure. However, this would require very accurate and carefully planned surveys.

B.3.5 Electromagnetic methods

Electromagnetic methods employ a source of electromagnetic energy, such as a coil carrying alternating current, which is introduced into the ground by inductive coupling. The receiver also detects its signal by induction.

Terrain-conductivity meter

This meter uses low-frequency electromagnetic waves to detect changes in the conductivity of the ground. It is commonly employed in engineering studios to assess the integrity of the ground in the same way as a resistivity survey (see Section B.3.2).

Ground conductivity is directly related to the presence of water within the ground and is thus appropriate for investigating walls which might be suffering from problems associated with the ingress of water. However, the presence of steel rods, metal fencing, etc. will seriously affect the results and thus care must be exercised in the selection of this technique (see the case history – Section B.6.2).

Ground-probing radar

Ground-probing radar uses the reflection of pulses of electromagnetic energy to detect interfaces in the medium which have contrasting physical properties. Its use is very similar to that of the seismic reflection technique described above (see Section B.3.1). It has been used to determine the thickness of permafrost, to detect fractures in rock salt and granite rock masses, to investigate shallow mine workings and in archaeological investigations.

This method has been used on a number of

waterfront walls in the UK. The most significant feature of the three case histories presented in Section B.6.1 is the practical problem of interpreting the radar records.

Several practical problems arise with ground-probing radar, such as overcoming the effects of reinforcing rods, and the difficulty of gaining sufficient penetration while still achieving good resolution. In particular, it should be noted that electromagnetic energy is highly attenuated in materials containing saline water. Lack of control at the interface with some of the targets has also been a problem.

In spite of these difficulties, the case histories mentioned, and other published work in the scientific literature, tend to confirm that ground-probing radar may become increasingly used in the investigation of waterfront walls. Of particular significance is the proposed introduction of high-power antennae and the development of advanced processing techniques in the analysis of radar records, which are now also being recorded in digital form.

B.4 Borehole methods

Boreholes give the geophysicist access to the mass being investigated and allow the deployment of sondes of various types. Since many site investigations include the installation of boreholes there are often opportunities for these to be considered for geophysical techniques.

B.4.1 Cross-hole methods

In the cross-hole method the seismic or electromagnetic source is lowered down one borehole and the receiver is lowered down another. The method has been used successfully, using a high-frequency seismic source, to identify voids, tunnels and other geological discontinuities. It has not been used in the investigation of waterfront walls although a similar technique has been employed for an investigation at Ramsgate, where the seismic measurements were made through a masonry pier (see Section B.6.3).

One problem which would be encountered in the detection of small discontinuities in a wall would be the number of boreholes required. Thus while this method might be feasible for the detailed study of a small extent of a wall it is unlikely to be justifiable for an extensive survey. Some improvement in effectiveness of cross-hole methods might be gained by the use of computerized tomographical modelling techniques, similar to those used in the medical field for X-ray imaging.

B.4.2 Surface-to-borehole methods

The surface-to-borehole method is used in an advanced form in the oil exploration field and is known as vertical seismic profiling. The seismic source is moved around the top of a borehole at several different radii and the signals received in the borehole give information on a series of cones in the mass being investigated. Although this method would appear to be fairly promising there is at present no record of it having been tried for the investigation of waterfront walls.

B.5 Well-logging methods

There are many tests that can be carried out on the surrounding ground from within a borehole. Most of these have been developed for the oil industry but there is an increasing interest in them for civil engineering applications. The three methods most likely to be of use in the investigation of waterfront walls are the natural gamma, gamma-gamma and neutron-neutron logs. However there are no records of their having been deployed yet for this purpose.

B.5.1 Natural gamma logging

The natural gamma log is a measurement of the natural radioactivity of a formation and is useful for lithological correlation.

B.5.2 Gamma-gamma logging

This log measures the intensity of gamma radiation, emitted from a radioactive source, as it is backscattered and attenuated within the surrounding mass. The recorded count rate is directly proportional to the density of the surrounding material. The sonde is lowered down a borehole. The log measures the effective density of the formation close to the borehole and this includes the effects of the material matrix and the contained fluids.

B.5.3 Neutron-neutron logging

The neutron-neutron sonde is used to measure formation porosity or moisture content since it responds primarily to the amount of hydrogen in the formation. It is employed in the same way as the gamma-gamma sonde. Results from the neutron-neutron log are sometimes used to give a Neutron Fracture Index.

B.6 Case histories

B.6.1 Ground-probing radar

Promenade between Boscombe and Southborne (Bournemouth)

In a survey of the promenade between Southborne and Bournemouth, Hampshire, a section of the sea wall had collapsed and the Council were concerned that similar subsidence might occur in other areas if voids were present beneath the promenade surface. A ground-probing radar survey was carried out over a 4.15 km length of the promenade and this was augmented by a drilling programme to examine anomalous areas identified in the radar records.

The ground-probing radar survey was carried out with a 500 MHz antenna. While greater penetration could be achieved with a 300 MHz antenna it was found that the radar records were affected by reflection from extraneous features such as the nearby fence, lamp posts and flagpoles. A prominent feature on the records was the identification of a conductive ash layer beneath the asphalt which gave rise to multiple reflections or 'ringing'. The bottom of the asphalt layer was also identifiable together with changes in the fill material along the promenade.

There was no evidence on the radar records of any reflection patterns normally associated with the presence of cavities. Subsequent drilling and sampling of anomalous areas which had been located did not find any voiding.

Promenade between Mablethorpe and Skegness (Anglian Water)

On the coast between Mablethorpe and Skegness, Lincolnshire, a large void, with a volume of some 300 m^3, had been discovered beneath the hard facing of the sea defences. This was thought to have undermined the wave return wall and much of the promenade, and it was considered necessary to identify the location of any other large voids. A ground-probing radar survey was carried out on the promenade using a 120 MHz antenna.

Two types of surfacing have been used on this promenade; tarmac and concrete slabs. In the areas of concrete slab surfacing it was found that the density of the reinforcing within the slabs was too great to allow delineation of any sub-surface features by ground-probing radar, and thus the survey was concentrated on the section of the promenade surfaced with tarmac.

From the radar records it was possible to identify a large number of features. Of particular interest was the record shown in Figure B.1, which includes the location of the large void already discovered. This record exhibits a number of elliptical features which usually result from diffraction of the electromagnetic energy by a point source, and are indicative of the presence of voids.

A trial was also carried out on the steeply sloping seaward face of the defences using both 120 MHz and 500 MHz antennae. In this

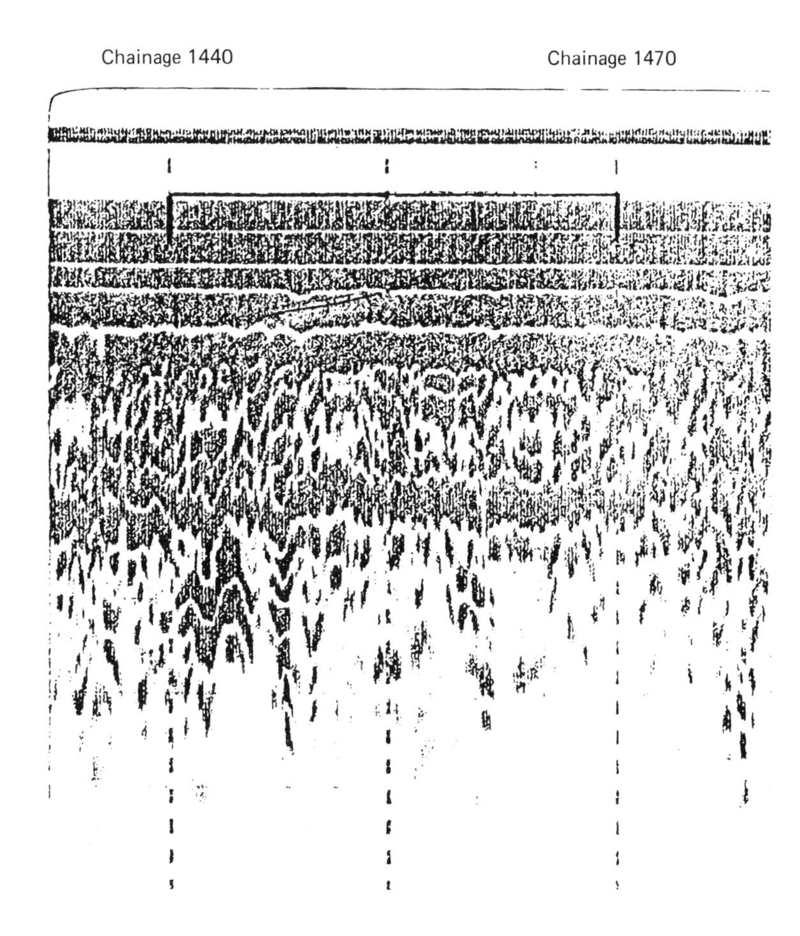

Figure B1 Radar reflection profile along 100 m of sea wall between Mablethorpe and Skegness showing classic void characteristics (Courtesy of Anglian Water).

location the radar records were complicated by the construction methods used, since in many cases new defences were built over the top of older ones and this resulted in complex radar characteristics and many interfering reflections. It was also difficult to sweep over the area of interest with the equipment supplied.

Sea defences at Aldeburgh

A ground-probing radar survey was carried out along the reinforced concrete sea defence wall at Aldeburgh, Suffolk, to investigate possible voiding. The concrete sea wall had collapsed due to the washout of the fill material from beneath it. The sea wall consisted of a number of reinforced concrete slabs in a stepped structure with a sloping concrete apron at the bottom of the steps extending to beach level.

The radar survey used a 500 MHz antenna and a number of survey lines were run along the sea wall. The presence of reinforcement within the concrete of the sea wall prevented the use of a lower-frequency antenna and the depth of penetration was limited to 1 to 2 m. It was anticipated that any voiding would occur directly beneath the 350 to 450 mm thick slabs but in the event no indications of voiding were observed. However, the presence of reinforcing bars within the concrete was observed on the records and also, in some cases, the interface between the concrete slab and the underlying shingle.

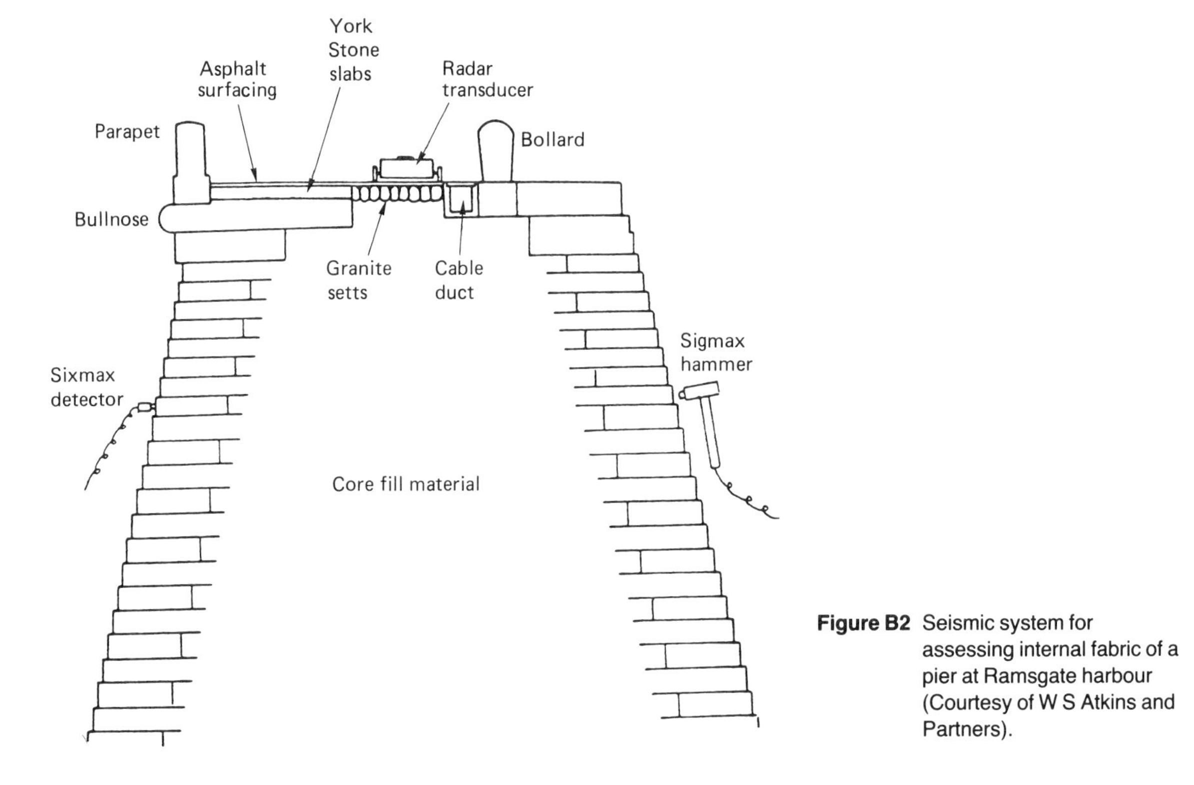

Figure B2 Seismic system for assessing internal fabric of a pier at Ramsgate harbour (Courtesy of W S Atkins and Partners).

B.6.2 Ground conductivity

This case is part of that described in Section B.6.1 above. Ground conductivity measurements were also made on the Mablethorpe to Skegness site and, as with the radar survey, the sections of promenade surfaced with reinforced concrete slabs adversely affected the performance of the electromagnetic equipment.

Conductivity readings were continuously monitored along the sea wall at 15 m intervals. Lateral inhomogeneities were delineated by recording both longitudinally and transversely oriented dipole positions at each survey point where the width of promenade allowed. The data was plotted on a continuous record of conductivity along the length of the promenade and two areas were covered with a 3 m × 2.5 m grid to delineate two large anomalous zones. These areas were later drilled and found to be underlain by a hard mass, which was probably concrete.

B.6.3 Seismic methods

A survey involving seismic methods was carried out on the East Pier at Ramsgate, Kent. The pier was believed to have been constructed of stone facing blocks, and a stone paving over a chalk fill material known as 'lime concrete'. At the time of the survey it was considered that there was a possibility of voiding existing in the core material as a result of wave washout through the open block joints.

The seismic method used was the transmission of a seismic pulse, generated by an instrumented hammer, through the structure to an accelerometer on the opposite face (see Figure B.2). Provided the distance between the source and the receiver is known, the compressional wave velocity can be calculated from the measured travel time at each survey position. The compressional wave velocity is a function of the state of the structure's core since any cracking or voiding attenuates and delays the seismic pulse to give a reduced velocity value. Readings were taken over a grid with a horizontal spacing of 1 m and a vertical spacing of two stone blocks.

Considerable variation in compressional wave velocity was observed over the length of the structure. This was attributed to three possible clauses:

1. variation in thickness of the facing material;
2. changes in the quality of the fill material;
3. deterioration of the fill material by erosion and the creation of voids.

It was concluded that all three factors were present. High-velocity values (> km/s) related to good-quality fill material. Lower-velocity values corresponded to a reduced fill quality. The presence of voids or major cracks resulted in the lowest velocities or a total loss of signal. It was concluded that the contour plots of apparent velocity were useful indications of the state of the internal portions of the pier.

This appendix has been compiled from information supplied by the British Geological Survey.

Performance of old waterfront walls – survey results

C.1 Introduction

Questionnaires were sent to wall owners, and those responsible for their upkeep, consultants, contractors, surveyors, suppliers of materials, site investigation firms and geophysical contractors. The objectives of the questionnaires were to assist in assessing the extent and types of old waterfront walls, the present performance of the walls, the nature and scale of the problems of maintenance and rehabilitation. The chief value of the answers to the questionnaires was in identifying a large number of examples of types of walls, the nature of problems which have occurred and the methods used for addressing the problems. The statistics on such items as numbers, length, percentages of different types and uses and costs were not particularly meaningful due to the following factors:

1. it was not practicable for those bodies who were responsible for a large number of walls to abstract the information;
2. in some cases such as bridge abutments the information was included with other items;
3. in other cases the data had not been recorded;
4. cost figures divided between maintenance and rehabilitation were sometimes not available;
5. the response to the questionnaire was uneven.

C.2 Summary of data

1. There were about 340 replies (including 'nil' returns) which was a return rate of just over 40%.
2. The numbers and types of individual walls reported were as follows:

	Number	%
Quay and dock	92	32
Breakwaters	23	8
Coast protection	57	20
Retaining	72	25
Flood defence	19	7
Bridge abutments	10	3
Other	16	5
	289	100

This distribution is not representative: in particular, bridge abutments were under-reported.

3. Reported amounts spent on maintenance and rehabilitation were as follows:

	Maintenance	Rehabilitation
	£ millions	
1983	1.2	2.3
1984	1.3	3.2
1985	1.6	8.2
1986	1.9	4.4
1987	2.8	5.5

4. Figures on state of repair, etc. were as follows:

	Metres
No significant defects	212 406
Requires significant maintenance	156 023
Requires rehabilitation	112 808
Rehabilitation in last 5 years	27 048*
Rehabilitation planned	32 834

* Average cost £878 per metre

5. The materials of wall construction were reported as follows:

Material	Number of walls	%
Masonry	107	45
Concrete/block	18	8
Brick	30	13
Mass concrete	83	34
	238	100

6. The circumstances affecting wall performance are summarized below.

Circumstances affecting wall performance	Quay or dock	Breakwater	Retaining	Coast protection	Flood defence	Bridge abutment	Other
Exceedence of original design loads							
excess horizontal or vertical loading	2	1	1	0	0	0	1
ship impact	8	1	2	1	0	0	1
Man-made changes to wall or environment							
excavation at toe	6	1	2	0	0	1	1
erosion due to propeller wash	8	1	2	1	0	0	1
effect of new adjacent structure	2	0	3	2	2	0	0
bomb damage (e.g. London Docks)	1	0	0	0	0	0	1
redevelopment of wall site	13	1	4	1	4	0	1
Natural changes in environment reduction in level in front of wall due to wave/current erosion, littoral drift, etc.	5	4	11	15	3	1	1
changes in water level in front of wall	2	0	0	0	0	0	0
increase in exposure to waves leading to overtoppping, loss of facing, etc.	0	0	1	1	0	0	0
forces imposed by vegetation	3	0	4	1	0	0	1
storm damage	5	3	3	7	1	0	1
Deterioration of the structure of the wall							
washout of hearting	7	7	3	3	1	1	2
washout of material behind wall	8	3	6	7	2	1	2
loss of wedges	0	1	0	1	1	0	0
cracking of concrete, masonry brickwork and concrete	9	3	7	10	2	5	1
abrasion	1	1	6	5	0	0	2
washout of pointing	6	4	4	4	1	3	3
acid water attack on mortar and concrete	1	0	1	1	0	0	1
decay of timber	2	1	1	1	1	1	0
corrosion of tie rods	1	2	1	1	1	0	1
concrete deterioration	1	1	6	9	1	0	0
Geotechnical							
softening of materials in founding layers	1	1	1	0	0	1	0
excess pore-water pressures	1	0	0	0	0	0	0
inadequate factors of safety in the original design	6	0	3	0	0	0	1

7. The various maintenance and rehabilitation methods are summarized below.

Maintenance and rehabilitation methods							
Modifying loads on structures							
redefining standard of service	3	0	0	0	0	0	0
reduction of load on back of wall, reinforced earth, etc.	4	1	3	1	1	0	1
Improving stability to wall toe							
rock revetment	2	1	2	3	1	2	0
rock loading of toe	0	0	0	0	0	0	1
concrete aprons	2	1	5	8	1	0	1

concrete toe	6	3	5	4	1	0	1
sheet-piling at toe	2	2	2	3	2	1	2
raising beach/seabed level	0	0	0	0	0	0	1
Increasing wall strength							
ground and rock anchors	11	2	4	1	1	1	2
piling behind, under and/or through wall	11	0	7	0	3	0	2
masonry bonding, stitching, dowelling and wedging	7	6	3	1	1	2	4
reconstruction whilst keeping original appearance	6	1	4	2	1	0	2
partial reconstruction	8	0	9	7	3	3	3
Filling voids							
repointing, including pressure pointing	14	4	12	10	0	2	5
grouting of wall	10	7	9	7	1	6	7
joint sealers	0	1	2	3	2	1	0
bagwork	3	1	1	0	0	2	0
voids filled with pumped concrete	4	2	1	1	0	0	1
Improvement to external surfaces of wall							
patching of concrete walls above water	3	0	4	2	0	1	0
patching of walls below water	3	0	4	2	0	1	0
replacement of brick and stone	5	1	6	5	0	1	1
additional skin of brickwork	5	1	1	1	1	0	0
waterproofing of top of wall	4	2	1	2	1	0	1
guniting	2	0	7	6	1	0	2
epoxy resin/polymers to face of wall	1	0	1	1	0	0	0
replacement of wall by new structure	4	1	7	5	2	0	0

Notes: Some walls have been classified as more than one type. Some walls have been recorded as having more than one type of problem or method of repair.

References

1. Haverfield, F. (1924) *The Roman Occupation of Britain*. Clarendon Press, Oxford.
2. Jackson, G. (1983) *The History and Archaeology of Ports*. Worlds Work, London.
3. Hadfield, C. (1974) *British Canals. An Illustrated History*. David and Charles, Newton Abbot.
4. Swann, D. (1959) *English Docks and Harbours 1660 to 1830*. PhD thesis, Leeds University.
5. Skempton, A.W. (1987) *British Civil Engineering 1640-1840: A Bibliography of Contemporary Printed Reports, Plans and Books*. Mansell Publishing Limited, London.
6. Milne, G. (1981) *Medieval riverfront reclamation in London*. Council for British Archaeology Research Report No. 41.
7. Marsden, P. (1981) *Early shipping and the waterfronts of London*. Council for British Archaeology Report No. 41.
8. Sowden, A.M. (1990) *Maintenance of Brick and Stone Masonry Structures*. E. and F.N. Spon, London.
9. CIRIA (1986) *Structural Renovation of Traditional Buildings*. Report 111.
10. Sganzin, J.-M. and Reibell, F. (eds) (1839) *Programme ou resumé des leçons d'un cours de construction, avec des applications tirées specialement de l'art de l'ingenieur des ponts et chaussées*, (4th edn.) Carilian-Goeury et V. Dalmont, Paris.
11. Kinipple, W.R. (1897) *Greenock Harbour. Proc. Inst. Civil Eng.*, **130**, 276–97.
12. Skempton, A.W. (ed.) (1981) *John Smeaton FRS*. Thomas Telford, London.
13. Gloucester and Berkeley Canal Co. (1795) *Mortar Specification*. Public Record Office, Kew, Ref 829/17.
14. Sutherland, R.J.M. (1988) How much did workmanship affect the robustness and load bearing capacity of old masonry walls? *Masonry Int.*, **Vol. 2.** No. 1, 1–34.
15. British Standards Institution. BS 5628: Part 1: 1978 (1985). *Structural use of unreinforced masonry*.
16. Chrimes, M. (1986) Pile foundations and piling: an historical introduction. Paper presented at a History Group meeting, Instn. Struct. Engnrs.
17. Creasy, E. (1856) *An Encyclopedia of Civil Engineering, historical, theoretical and practical*. Longman, Brown, Green and Longmans, London.
18. Stoney, B.B. (1874) On construction of harbour and marine works with artificial blocks of large size. *Proc. Instn. Civil Engnrs*, **XXXVII**, 332–80.
19. Buchan, A.R. (1985) Peterhead, Scotland's 100-year harbour of refuge. Discussion. *Proc. Instn. Civil Engrs*, Part 1, **78**, Oct., 1237–9.
20. Posford, Pavry & Partners and Lewis & Duvivier (1986) *Sea Walls. Survey of Performance and Design Practice*. CIRIA, Technical Note 125.

21. Home, G. (1931) *Old London Bridge*, John Lane, The Bodley Head, London.
22. Labelye, C. (1739) *Short Account for Building a Bridge at Westminster.*
23. Salman, T.R. (1878) The River Lagan and Harbour of Belfast. *Proc. Instn. Civil Engnrs.*, **LV**, Session 1878–9, Part 1.
24. Hall, W.J. (1890) On the failure of the Limerick dock walls and the methods adopted for reconstruction and repairs. *Proc. Instn. Civil Engnrs.* **Vol. CIII**, Session 1890–1, Part 1.
25. Savile, L.H. (1903) Lowering the sill of the Ramsden Lock, Barrow in Furness. *Proc. Instn. Civil Engnrs.*, **CLVIII**, Session 1903–4, Part IV.
26. Prosser, M.J. (1986) *Propeller Induced Scour.* BPF/BHRA Fluid Engineering Report No. 2570.
27. Pease, J. (1989) *New Civil Engineer*, 21 September, p. 7.
28. Whittle, I.R. (1989) The Greenhouse Effect. Lands at risk. An assessment. *MAFF Conference of River and Coastal Engineers*, Loughborough.
29. British Standards Institution. BS 3921:1985 *Specification for clay bricks.*
30. CIRIA. (In preparation) *Seawall Design Guidelines.* Research Project 353.
31. British Standards Institution, BS 8002. *Code of practice for earth retaining structures.* (In preparation).
32. Baker, B. (1880) The actual lateral pressure of earthwork. *Proc. Inst. Civil Engnrs.*, **LXV**, 1881, Paper 1759. Session 1880/1881 Part III.
33. Romilly, Allen, J. (1876) *Design and Construction of Dock Walls.* E and FN Spon, London.
34. Simmons, H.P. (1987) The maintenance of harbour structures. Paper presented at the seminar on Coastal and Port Engineering in Developing Countries, Beijing, China, Ocean Press, Nanjing.
35. British Standards Institution, BS 8210:986. *Building Maintenance Management.*
36. Council of the European Communities. *Directive of 27 June 1985 on the assessment of certain public and private projects on the environment.* 85/337/EEC; Official Journal of the European Communities L 175/40, 5th July 1985.
37. Ministère des Transports (1980) *Fondations de ponts en site aquatique en état précaire, Guide pour la surveillance et le confortement.* Paris, Le Laboratoire Central des Ponts et Chaussées/Le Service d'Etudes Techniques des Routes et Autoroutes.
38. Penning-Rowsell, E.C. and Chatterton, J.B. (1977) *The Benefits of Flood Alleviation: a manual of assessment techniques.* Saxton House.
39. *Investment appraisal in the public sector. A management guide.* HMSO.
40. *Investment in the public sector. A technical guide.* HMSO.
41. OECD (1976). *Bridge Inspection.* A report prepared by an OECD Road Research Group. OECD, Paris.
42. Department of Transport (1983) *Bridge Inspection Guide.* HMSO, London.
43. Ministère des Transports (1982) *Défauts apparents des ouvrages d'art en maconnerie.* Laboratoire Central des Ponts et Chauseées, Paris.
44. Rendel Palmer and Tritton. (1983) *Final Report to London Docklands Development Corporation on Dock and River Wall Inspections, Southwark Site.*
45. Berry, J.G. (1983) Subsea NDT and CCTV techniques for piling and quay wall inspection. *Subsea Challenge Conference*, Amsterdam.

46. British Standards Institution. BS 1377:1975. *Methods of testing soils for civil engineering purposes.*
47. Dunnicliff, J. and Green, G.E. (1988) *Geotechnical Instrumentation for Monitoring Field Performance.* Wiley Interscience, New York.
48. Hume, I. (1983) *The Monitoring of Structures and Landscapes.* English Heritage, Civil and Structural Engineering Section.
49. Kalaugher, P.G. (1987) Structural inspection using photographic colour transparencies: planning a monitoring exercise. Structural Faults & Repairs 87, *Proc. Int. Conf. on Structural Faults and Repairs*, London, Edinburgh, Engineering Technics Press.
50. British Ports Federation (1988) *Evaluation of echo sounders for hydrographic surveying in ports.*
51. British Ports Federation (1981) *Hydrographic surveying in small ports.* Guidance note.
52. ABP Research & Consultancy Ltd. (1987) *Dock wall integrity profiling, Prince of Wales Dock, Swansea.* Research Note No. R.640.
53. Komeyl-Birjandi, F., Forde, M.C. and Batchelor, A.J. (1987) Sonic analysis of masonry bridges. *Proc. Int. Conference on Structural Faults and Repairs*, Engineering Technics Press, London.
54. British Standards Institution. BS 6349:Part 1:1984 *General criteria. Code of practice for maritime structures.*
55. Bureau of Ports and Harbours, Ministry of Transport, Tokyo, 1980. *Technical Standards for Port and Harbour Facilities in Japan.*
56. British Standards Institution. BS 6031:1981. *Code of practice for earthworks.*
57. British Standards Institition. BS 8004:1986. *Code of practice for foundations.*
58. Cochrane, D.J. (1986) Caledonian Canal – repairs to locks at Fort Augustus. *Proc. Instn. Civil Engnrs.*, Part 1, 80, Oct., 1363–83.
59. Barber, P.C. and Davies, C.D. (1985) Offshore breakwaters – Leasowe Bay. *Proc. Instn. Civil Engnrs*, Part 1, 77, Feb., 85–109.
60. Allsop, N.W.H., Powell, K.A. and Bradbury, A. (eds.) (1986) *The use of rock in coastal structures.* Summary of a seminar, 16 January 1986, Hydraulics Research Ltd, Wallingford.
61. British Standards Institution. BS 8081:1989. *Ground Anchorages.*
62. Beckett, T.K.H. (1989) Private communication.
63. Allen, R.T.L. and Edwards, S.C. (1989) *Repair of Concrete Structures.* Blackie, Glasgow.
64. Perkins, P.H. (1986) *Repair, Protection and Waterproofing of Concrete Structures.* Elsevier Applied Science Publishers, Barking.
65. Le Franc, M. and Virlogeux, M. (1982) *Symposium on Civil Engineering Structure Management Brussels – Paris, 1981, and An Example of an Ancient Bridge Rehabilitation.* Final Report IABSE Symposium, Washington DC.
66. Concrete Society. (1979) *Specification for sprayed concrete.*
67. UEG/The Association of Offshore Diving Contractors/The UK Department of Energy. (1984) *The principles of safe diving practice.* Report UR 23, MTD, London.
68. Underwater Engineering Group (1983) *Handbook of Underwater Tools.* 2nd edn. UR18.

Bibliography of geophysical methods of investigation

Annan, A. P. and Davis, J.L. (1976) Impulse radar sounding in permafrost. *Radio Science*, **11**, 383–94.

Anon. (1987) Engineering geophysics. Report by the Geological Society Engineering Group Working Party, *Quarterly Journal of Engineering Geology*, **21**, No. 3, 207–71.

Arrowsmith, D. J. and Rankilor, P. R. (1981) Dalton by-pass: site investigation in an area of abandoned haematite mine workings. *Quarterly Journal of Engineering Geology*, **14**, 207–18.

Conway, B. W., McCann, D. M., Sorginson, M. and Floyd, R. A. (1984) A geophysical survey of the Crouch/Roach river system in south Essex with special reference to buried channels. *Quarterly Journal of Engineering Geology*, **17**, Part 3, 269–82.

Cook, J. C. (1975) Radar transparencies of mine and tunnel rocks. *Geophysics*, **40**, No. 5, 865–85.

Cornwall, J. D. and Carruthers, R. M. (1986) Geophysical studies of a buried valley system near Ixworth, Suffolk. *Proceedings of the Geologists' Association*, **97**, 357–64.

Cratchley, C. R., McCann, D. M. and Ates, M. (1976) Application of geophysical techniques to the location of weak tunnelling ground with an example from the Foyers hydroelectric scheme, Loch Ness. *Transactions of the Institution of Mining and Metallurgy*, Section A, **85**, A127–A135.

Dearman, W. R., Baynes, F. J. and Pearson, R. (1977) Geophysical detection of disused mineshafts in the Newcastle upon Tyne area, N.E. England. *Quarterly Journal of Engineering Geology*, **10**, 257–69.

Early, K. R. and Dyer, K. R. (1964) The use of a resistivity survey on a foundation site underlain by karst dolomite. *Geotechnique*, **14**, 341–8.

Gallagher, C. P., Henshaw, A. C., Money, M. S. and Tarling, D. H. (1978) The location of abandoned mineshafts in rural and urban environments. *Bulletin of the International Association of Engineering Geology*, No. 18, 179–85.

Godson, R. H. and Watkins, J. S. (1968) Seismic resonance investigation of a near-surface cavity in Anchor Reservoir, Wyoming. *Bulletin of the Association of Engineering Geologists*, **5**, 27–36.

Grainger, P. and McCann, D. M. (1977) Inter-borehole acoustic measurements in site investigation. *Quarterly Journal of Engineering Geology*, Geological Society, London, **10**, No. 3, 241–55.

Grainger, P., McCann, D. M. and Gallois, R. W. (1973). The application of the seismic refraction technique to the study of the fracturing of the Middle Chalk at Mundford, Norfolk. *Geotechnique*, **28**, No. 2, 219–32.

Greenfield, R. J. (1979) Review of geophysical approaches to the detection of Karst. *Bulletin of the Association of Engineering Geologists*, **XVI**, 393–408.

Hadley, L. M. (1982) A geophysical method of evaluating existing earth embankment. *Proceedings of the 25th Annual Meeting of the Association of Engineering Geologists*. Montreal, Canada.

Hasselstrom, B. (1969) Water prospecting and rock investigation by the seismic refraction method. *Geoexploration*, 7, 113–32.

Hunter, J. A., Burns, R. A., Gagne, R. M., Good, R. L., Macaulay, H. A. and Pullan, S. E. (1982) Field experience with the 'optimum window' hammer seismic reflection technique. Paper presented at the 52nd Annual International Meeting of the Society of Exploration Geophysicists in Dallas, Texas, USA.

Imai T., Sakayoma, T. and Kanemori, T. (1987) Use of ground probing radar and resistivity surveys for archaeological investigations. *Geophysics*, **52**, 137–150.

Jackson, P. D., Suddaby, D. L. and Baria, R. (1978) Electrical measurements on the proposed route of the M4 near Bridgend, South Wales. Report of the Engineering Geology Unit, Institute of Geological Sciences, No. 78/5.

Keller G. V. and Frischknecht, F. C. (1966) *Electrical methods in Geophysical Prospecting*. Pergamon Press, Oxford.

Kennett, P., Ireson, R. L. and Conn, P. J. (1980) Vertical seismic profiles: their application in exploration geophysics. *Geophysical Prospecting*, **28**, 676–9.

Knill, J. L. (1970) The application of seismic methods in the prediction of grout take in rocks. In *Proceedings of the Conference on In Situ Investigation in Soils and Rocks*. British Geotechnical Society, London, pp.93–100.

Kunetz, G. (1966) *Principles of Direct-current Resistivity Prospecting*. Geoexploration Monographs. Series 1, No.1, Gebrüder Borntaiger, Berlin

Leggo, P. J. and Leech, C. (1983) Subsurface investigation for shallow mine workings and cavities by the ground impulse radar technique. *Ground Engineering*, **16**, January, 20–3.

McCann, D. M. and Jackson, P. D. (1988) Seismic imaging of the rock mass. *University of Wales Review*, No. 4, 21–8.

McCann, D., Grainger, P. and McCann, C. (1975) Interborehole acoustic measurements and their use in engineering geology. *Geophysical Prospecting*, **23**, 50–69.

McCann, D. Baria, R., Jackson, P. D., Culshaw, M. G., Green, A. S. P., Suddaby, D. L. and Hallam, J. R. (1982) The use of geophysical methods in the detection of natural cavities, mineshafts and anomalous ground conditions. Report of the Engineering Geology Unit, Institute of Geological Sciences, No. EG/82/15.

McCann, D. M., Baria, R., Jackson, P. D. and Green, A. S. P. (1986) Application of cross-hole seismic measurements to site investigations. *Geophysics*, **51**, No. 4, 914–25.

McCann, D. M., Jackson, P. D. and Culshaw, M. G. (1987) The use of geophysical surveying methods in the detection of natural cavities and mineshafts. *Quarterly Journal of Engineering Geology*, **20**, 50–73.

McDowell, P. (1975) Detection of clay-filled sink holes in the chalk by geophysical methods. *Quarterly Journal of Engineering Geology*, **8**, 303–10.

McDowell, P. W. (1981) Recent developments in geophysical techniques for the rapid location of near-surface anomalous ground conditions. *Ground Engineering*, **14**, 20–3.

Neumann, R. (1977) Microgravity methods applied to the detection of cavities. *Proceedings of the Symposium on Detection of Subsurface Cavities*, Soils and Pavements Laboratory, US Engineering, Waterways Experiment Station, Vicksburg, Mississippi.

Olsson, O., Sondberg, E. and Nilsson, B. (1983) Cross-hole investigations – The use of borehole radar for the detection of fracture zones in crystalline rock. Stripa Project Internal Report 83-06. Swedish Nuclear Fuel Supply Co/Division KBS.

Parasnis, D. S. (1986) *Principles of Applied Geophysics* (4th edn). Chapman and Hall, London.

Richtien, R. D. (1975) A high frequency wave generator for application to cavity delineation. *Proceedings of the 3rd Symposium on Salt*, **2**, 357–66. J. L. Ran, and L. F. Deilwig (eds.) The Northern Ohio Geological Society, Cleveland, Ohio.

Richtien, R. D. and Stewart, D. M. (1975) A seismic investigation over a near-surface cavern. *Geoexploration*, **13**, 235–46.

Scalabrini, M., Carrugo, A. and Carati, L. (1964) Determination in situ of the Frera Dam Foundation by the sonic method, its improvement by consolidation grouting, and verification of the result by again using the sonic method. *8th International Congress on Large Dams*, Edinburgh, **Volume 1**, 585–600, The International Commission on Large Dams.

Telford, W. M., Geldart, L. P., Sherriff, R. E. and Keys, D. A. (1976) *Applied Geophysics*. Cambridge University Press.

Unterberger, R. R. (1978) Radar propagation in rock salt. *Geophysical Prospecting*, **26**, 312–28.

Location index

Subject index